THE SOURCE REGION OF THE SOLAR WIND

The Source Region
of the Solar Wind

(IX Lindau Workshop, November 1981)

(Invited Review Papers)

Edited by

W. K. H. SCHMIDT and H. GRÜNWALDT

Reprinted from

Space Science Reviews, Vol. 33, Nos. 1/2

D. Reidel Publishing Company

Dordrecht : Holland / Boston : U.S.A.

ISBN-13:978-94-009-7955-0   e-ISBN-13:978-94-009-7953-6
DOI: 10.1007/978-94-009-7953-6

TABLE OF CONTENTS

INTRODUCTORY REMARKS

This issue of *Space Science Reviews* contains the written versions of a number of invited/review lectures that were presented at the IX-th Lindau Workshop under the theme of 'The Source Region of the Solar Wind'. Contributed papers to this workshop meeting will be published in the companion journal *Solar Physics*. We separated the publications of review and original research papers because we felt they were worth publishing in established journals, and such journals will generally not accept a mixture of the two kinds of articles. At this point we want to thank very much the managing editors of *Space Science Reviews* and *Solar Physics*, C. de Jager and Z. Švestka, for helping us to work out this procedure, and for being very cooperative in implementing it.

In the sense that Solar Wind research and Coronal research have developed into different branches of research in the past because of their differing experimental techniques, our meeting could be called an interdisciplinary one. We had heard several suggestions that Solar physicists and Solar Wind physicists should be talking to each other much more than they have done in the past, and that a prerequisite to that is that both groups learn to understand each others' languages. With our meeting we tried to contribute to the communication between scientists of different special branches in a research field where different specialties should not have developed in the first place.

In the workshop we tried to emphasize research results rather than instrumentation. At the beginning of the conference, though, we had a series of lectures on observational methods, which were intended to give an overview over the large variety of methods that can contribute data to Coronal and Solar Wind research. At the same time these presentations set the stage for later discussions.

The articles in this volume are arranged in the same sequence as they were during the workshop. Naturally, such a short workshop cannot accomplish exhaustive coverage of such a variety of phenomena. But although these proceedings are particularly incomplete without the inclusion of the contributed papers, and also because written versions of some presentations unfortunately could not be prepared by their authors for one reason or another, we feel that this sequence of papers forms an instructive selection of present knowledge in this field, and we thank all authors for their contributions.

WOLFGANG K. H. SCHMIDT
HEINER GRÜNWALDT

Space Science Reviews **33** (1982) 5.   0038–6308/82/0331–0005$00.15.

LIST OF PARTICIPANTS

W. Allan	Max-Planck-Institut für Aeronomie, Lindau
R. C. Altrock	Sacramento Peak Observatory, Sunspot
W. I. Axford	Max-Planck-Institut für Aeronomie, Lindau
M. Bird	Universität Bonn, Bonn
P. Bochsler	Universität Bern, Bern
A. Bürgi	Universität Bern, Bern
C. Chiuderi	Osservatorio Astrofisico di Arcetri, Firenze
P. Couturier	Meudon Observatory, Meudon
P. W. Daly	Max-Planck-Institut für Aeronomie, Lindau
K. U. Denskat	Technische Universität Braunschweig, Braunschweig
M. Dryer	NOAA, Boulder
G. Elwert	Universität Tübingen, Tübingen
J. Fejer	Max-Planck-Institut für Aeronomie, Lindau
R. Fisher	National Center for Atmospheric Research, Boulder
J. Geiss	Universität Bern, Bern
W. M. Glencross	University College, London
H. Goldstein	Max-Planck-Institut für Aeronomie, Lindau
J. T. Gosling	Los Alamos Scientific Laboratory, Los Alamos
H. Grünwaldt	Max-Planck-Institut für Aeronomie, Lindau
S. Habbal	Harvard College Observatory, Cambridge
E. Haug	Universität Tübingen, Tübingen
T. E. Holzer	National Center for Atmospheric Research, Boulder
D. Hovestadt	Max-Planck-Institut für Extraterr. Physik, Garching
R. A. Howard	Naval Research Laboratory, Washington
N. Kömle	Institute für Weltraumforschung, Graz
S. Kunz	Universität Bern, Bern
G. F. Krymsky	Institute of Cosmophysical Research and Astronomy, Yakutsk
E. Leer	National Center for Atmospheric Research, Boulder
S. Livi	Max-Planck-Institut für Aeronomie, Lindau
M. Malinovsky-Arduini	Laboratoire de Physique Stellaire at Planetaire, Verrieres-le-Buisson
B. C. Monsignori-Fossi	Osservatorio Astrofisico di Arcetri, Firenze
E. Marsch	Max-Planck-Institut für Aeronomie, Lindau
M. Neugebauer	Jet Propulsion Laboratory, Pasadena
K. W. Ogilvie	Goddard Space Flight Center, Greenbelt
V. Pizzo	National Center for Atmospheric Research, Boulder

Space Science Reviews **33** (1982) 7–8. 0038–6308/82/0331–0007$00.30.
Copyright © 1982 by D. Reidel Publishing Co., Dordrecht, Holland, and Boston, U.S.A.

A. K. Richter	Max-Planck-Institut für Aeronomie, Lindau
B. J. Robertson	University of London, London
H. Rosenbauer	Max-Planck-Institut für Aeronomie, Lindau
W. K. H. Schmidt	Max-Planck-Institut für Aeronomie, Lindau
M. Schüssler	Universitäts-Sternwarte, Göttingen
R. Schwenn	Max-Planck-Institut für Aeronomie, Lindau
N. R. Sheeley, Jr.	Naval Research Laboratory, Washington
R. N. Smartt	Sacramento Peak Observatory, Sunspot
R. Steinitz	Universität Ulm, Ulm
V. M. Vasyliunas	Max-Planck-Institut für Aeronomie, Lindau
G. M. Webb	Max-Planck-Institut für Aeronomie, Lindau
E. Weisshaar	Universitäts-Sternwarte, Göttingen
G. L. Withbroe	Center for Astrophysics, Cambridge
S. T. Wu	The University of Alabama, Huntsville

OPTICAL OBSDERVATIONS OF THE SOLAR CORONA*

R. R. FISHER

*High Altitude Observatory National Center for Atmospheric Research** Boulder, CO 80303, U.S.A.*

Abstract. The history of optical observations of the solar corona is traced through the development of lour topics: the observation of naturally occurring total solar eclipses, the discovery of forbidden emission lines in the sclar corona, the development of narrow spectral bandpass observing techniques, and the invention of the coronagraph. These events occurred over a span of time from the middle of the last century until the present.

1. Introduction

Prior to the middle of the nineteenth century, eclipses were observed as astronomical events to be used to fix either time or location. One of the most famous expeditions of the eighteenth century actually centered upon a different sort of solar astronomical event, namely the attempt, in 1769, to set a value, in terrestrial units, to the distance between the Earth and Sun by observing the transit of the planet Venus across the disk of the Sun. This attempt was led by James Cook, R. N.; a young officer thought to be well qualidied for this most important task, this was not the first time that Cook had undertaken astronomical observations. On 5 August, 1766, Cook observed the altitude and azimuth of a solar eclipse from a shore station on the coast of the headlands of Newfoundland. Dr John Bevis, a physician and member of the Royal Society, used these data to calculate the longitude of one of the Burgeo Islands; the value he inferred was only six minutes of arc smaller than the modern, accepted value of $57° 37'$ W (Beaglehole, 1974). We shall return briefly to Cook and his travels a little later. The history of the development of the astronomy and astrophysics of the solar corona may be traced through the work performed over the last century in four areas of observation of the solar corona. These topics are not exclusive of each other and did not lake place as separate events occurring neatly in a linear progression, even though it is possible to view them as such from our present perspective. These four important observational topics are: the observation of naturally occurring total solar eclipses, the dicovery of forbidden emission lines in the solar corona, the development of narrow spectral bandpass observing techniques, and the invention of the coronagraph. In the remainder of this discussion, these subjects will be reviewed briefly.

2. Eclipse Observations

Changes in the way the Sun was observed, along with changes in the way in which men viewed the nature of the Sun, proceeded slowly from the eighteenth to nineteenth century.

* Paper presented at the IX-th Lindau Workshop 'The Source Region of the Solar Wind'.
** The National Center for Atmospheric Research is sponsored by the National Science Foundation.

Space Science Reviews **33** (1982) 9–16.   0038–6308/82/0331–0009$01.20.

In Cook's time, it was the view that the duty of science was to yield useful information about the earth, information that might aid in the navigation of the Earth's oceans for example. The corona, during this period was regarded as likely to be an artifact produced in the atmosphere of the Earth, a view held by the man who was most influential in the organization of Cook's first voyage to the South Pacific, Edmund Halley.

In 1869, observing from a station in the middle of the Nort American continent, C. A. Young viewed the solar corona through a modest prism spectrograph. He was able to find a single, moderately bright, emission line in the radiance emitted by the corona. The emission line was observable to a height of five to eight arc min above the limb at the 1869 eclipse, but subsequently was seen to a height of 20 arc min above the limb at the eclipse of 1870 (Young, 1881). This emission line was located in the green region of the spectrum, the wavelength was given as 1474 in the Kirchoff scale, $\lambda 5316$ Å in modern terminology. This finding strongly suggested an extraterrestrial origin for the solar corona to Young (1881), although this view was not widely accepted until a few years later. Writing after Young's announcement of his dicovery, Lockyer (1874) speculated that without confirmation in the near future, it would be reasonable for Professor Young to withdraw his suggestion of a solar origin for the emission line, since a homogeneous solar atmosphere would tend to scatter continuum radiation from the photosphere rather than an isolated emission feature. This variance of opinion was to be resolved by the use of a new nineteenth century technology, capturing eclipse images on photographic plates for quantitative analysis.

Total solar eclipses have been photographed from the early 1850s onward, and improved, dry photographic emulsion became available for use later about 1885. In his general astronomy text, Young (1895) states that photographic images of a solar eclipse obtained from India, Ceylon, and Java were found to be identical after the 1871 eclipse. As late as 1883 somme doubt must have remained, since at this time Yale investigators interpreted their data as being the result of diffraction around the edge of the moon. The issue was resolved by a classic set of large scale coronal images, secured by the Lick Observatory between 1889 and 1922, from that observatory's program of eclipse photography (Eddy, 1971).

The major achievement of optical observations during the nineteenth century was the identification of the corona, or at least some component of the corona, as a solar phenomenon. This was done by careful study of the eclipse images, which revealed that the general morphology was time dependent and specifically related to the phase of the solar cycle. By the technique of low resolution spectrocopy it was discovered that there was a kind of permanent component of the upper atmospheric layers of the Sun which emitted a kind of solar aurorae. Young attributed this radiance to an unrecognized element:

We have as yet been unable to identify with any terrestrial element the substance to which the line is due, but the provisional name coronium has been proposed for it.

3. Forbidden Lines of the Solar Corona

After the discovery of the green line of the solar corona in 1869, little further understanding of the origin, or the significance, of this emission feature was gained during the rest of the nineteenth century. As techniques improved for spectroscopic analysis, it was learned that there was not a single emission feature but, in fact, several emission features are present in the solar corona. Two important observational facts predated the identification of the solar coronal line – these were the spectral observations of the novae RR Pictoris in 1928 and Rs Ophiuchi (1933). H. Spencer Jones (1932) described the time variation of the spectrum of the former of these two events for the period 1928–1931, showing that in the post-maximum phase there were a number of emission features in the spectrum which could not be identified with existing line lists obtained from laboratory spectra or from the known spectra of nebulae. One year after the publication of this observation, both the red and green coronal lines were detected in the spectrum of the nova RS Ophiuchi by Adams and Joy (1933). Thus, for the first time, it was clear that the material responsible for radiation at these wavelengths was not unique to the solar corona. Complete identification of the emission lines seen in this nova were later provided by Joy and Swings (1945). Bowen (1935) was able to identify three of these features, $\lambda 4968$, $\lambda 5146$, and $\lambda 5177$ as lines originating from the $^4F-^2G$ transistion of Fe VI. Popper (1939) noticed that the same emission features was observed in yet a third nova, DQ Herculis of 1934, and Bowen and Edlen (1939) successfully identified lines from $^3F-^1G$ and $^3F-^1D$ transitions of Fe VII in spectra of this nova.

As early as 1932, H. Spencer Jones (1932b) speculated that at least some of the lines appearing in the spectrum of yet another nova, η Carinae, probably could be attributed to forbidden transistions of ionized iron. In his letter of 13 February, 1937 to B. Edlen, W. Grotian pointed out that the separation of the ground terms in Fe X and Fe XI coincided with the wave numbers of the coronal emission lines $\lambda 6374$ and $\lambda 7892$ (the later emission line having been recently discovered by B. Lyot in coronagraph observations), and the identities of two of the emission features found in the solar corona were established. By considering the Al I isoelectronic sequence, Edlen (1943) was able to identify the green line, $\lambda 5303$, as originating from the ion Fe XIV, and further he was able to identify most of the remaining lines seen in the coronal spectrum. Reviewing his work, Edlen, in the Darwin Lecture of 1945 (Edlen, 1945), pointed out that the solar corona was finally understood to be composed of free electrons and ions of highly ionized elements, ions which could only exist at extremely high temperatures.

4. Monochromatic Observations and Coronagraphs

By the third decade of the 20th century, the general properties of the plasma found in the solar corona were understood from spectroscopic observations and eclipse images. Because of the short duration of eclipses, and their relative infrequency, the main temporal variation of the corona and the threedimensional distribution of coronal material remained obscure. Prominences were first observed outside of eclipse by

Janssen and Lockyer in 1868, both using a slitless spectograph technique and looking above the limb. Hale and Ellerman (1903) perfected a technique of simultaneously scanning the solar image with a spectrograph slit and focal plane bandpass slit – the spectroheliograph, which allowed the observation of prominences above the limb as well as against the disk. While not directly observing the corona, spectroheliograms taken in the cores of strong chromospheric lines were essential for following the temporal changes in prominences. Lockyer, (1903) showed that there was a close correlation between the form of the corona seen at eclipse, and the latitude distribution of prominences. Further, the existence of electromagnetic forces acting above the photosphere were made obvious to observers of prominences such as Pettit (1924) using the spectroheliograph technique.

Spectroheliograph observations of the distribution of prominences were initiated by L. D'Azambuja, beginning in March 1919, and have been continued by others up to the present. Hale, and many others, attempted to observe the solar corona outside of total eclipse, but without success.

The problem was finally solved by B. Lyot, and his approach was founded on his analysis of the origin of scattered photospheric light in the final image plane. The necessary reduction was accomplished by the elimination of sources of scattered light within the telescope and the selection of an observing site relatively free from the effects of atmospheric scattering. Lyot (1930) identified the major sources of scattering within the telescope as the imperfections within the glass of the objective lens, the quality of the polish on the surfaces of the objective, multiple reflections between the surfaces of the objective and finally light diffracted arond the entrance aperture of the telescope. The brightness of the corona is usually between 1 to $3 \times 10^{-6} B_{\odot}$, a few are min above the limb, and the light scattered by the earth's atmosphere is usually far above this value. Lyot (1931, 1935) however, was able to secure photographic coronal images outside of eclipse on several occasions using red glass filters and a coronafraph. Since emission lines of the corona have an equivalent width which is relatively large compared to a 1 Å slice of continuim radiation, it is much easier to detect coronal line radiation against the background of the atmospheric and instrumentally scattered light. Emission lines at $\lambda 7892$, (Fe XI), $\lambda 8024$ (N XV), $\lambda 10747$, and $\lambda 10798$ (Fe XIV) were discovered using coronagraph observations. The combination of a coronagraph-spectrograph was used daily by Waldmeier to study the emission line corona, beginning in 1939. Using emission line data from a variety of observatories, this study extended over a solar cycle, and it became the landmark for synoptic studies of the corona. From this work came the first clear recognition that long-lived, low density structures exist in the solar corona, (Locher, or holes) now commonly thought of as sources of high speed solar wind plasma (Waldmeier, 1957).

In order to observe the details of the coronal emission line structures, a final development in optical instrumentation was required. The birefingent filter was proposed by Lyot (1933) and later the same invention was proposed by Ohman (1938) who produced the first instrument of this type. Later Lyot (1944) and Evans (1949) both gave more extensive discussion to various improvements in filter design, such as increased transmission, reduction in off-band light, enlarged fields of view, and continuous wavelength adjustment. In recent times, this type of filter has been improved so that the spectral resolution rivals that of the spectroheliograph, and it is now possible to construct filter magnetographs for the detection of photospheric magnetic fields (Title, 1975a, b). With the availability of filters for operation at the brightest coronal emission lines, usually 5303 of Fe XIV and 6374 of Fe X, it was possible finally to obtain images of the solar corona with good dimensional spatial resolution, and when used in concert with a coronagraph, observations could be extended over a number of hours to reveal the details of the evolution of various coronal structures. The relationship between the active regions of the chromosphere and the emission line coronal structures, loops and transient changes, is a subject investigated by a combined coronagraph-filter technique by Lyot and Dollfus (1971). A comprehensive study of coronal evolution has been given by Dunn (1971), who used a $\lambda 5303$ filter for his observing program.

Recent optical observations tend to extend the techniques discussed above or process the data in a more economical and efficient manner. Some recent developments deserve mention. The spectroheliograph technique was used, with some success, to follow the development of coronal hole structures over the last part of sunspot cycle 20 and the first portion of cycle 21 by Sheeley and Harvey (1981). This is possible to do using the weak chromospheric line $\lambda 10830$ of He I, and efforts are taken enhance the contrast in the image obtained. This technique is now used daily at Kitt Peak National Observatory. Two recent developments concerning coronagraphics which deserve further discussion. The polarized component of the electron scattered corona may be detected rather easily by using a conventional white-light coronagraph and analyzing the observed radiation for linear polarization. An instrument intended for coronal observations which takes adventage of this fact was developed by Wlerick and Axtell (1957) for the old HAO observatory at Climax, Colorado; later this instrument was moved to Mauna Loa, Hawaii. At present there are K-coronameters operating at Pic du Midi, France, and Kazakh, U.S.S.R. Considerable progress in understanding the variation in coronal morphology over a solar cycle, from 1966 to 1978, has been made by Hundhausen, Hansen and Hansen (1981) using this type of instrument. A third version of the K-coronameter was recently installed at the Mauna Loa site which produces digital images of the white light corona with a time resolution of about one observation per minute. This is described by Fisher et al. (1981); by a curious coincidence, this new K-coronameter is located near Kealakekua Bay, the place where James Cook lost his life in early 1779.

At extremely high altitudes, and at the altitudes, that spacecraft operate, atmospheric scattering effects are reduced or non-existent. A new coronagraph technique isused for instruments operating in these favorable circumstances, the externally occulted coronagraph, in which the scattered light is reduced to extremely low levels. The method for producing very low scattered light in an externally occulted device was described by Newkirk and Bohlin (1963). This type of coronagraph was used by the Naval Research Laboratory on OSO-7 and more recently on P78–1, the latter instrument being described by Sheeley et al. (1980). The NRL version of this instrument is capable of observing the solar corona at latitudes of nearly $10R_\odot$. Externally occulted instruments with smaller

Fig. 1a. The total solar eclipse of 18 July 1960, as drawn by G. Tempel from a site near Torreblanca Spain. North is at the top and east to the left.

fields of view were carried on ATM skylab and the SMM spacecraft, the former experiment was discussed by MacQueen, *et al.* (1974), the latter by House *et al.* (1980). These externally occulted devices have been quite successful for the study of coronal transient events. White light coronal transient activity was initially noted by Tousey (1973) in the OSO-7 data, and soon afterward the HAO instrument on ATM Skylab was used by Munro *et al.* (1975) to measure the physical parameters of a large number of these events. An interesting question following these developments was addressed by Eddy (1974) who surveyed available eclipse records to see if it might be possible to defect coronal transient activity in historical records. The most likely candidate for such an observation is the one shown in Figure 1a, the drawing of the eclipse 1860 by Gughemo Temple at Torreblanca Spain. This peculiar image may be compared with the image of

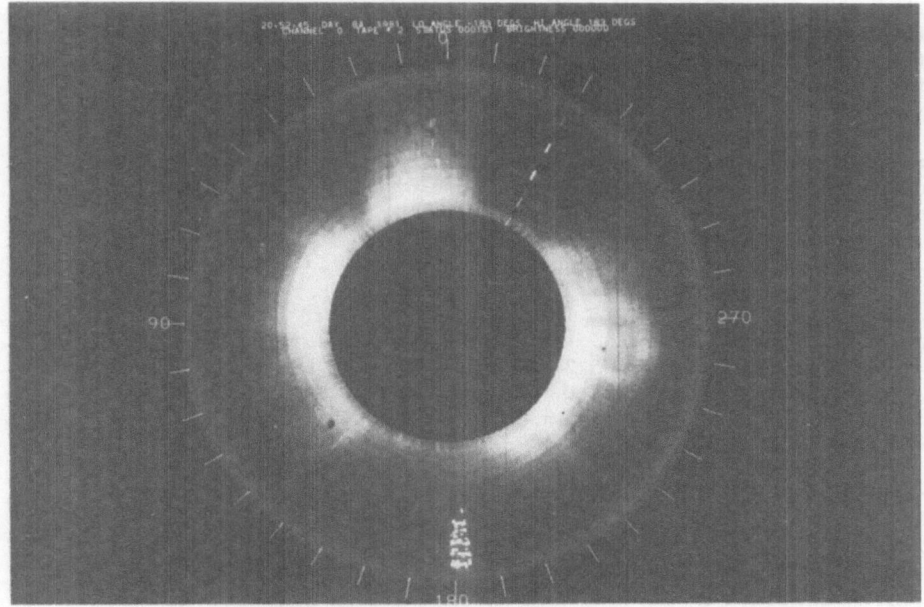

Fig. 1b. A flare-associated coronal transient observed with the Mk-III K-coronameter located at Mauna Loa Hawaii on 25 March, 1981. (Position angle 260°). The scale and orientation are the same as in Figure 1a.

a flare induced transient, Figure 1b, which was observed with the Mk-III K-coronameter at the Mauna Loa, site on 25 March 1981. The observed velocity was 800 km s⁻¹ for the leading edge of the transient, a high enough speed so that it remained in the field of view of the instrument for less than 15 min. The similarity is striking in the extreme; it remains a matter of conjecture for the reader as to the true nature of Temple's object.

Acknowledgement

The author gratefully acknowledges the careful reading of this manuscript by Dr R. Munro of the High Alltitude Observatory. This work was supported by the High Altitude Observatory, a division of the National Center for Atmospheric Research, under contract with the National Science Foundation.

References

Adams, W. S. and Joy, A. H.: 1933, *Publ. Astron. Soc. Pacific* **45**, 301.
Beaglehole, J. C.: 1974, *The Life of Captain James Cook*, Stanford University Press, Stanford, California.
Bowen, I S.: 1935, *Phys. Rev.* **437**, 924.
Bowen, I S. and Edlen, B.: 1939, *Nature*, 4 March.
Dollfus, A.: 1971, in C. J. Macris (ed.), *Physics of the Solar Corona*, Athens, Greece, p. 97.
Dunn, R. B.: 1971, in C. J. Macris (ed.), *Phisics of the Solar Corona*, Athens, Greece, p. 114.

Eddy, J.: 1971, *J. Hist. Astron.* **2**, 1.
Eddy, J.: 1974, *Astron. Astrophys.* **34**, 235.
Edlen, B.: 1943, *Z. Astrophys.* **22**, 30.
Edlen, B.: 1945, *Monthly Notices Roy. Astron. Soc.* **105**, 323.
Evans, J. W.: 1949, *J. Opt. Soc. Am.* **39**, 229.
Fisher, R., Lee, R., MacQueen, R., and Poland, A.: 1981, *Appl. Opt.* **20**, 1094.
Hale, G. E. and Ellerman, F.: 1903, *Publ. Yerkes Obs.* **3**, 3.
House, L. L., Wagner, W., Hildner, E., Sawyer, C., and Schmidt, H.: 1981, *Astrophys. J.* **244**, L117.
Hundhausen, A. J., Hansen, R. T., and Hansen, S. F.: 1981, *J. Geophys. Res.* **86**, 2079.
Joy, A. H. and Swings, P.: 1945, *Astrophys. J.* **102**, 353.
Lockyer, J. N.: 1874, *Contributions to Solar Physics*, MacMillan and Co., London.
Lockyer, W. J. S.: 1903, *Monthly Notices Roy. Astron. Soc.* **63**, 481.
Lyot, B.: 1931, *Compt. Rend. Acad. Sci.* **193**, 1169.
Lyot, B.: 1933, *Compt. Rend. Acad. Sci.* **197**, 1593.
Lyot, B.: 1935, *Compt. Rend. Acad. Sci.* **200**, 219.
Lyot, B.: 1944, *Ann. Astrophys.* **7**, 31.
MacQueen, R., Eddy, J., Gosling, J., Hildner, E., Munro, R., Newkirk, G., Poland, A., and Ross, G.: 1974, *Astrophys. J.* **187**, L85.
Munro, R., Gosling, J., Hildner, E., MacQueen, R., and Ross, C.: 1979, *Solar Phys.* **61**, 201.
Newkirk, G. and Bohlin, D.: 1963, *Appl. Opt.* **2**, 131.
Ohman, Y.: 1938, *Nature* **141**, 157.
Pettit, E.: 1925, *Publ. Yerkes Obs.* **3**, 205.
Popper, D. M.: 1939, *Publ. Astron. Soc. Pacific* **51**, 168.
Sheeley, N. R. and Harvey, J.: 1981, *Solar Phys.* **12**, 23.
Sheeley, N. R., Michels, D., Howard, R., and Kooman, M.: 1980, *Astrophys. J.* **237**.
Spencer-Jones, H.: 1932a, *Monthly Notices Roy. Astron. Soc.* **91**, 794.
Spencer-Jones, H.: 1932b, *Monthly Notices Roy. Astron. Soc.* **92**, 728.
Title, A. M.: 1975a, *Appl. Opt.* **14**, 229.
Title, A. M.: 1975b, *Appl. Opt.* **14**, 445.
Tousey, R.: 1973, in M. Rycroft and S. Runcorn (eds.), *The Solar Corona in Space Research*, *XIII*, Akademie-Verlag, Berlin.
Waldmeier, M.: 1957, *Die Sonnenkorona* **2**, Birkhäuser Verlag, Basel.
Wlerick, G. and Axtell, J.: 1957, *Astrophys. J.* **126**, 253.
Young, C. A.: 1881, *The Sun*, D. Appelton and Co., New York.
Young, C. A.: 1895, *General Astronomy*, Ginn and Co., Boston.

PROBING THE SOLAR WIND ACCELERATION REGION USING SPECTROSCOPIC TECHNIQUES*

GEORGE L. WITHBROE, JOHN L. KOHL, and HEINZ WEISER

Harvard-Smithsonian Center for Astrophysics, Cambridge, Mass., U.S.A.

and

RICHARD H. MUNRO

High Altitude Observatory, Boulder, Colo., U.S.A.

Abstract. Measurements of the intensities and profiles of UV and EUV spectral lines can provide a powerful tool for probing the physical conditions in the solar corona out to $8R_\odot$ and beyond. We discuss here how measurements of spectral line radiation in conjunction with measurements of the white light K-corona can provide information on electron, proton and ion temperatures and velocity distribution functions; densities; chemical abundances and mass flow velocities. Because of the fundamental importance of such information, we provide a comprehensive review of the formation of coronal resonance line radiation, with particular emphasis on the H I Lα line, and discuss observational considerations such as requirements for rejection of stray light and effects of emission from the geocorona and interplanetary dust. Finally, we summarize some results of coronal H I Lα and white light observations acquired on sounding rocket flights.

1. Introduction

Improved knowledge of the physical conditions (temperatures, densities and mass flow velocities) in the solar corona is critical to the development of an understanding of the physical state of the corona and the physical mechanisms responsible for coronal heating, solar wind acceleration and the transport of mass, momentum and energy. In this paper we review recent developments in UV and EUV coronagraphic spectroscopy which is becoming a powerful tool for acquiring information on the physical conditions in coronal regions out to 8 solar radii $(R_\odot)$ or more from Sun-center. Of particular interest is the solar wind acceleration region beyond $r = 1.5 R_\odot$ where present empirical information is extremely limited.

Until recently coronal spectroscopy has been confined primarily to the low corona within a few tenths of a solar radius from the solar surface. At higher levels measurements made through remote sensing techniques have consisted primarily of broad-band measurements of the electron-scattered white light corona, limited observations of forbidden emission lines and limited measurements at radio wavelengths. As a consequence in the solar wind acceleration region for heliocentric distances $r \gtrsim 1.3 R_\odot$ there has been very little information on temperatures, flow velocities, the chemical composition and the spatial and temporal variations of these parameters. The available information has been insufficient to place significant empirical constraints upon the identities and functions of:

- plasma heating processes;
- solar wind acceleration processes;

---

* Paper presented at the IX-th Lindau Workshop 'The Source Region of the Solar Wind'.

Space Science Reviews **33** (1982) 17–52.   0038–6308/82/0331–0017$05.40.

— processes for transport of energy and momentum; and

— processes for producing variations in chemical abundances; or to determine

— the role of different regions in contributing to the outward flow of plasma in the solar wind;

— the dominant physical processes controlling the evolution of transients and the detailed effects of transients on the ambient medium.

This situation is changing rapidly due to the development of coronagraphic techniques employing reflecting optics that permit measurements at large distances above the solar surface in the UV and EUV regions of the spectrum (Kohl *et al.*, 1978, 1981). Access to these spectral regions is important because there are a number of strong permitted coronal lines located there, lines which can provide critical plasma diagnostic information.

Observations of the low corona $r \lesssim 1.3 R_\odot$ acquired with ground-based, rocket and satellite experiments operating in the visible, UV, EUV, and soft X-ray regions of the spectrum have been discussed extensively in the literature and in recent review papers (cf. Withbroe and Noyes, 1977; Vaiana and Rosner, 1978; Zirker, 1981; and references cited therein). We will restrict the present review to the higher layers which, by comparison, have been subjected to only limited spectroscopic probing and therefore offer rich opportunities for empirical investigations. We discuss (1) spectroscopic plasma diagnostic techniques that can be used in the solar wind acceleration region and (2) results of the application of some of these techniques to observations from the Center for Astrophysics/High Altitude Observatory rocket coronagraph program.

2. Coronal Ionization Balance and Particle Velocity Distributions

UV and EUV spectroscopic diagnostics make use of information contained in measurements of the intensities and profiles of spectral lines. These observable parameters depend upon the properties of the plasma emitting the measured radiation. In the upper regions of the corona the densities are sufficiently low that various particle species such as electrons, protons and other ions may have velocity distribution functions characterized by different temperatures. In addition, the ionization balance for a given atomic species may be decoupled from the local electron temperature. One method of illustrating where these effects become important is through comparison of the characteristic time for coronal expansion with the relevant thermalization and ionization equilibrium times. Figure 1 does this for a typical equatorial region (upper graph) and polar region (lower graph). The densities used for these regions are from respectively Saito (1970) and Munro and Mariska (1977), supplemented by values from Allen (1963). The Munro/Mariska model is an updated version of a model by Munro and Jackson (1977) for a large polar coronal hole with rapidly diverging geometry observed during solar minimum in 1973.

For estimating the thermalization times we assumed $T_e = T_p = 2 \times 10^6$ K in the equatorial region and $T_e = T_p = 10^6$ K in the polar region. The expansion times (heavy solid line) $\tau_{\exp} = [(V/n) \, dn/dr]^{-1}$ were calculated assuming a particle flux $nVA = $ constant with the area $A \sim r^2$ in the equatorial region and the constant evaluated so as to give a value for the particle flux at 1 AU comparable to that typically observed, 3.8×10^8 protons $cm^{-2} s^{-1}$ (Feldman *et al.*, 1977). For the polar region we used the flow velocities V given by Munro and Mariska who suggested that the outflow in the observed region would be supersonic at low heights ($r \gtrsim 2.3 R_\odot$) *if* the mass flux in polar coronal holes is as large as measured in high speed solar wind streams observed in the ecliptic. The times given in Figure 1 were calculated using H I collisional cross-sections from Percival (1966), hydrogen/proton charge-exchange cross-sections from Fite *et al.* (1962), and the relations for thermalization times given by Spitzer (1962). The ionization equilibrium times for heavy ions vary with the element and stage of ionization. For a representative value we used $\tau_{ions} = 10^{10}/N_e$ (cf. Bame *et al.*, 1974; Owocki, 1982). The photoionization of H I was determined using measurements of the solar EUV spectrum made using OSO-6 (Dupree *et al.*, 1973) and Skylab (Vernazza and Reeves, 1978).

Because of the long lifetimes of ions in the low density coronal plasma, the ionization balance of heavier elements (e.g. O, Si, Fe) is expected to 'freeze in' within several solar radii of the surface where the lifetimes of the ions become comparable with the coronal expansion time (cf. Hundhausen, 1972; Bame *et al.*, 1974; Owocki, 1982). The ionization equilibrium time for hydrogen $\approx (1/\tau_v + 1/\tau_c)^{-1}$ is much shorter than that of minor ions (see Figure 1), hence the fractional amount of H I freezes in at much greater distances from the Sun. In equatorial regions where the H I collisional ionization times τ_c are much smaller than the corresponding times for photoionization τ_v and coronal expansion (Figure 1a), the ionization balance of H I depends primarily on the electron temperature for $r \lesssim 5 R_\odot$. In low density coronal holes where the H I ionization time may equal or exceed the coronal expansion time at $r \approx 3$ to $4 R_\odot$ (Figure 1b), the H I ionization balance may 'freeze in' within several solar radii of the Sun.

The lifetime of H I atoms $\tau_{HI} = (1/\tau_e + 1/\tau_v + 1/\tau_{hp})^{-1}$ is much shorter than the coronal expansion time for $r \lesssim 8 R_\odot$ in equatorial regions and $r < 3 R_\odot$ in low density coronal holes. The quantity τ_{hp} is the characteristic time for charge exchange between hydrogen atoms and protons. Thus, in these layers the coupling between the hydrogen atoms and protons is sufficiently strong that the proton and hydrogenic temperatures should be equal. As we shall see in Section 4, this means that the profile of the resonantly scattered H I Lα line can be used to measure proton temperatures in these layers.

Although the neutral hydrogen and proton temperatures are expected to be equal throughout most of the solar wind acceleration region, this is not true for other particles. At $r \gtrsim 2 R_\odot$ the electron/proton thermalization time τ_{ep} is comparable to or larger then the coronal expansion time. Electron/ion thermalization times are also long. Consequently, the temperatures of electrons, protons and ions can differ significantly in this region as shown by theoretical multi-fluid solar wind models (see Hartle and Sturrock, 1968; Nerney and Barnes, 1977; Joselyn and Holzer, 1978; Hollweg, 1978). Thus it is important to develop a technique for distinguishing between the temperatures of different particles species. As shown below, measurements of the profiles of the electron and resonantly scattered components of the coronal Lα line provide a potential means of accomplishing this for electrons and protons. Measurements of spectral lines from heavier ions can provide information on temperatures of other particle species.

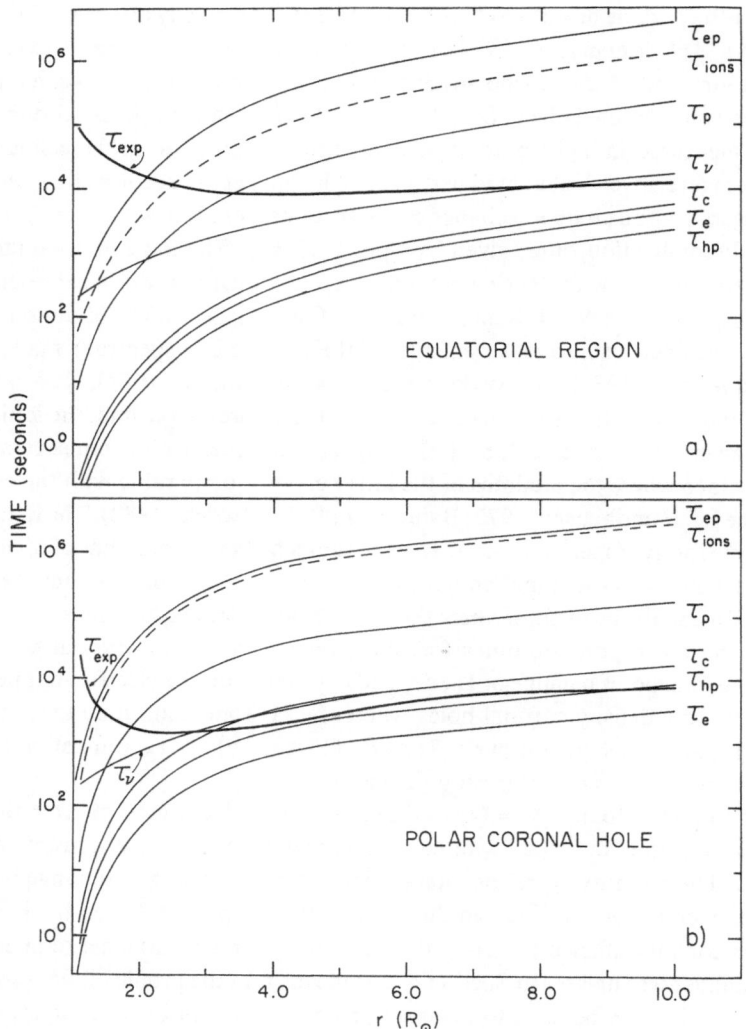

Fig. 1.   (a) Characteristic times plotted as a function of heliocentric distance in a typical equatorial region; coronal expansion time τ_{exp}, ionization equilibrium time for typical ions τ_{ions}, H I collisional ionization time τ_c, H I photoionization time τ_ν, electron thermalization time τ_e, proton thermalization time τ_p, time for hydrogen-proton charge exchange τ_{hp} and time for the electron and proton temperatures to equalize τ_{ep}. (b) Same for a low density polar coronal hole.

3. Determinations of Coronal Temperatures and Particle Velocity Distribution

The width of an optically thin spectral line in a low density plasma depends on the kinetic temperature $T_{kinetic}$ of the plasma where the line originates. If the line has a shape which is Gaussian or nearly Gaussian, one can define the kinetic temperature by the relation

$$\Delta\lambda_{\text{FWHM}} = \frac{2.4\lambda}{c}\left(\frac{2kT_{\text{kinetic}}}{M}\right)^{1/2},\tag{1}$$

where T_{kinetic} includes the effect of both thermal and nonthermal motions and M is the mass of the particle species producing the spectral line. The kinetic temperature is a quantity determined directly from the line width. If one wishes to separate the effects of thermal and nonthermal line broadening, then one can define

$$\Delta\lambda_{\text{FWHM}} = \frac{2.4\lambda}{c}\left(\frac{2kT_{\text{thermal}}}{M} + \xi^2\right)^{1/2},\tag{2}$$

where ξ is the rms velocity component due to plasma motions that occur on a spatial scale much larger than the particle mean free path, but smaller than the path length over which the spectral line is formed. (In the lower solar atmosphere motions of this type are often referred to as microturbulence.) For example, Alfvén waves propagating through the corona (e.g., Hollweg, 1978) may cause plasma motions of sufficient amplitude to significantly broaden spectral lines, perhaps even be the dominant source of broadening for spectral lines of heavy ions. By observing lines from ions of different masses one can obtain empirical constraints on the magnitude of mass-dependent and mass-independent motions. (Line profiles can also be affected by solar wind outflow velocities as will be discussed in Section 4.)

'Temperature' is a convenient parameter, but the line profile gives much more information, namely a direct determination of the velocity distribution of the particles along the line of sight. A static isothermal plasma with a Maxwellian particle velocity distribution produces Gaussian profiles, while a multi-temperature or nonthermal plasma with a non-Maxwellian particle velocity distribution function will yield non-Gaussian profiles. As indicated in Section 2, thermalization times in the solar wind acceleration region are long. As a result, plasma heating, acceleration and/or transport processes that are mass- or charge-to-mass dependent can produce differences in thermal temperatures and/or non-Maxwellian velocity distributions among different species of particles. Consequently, measurements of spectral line profiles can provide constraints on mechanisms for these processes, particularly if measurements of spectral lines from ions of different masses can be acquired.

4. Resonantly Scattered Component of Lα and Determination of T_H and T_p

The H I Lα line provides one of the primary tools for studying the upper corona. There are two components of interest, the resonantly scattered component (FWHM ≈ 1 Å) and a much weaker (by about 3 orders of magnitude) electron scattered component (FWHM ≈ 50 Å). The resonantly scattered component depends upon the scattering of chromospheric Lα photons by neutral hydrogen atoms in the corona. Even though only about one proton in 10^7 is tied up in neutral hydrogen at coronal temperatures and densities (Gabriel, 1971), the large coronal proton abundance coupled with the high intensity of the chromospheric Lα radiation gives rise to a coronal resonantly scattered component of Lα strong enough to be measured out to large distances above the solar surface. Resonantly scattered Lα radiation was first observed during the 1970 solar eclipse (Gabriel *et al.*, 1971) and has subsequently been measured by a UV coronagraphic instrument flown on a sounding rocket (Kohl *et al.*, 1980; Weiser *et al.*, 1981). In this section we discuss the formation of this component of Lα. The next section discusses the electron scattered component.

Gabriel (1971) and Beckers and Chipman (1974) have derived equations for the intensity and profile of the resonantly scattered Lα line. The number of coronal hydrogen atoms with velocities between **v** and **v** + d**v** excited per second from level 1 (the ground level) to level 2 (the first excited level) by a beam of chromospheric radiation with wavelengths between λ' and $\lambda' + d\lambda'$ and angular direction between ω and $\omega + d\omega$ is

$$dN_2(\mathbf{v}) = N_1(\mathbf{v})hB_{12}\lambda_0^{-1}I(\lambda', \omega) \times$$

$$\times \delta\left(\lambda' - \lambda_0 - \frac{\lambda_0}{c}\,\mathbf{v}\cdot\mathbf{n}'\right) d\omega\, d\lambda'\, d\mathbf{v}, \tag{3}$$

where B_{12} is the Einstein coefficient, h is Planck's constant, and $I(\lambda', \omega)$ is the intensity of the chromospheric radiation at wavelength λ'. The only photons that can be scattered by a hydrogen atom moving with a velocity **v** are those with $\lambda' = \lambda_0 + (\lambda_0/c)\mathbf{v}\cdot\mathbf{n}'$ where λ_0 is the central wavelength of the Lα transition and $\mathbf{n}'$ is the vector describing the direction of the incident chromospheric radiation; hence a Dirac delta function has been introduced in Equation (3). (The effect of the natural Lα line width can be ignored for scattering at coronal temperatures.)

The number of photons scattered per second in the direction **n** toward an observer is

$$dN = \frac{dN_2(\mathbf{v})}{4\pi}\,\frac{1}{12}\left(11 + 3(\mathbf{n}\cdot\mathbf{n}')^2\right)\delta\left(\lambda_0 - \lambda + \frac{\lambda_0}{c}\,\mathbf{v}\cdot\mathbf{n}\right), \tag{4}$$

where $(11 + 3(\mathbf{n}\cdot\mathbf{n}')^2)/12$ is the angular dependence of the Lα scattering process (see Beckers and Chipman, 1974) and the Dirac delta function transforms the scattered wavelength from the atom's frame to the observer's frame. Now

$$N_1(\mathbf{v})\,d\mathbf{v} = \frac{N_1}{N_{\mathrm{HI}}}\,\frac{N_{\mathrm{HI}}}{N_p}\,\frac{N_p}{N_e}\,N_e f(\mathbf{v})\,d\mathbf{v}$$

$$= 0.8 N_e R f(\mathbf{v})\,d\mathbf{v} \tag{5}$$

where $N_1/N_{\mathrm{HI}} = 1$ (because of the low coronal density), $R = N_{\mathrm{HI}}/N_p$, $N_p/N_e = 0.8$ for a fully ionized plasma with 10% helium and $f(\mathbf{v})$ is the velocity distribution function of the hydrogen atoms.

If we employ a rectangular coordinate system with the observer's line of sight being the x-axis, we have

$$I_s(\lambda) = \frac{0.8hB_{12}}{48\pi\lambda_0} \int\limits_{-\infty}^{\infty} N_e R \, dx \int\limits_{\omega} (11 + 3(\mathbf{n}\cdot\mathbf{n})^2) \, d\omega \times$$

$$\times \int\limits_{-\infty}^{\infty} I(\lambda', \omega) \, d\lambda' \int\limits_{-\infty}^{\infty} f(\mathbf{v})\delta\left(\lambda' - \lambda_0 - \frac{\lambda_0}{c} \mathbf{v}\cdot\mathbf{n}'\right) \times$$

$$\times \delta\left(\lambda_0 - \lambda + \frac{\lambda_0}{c} \mathbf{v}\cdot\mathbf{n}\right) d\mathbf{v}, \qquad (6)$$

where $I_s(\lambda)$ is the intensity of the scattered radiation and $\mathbf{v}\cdot\mathbf{n} = v_x$. If a Maxwellian

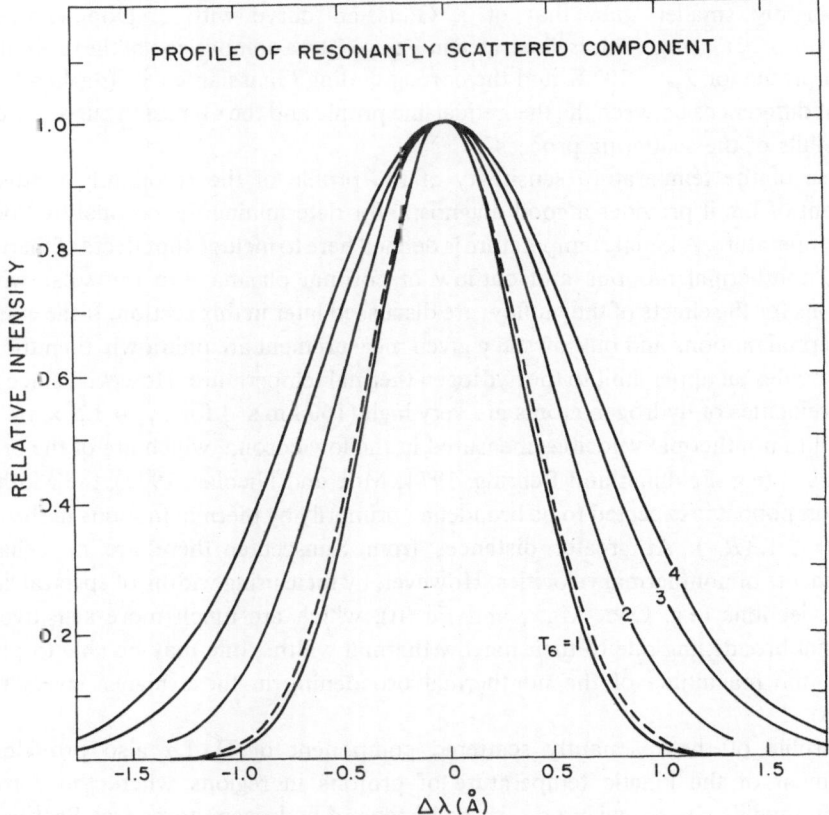

Fig. 2. Profiles of the resonantly scattered component of Lα calculated with a spheric isothermal corona for a line-of-sight intersecting the plane of the solar disk at $\rho = 3\,R_\odot$ from Sun-center. Profiles for 4 coronal temperatures (in units of 10^6 K) are shown. For comparison a pure Gaussian with a width corresponding to the thermal width is also plotted (dashed line).

velocity distribution is assumed, the integral over $\mathbf{v}$ can be integrated analytically to yield an expression for $I_s(v)$ which can then be integrated numerically over x, ω, and λ'.

Figure 2 presents profiles for the resonantly scattered component of $L\alpha$ for several values of the hydrogen temperature T_H (in units of 10^6 K). These profiles were calculated using an isothermal spherically symmetric corona with a radial density distribution given by Saito (1970). For the profile of the incident chromospheric radiation we used the measurements of Gouttlebroze et al. (1978). We assumed that the intensity and shape of the chromospheric profile are constant across the solar disk. The chromospheric $L\alpha$ line shows little center-limb variation (Prinz, 1974; Basri et al., 1979; and unpublished data from the Harvard Skylab experiment). If the effects of nonthermal motions can be ignored, then the temperatures given in Figure 2 are the hydrogen thermal temperatures T_H. If isotropic nonthermal motions are present which yield a Maxwellian velocity distribution for hydrogen atoms, then the hydrogen temperatures given in Figure 2 correspond to a kinetic temperature $T_k = T_H + \xi^2 m_H/2k$, where ξ is the rms nonthermal velocity. The theoretical resonantly scattered profiles are very nearly Gaussian with widths slightly smaller than that of a Gaussian curve with a Doppler width $\Delta\lambda_0 = (\lambda/c)\sqrt{2kT_H/m_H}$. This is illustrated in Figure 2 by a comparison of the resonantly scattered profile for $T_H = 10^6$ K and the corresponding Gaussian curve (dashed line). The slight differences between the theoretical line profile and the Gaussian curve are due to the details of the scattering process.

Because of the temperature sensitivity of the profile of the resonantly scattered component of $L\alpha$, it provides a good diagnostic for determining the coronal hydrogen kinetic temperature. A kinetic temperature is defined here to include the effects of thermal motions, nonthermal motions and outflow of coronal plasma into the solar wind. Corrections for the effects of the outflow are discussed later in this section. If the effects of nonthermal motions and outflows in a given measurement are unknown, then the $L\alpha$ profile provides an upper limit to the hydrogen thermal temperature. However, since the thermal velocities of hydrogen atoms are very high (160 km s^{-1} for $T_H = 1.5 \times 10^6$ K) compared to nonthermal velocities measured in the low corona, which are of the order of 30 km s^{-1} (e.g. Feldman and Behring, 1974; Moe and Nicolas, 1977), the width of this $L\alpha$ component is expected to be broadened primarily by thermal motions in the low corona ($r \lesssim 1.3 R_\odot$). At greater distances from Sun-center there are no reliable measurements of nonthermal velocities. However, by measuring widths of spectral lines from heavier ions (e.g. O VI, Mg X, and Fe XII), which are much more sensitive to nonthermal broadening due to their narrow thermal widths, one may be able to place limits on the magnitude of the nonthermal broadening in these higher layers (see Section 6).

The profile of the resonantly scattered component of H I $L\alpha$ also provides a measurement of the kinetic temperature of protons in regions where the coronal expansion time $\tau_{exp} \gg \tau_{H I}$ where $\tau_{H I}$ is the lifetime of hydrogen atoms (see Section 2). For most solar regions this means $T_p \approx T_H$ for $r \lesssim 5$ to $10 R_\odot$ (see Figure 1a). In low density coronal holes the solar wind may reach supersonic velocities at low heights. In that case $T_p \approx T_H$ can be assumed only for $r \lesssim 3 R_\odot$ (see Figure 1b). Because of the weak coupling between protons and electrons, the electron temperature may differ significantly from the hydrogen and proton temperatures as indicated in Section 2.

It is important to recognize the advantages of measuring not only the width of the resonantly scattered component, but also its shape, since the shape of the profile contains information on the velocity distribution of the hydrogen atoms (and protons) along the line of sight. In coronal layers more than a few tenths of a solar radius from the solar surface, the thermalization times for protons (and hydrogen atoms) are long as discussed in Section 2. Hence one might expect that solar wind acceleration processes, wave motions and/or coronal heating processes (in particular any which dissipate nonradiative energy by heating protons) may produce non-Maxwellian hydrogen velocity distributions which are directly measurable from the shape of the resonantly component of the Lα profile.

Figure 3 shows how the width of the profile scattered from a plasma element in the line of sight depends on the angle ψ between the vector from Sun-center to the plasma element and the vector defined by the line of sight. Because of the non-negligible effect of the scattering geometry on the profile (particularly for a bright feature such as a streamer far from the plane of the solar disk) the magnitude of the uncertainty in the

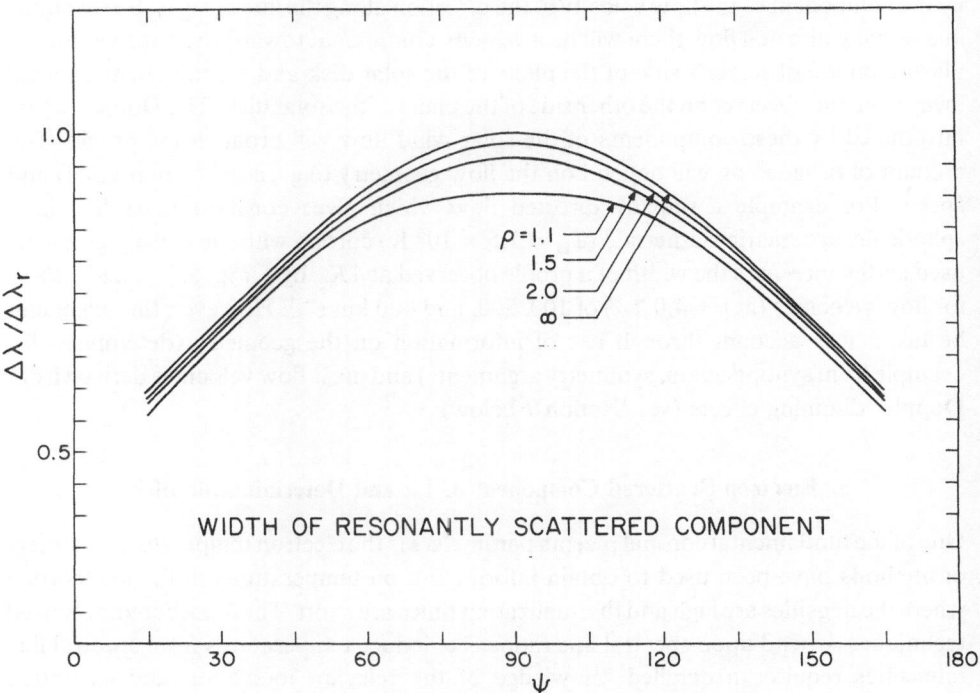

Fig. 3. Width of the resonantly scattered profile from plasma elements at different angular distances ψ where $\psi = 90°$ for a plasma element in the plane of the solar disk. The widths are given in units of the thermal width $\Delta\lambda_r = (\lambda_0/c)(2kT_H/m_H)^{1/2}$.

temperature (and hydrogen velocity distribution) derived from a measurement of the profile of the resonantly scattered component of Lα will depend on the uncertainty in the geometry assumed in analyzing the measurements. The greater the amount of information that is available concerning the geometry, the lower the uncertainty in the derived temperatures and velocity distribution. Observations of the total intensity and polarization of coronal white light radiation can be used to obtain the required geometrical information for streamers (e.g. Newkirk *et al.*, 1970). For stable features such as coronal holes and many streamers, knowledge of the geometry can be derived from synoptic observations (measurements made periodically over several days) where coronal features can be viewed from different angles as solar rotation carries them around the Sun (e.g. Perry and Altschuler, 1972; Wilson, 1977). Alternatively, or in addition, one can make use of information provided by measurements of features observed on the disk (e.g. He I 10830 observations of coronal holes) or geometries inferred from coronal magnetic field configurations calculated from photospheric magnetograms. For typical observations the uncertainty in T_p resulting from line-of-sight effects should be 10% or less.

There is one additional factor that can affect the profile of the resonantly scattered component of Lα. If one is observing radiation from a region with high mass flow velocities resulting from the outward flow of the solar wind, the profiles can be broadened by the component of the flow velocity in the direction along the line of sight. For example in a radially directed flow there will be a velocity component toward the observer for the plasma on the observer's side of the plane of the solar disk and a component directed away from the observer on the other side of the plane of the solar disk. The Doppler shifts introduced by these components of the solar wind flow will broaden the profile. The amount of broadening will depend on the flow geometry (e.g. radial or nonradial) and speed. For example a radially directed flow which gives constant mass flux in a spherically symmetric isothermal ($T_H = 1.5 \times 10^6$ K) corona with the density gradients used earlier increases the width of a profile observed at $4 R_\odot$ by 1.7%, 5.9%, and 12.8% for flow velocities (at $r = 4.0 R_\odot$) of 100, 200, and 400 km s^{-1}. However, this effect can be taken into account through use of information on the geometry (determined for example from synoptic data, symmetry arguments) and mass flow velocities derived from Doppler dimming effects (see Section 7 below).

5. Electron Scattered Component of Lα and Determination of T_e

One of the fundamental coronal plasma parameters is the electron temperature. A variety of methods have been used to obtain information on temperatures in the low corona where the densities are high and thermalization times are short. The most commonly used techniques depend upon spectral line ratios. To deduce temperatures from spectral line intensities requires a detailed knowledge of the relevant ionization and excitation processes and in some cases chemical abundances. Ideally, these methods utilize observations of several spectral lines and a large body of atomic data. Such methods are more difficult to apply in the tenuous outer corona; both because of the limited number of observable lines and the uncertainties associated with the long thermalization times. For these reasons, indirect determination of electron temperatures from intensities should not be relied upon, exclusively, for electron temperatures is the solar wind acceleration region, although they do provide some information. Widths of spectral lines from heavy ions are also unreliable indicators of electron temperatures in the solar wind acceleration region due to (1) their sensitivity to broadening by nonthermal or turbulent motions and (2) the reduced collisional coupling between ions and electrons in this atmospheric region which can lead to large differences between electron and ion kinetic temperatures. Temperatures based on radio measurements are affected by uncertainties due to the possible presence of nonthermal sources, optical depth effects, the dispersive, inhomogeneous nature of the coronal plasma, and uncertainties in instrumental calibration. Electron temperatures deduced from electron density gradients are unreliable because of the effects of flows and fact that the electron density gradient depends on the sum of the electron and proton temperatures whose relative values are unknown.

Cram (1976) has advocated use of the shape of the electron scattered photospheric spectrum near 4000 Å as an electron temperature diagnostic. However, the requirements on photospheric precision are extremely tight. The uncertainty in coronal temperature corresponding to $\pm 1\%$ accuracy in the ratio of the intensities at 4100 and 3900 Å is $\pm 0.2 \times 10^6$ K. The problem of rejecting stray light in the visible and separating the contributions of the K-corona (electron scattered corona) and F-corona (dust scattered corona) pose additional difficulties. Corrections for scattering geometry for this method, which are similar to those required for temperature measurements based on electron scattered H I Lα, are discussed later in this section. Given the above considerations it is useful to consider a means of directly measuring coronal electron temperatures, a method that depends upon measurements of the electron scattered component of H I Lα. This means of measuring coronal temperatures was first suggested by Hughes (1965).

The electron scattered component of the coronal Lα radiation is produced by Thomson scattering of chromospheric Lα emission. The problem of determining the scattering of monochromatic radiation by coronal electrons has been addressed by van Houten (1950) and others concerned with the scattering of the photospheric white light spectrum. Consider the scattering of a beam of chromospheric Lα radiation with intensity $I(\lambda', \omega)$, wavelength between λ' and $\lambda' + d\lambda'$ and angular direction between ω and $\omega + d\omega$. The number of incident Lα photons $cm^{-3}\,s^{-1}$ scattered by coronal electrons with velocities between v_e and $v_e + dv_e$ is

$$dN = N_e(v_e)\sigma I(\lambda', \omega)\,d\lambda'\,d\omega\,dv_e\,, \tag{7}$$

where c is the Thomson scattering cross-section. The fraction of these photons scattered toward an observer is

$$dN(\lambda) = \frac{dN}{4\pi}\frac{3}{4}(1 + (\mathbf{n}\cdot\mathbf{n}')^2)\delta\left[\left(\lambda' - \frac{\lambda'}{c}\,\mathbf{v}_e\cdot\mathbf{n}'\right) - \left(\lambda - \frac{\lambda}{c}\,\mathbf{v}_e\cdot\mathbf{n}\right)\right], \tag{8}$$

where $3((1 + (\mathbf{n} \cdot \mathbf{n}')^2)/4$ is the angular dependence of Thomson scattering and $\mathbf{n}'$ and $\mathbf{n}$ are the directions of the incident and scattered photons. Since these photons must have the same wavelength in the rest frame of the electrons, the Dirac delta function specifies which incident photons will be scattered at wavelength λ measured by the observer. Hence the intensity (photons $cm^{-2} s^{-1} sterad^{-1}$) of the electron scattered component of Lα is given by

$$I_e(\lambda) = \frac{3\sigma}{16\pi} \int_0^\infty N_e \, dx \int_\omega^\infty d\omega (1 + (\mathbf{n}' \cdot \mathbf{n})^2) \int_{-\infty}^\infty I(\lambda', \omega) \, d\lambda' \times$$

$$\times \int_{-\infty}^\infty f(\mathbf{v}_e) \delta\left[\left(\lambda' - \frac{\lambda'}{c} \mathbf{v}_e \cdot \mathbf{n}'\right) - \left(\lambda - \frac{\lambda}{c} \mathbf{v}_e \cdot \mathbf{n}\right)\right] d\mathbf{v}_e, \qquad (9)$$

where x is the distance from the observer along the line of sight. Because of the high thermal velocity of electrons at coronal temperatures, nearly 7000 km s^{-1} for $T_e = 1.5 \times 10^6$ K, we will neglect the effects of solar wind flows and turbulence on the

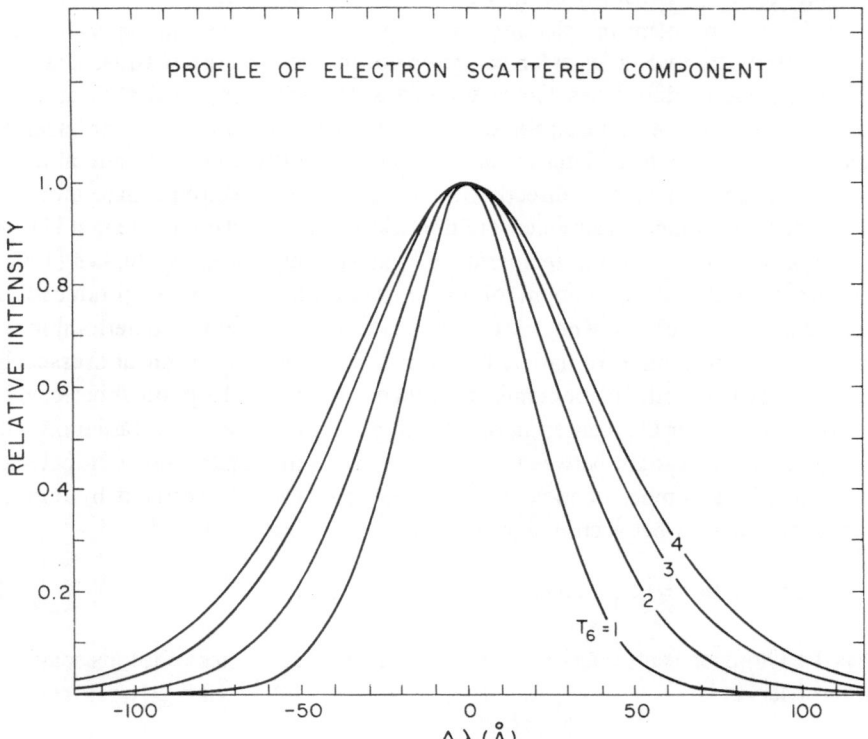

Fig. 4.   Profiles of the electron scattered component of Lα calculated with a spherically symmetric isothermal corona for $\rho = 3$. Profiles for 4 coronal temperatures (in units of 10^6 K) are shown.

electron velocity distribution. If one further assumes that the coronal electrons have a Maxwellian velocity distribution, one can integrate over $\mathbf{v}_e$ analytically and obtain an expression for the scattered radiation which can be numerically integrated over x, ω, and λ'.

Figure 4 contains plots of profiles for the electron scattered component of coronal Lα radiation computed using Equation (9) for a spherically symmetric isothermal corona with the same radial density distribution used above for calculating resonantly scattered profiles. These profiles show the sensitivity of the width of the electron scattered component to the magnitude of the coronal electron temperature. Due to the low mass of the scattering particles, the width of this component is much larger than the resonantly scattered component of Lα. The sensitivity of the profile of the electron scattered component to variations in the coronal electron temperature means that it can provide a method for spectroscopic determination of this critical coronal parameter. It is particularly important to note that not only can one obtain information on the electron temperature, but also information on the velocity distribution of the electrons along the line of sight (which determines the shape of the profile). As shown in Figure 1 electron thermalization times in the corona increase rapidly with height. Consequently, solar wind acceleration processes or nonthermal processes which heat the coronal electrons may produce departures from a Maxwellian velocity distribution that are detectable as distortions in the shape of the electron scattered Lα profile.

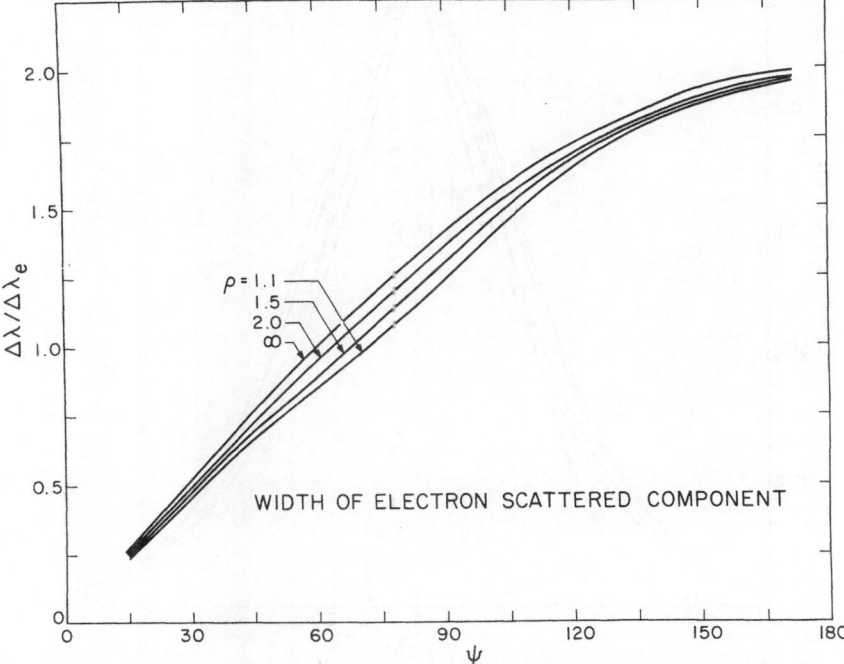

Fig. 5.   Width of the electron scattered profile for plasma elements at different angular distances ψ where $\psi = 90°$ for a plasma element in the plane of the solar disk. The widths are given in units of the thermal width
$$\Delta\lambda_e = (\lambda_0/c)(2kT_e/m_e)^{1/2}.$$

It is important to note that the shape of the electron scattered component of Lα depends on the scattering geometry and is not simply a Gaussian with a Doppler width $\Delta\lambda_0 = (\lambda_0/c)\sqrt{2kT_e/m_e}$. The width $\Delta\lambda_e$ of the scattered radiation from a coronal plasma element depends on the angle ψ between the vector from Sun-center to the plasma element and the vector defined by the line of sight. One can show that $\Delta\lambda_e \approx 2\Delta\lambda_0 \sin(\psi/2)$ (van Houten, 1950). For a plasma element in the plane of the solar disk ($\psi = 90°$) the profile is Gaussian with a characteristic width that is a factor of $\sqrt{2}$ larger than the Doppler width. For plasma elements on the observer's side of the plane of the disk the width $\Delta\lambda_e$ decreases (as $\sin \psi/2$) with increasing angular distance (measured from Sun-center) from the plane of the disk, while on the opposite side the width increases with increasing angular distance. This is illustrated in Figure 5 which presents curves showing the variation of the width of profiles from plasma elements at different angular distances ψ from the plane of the disk. Curves (calculated with Equation (9)) are plotted for several values of ρ where ρ is the distance in solar radii from

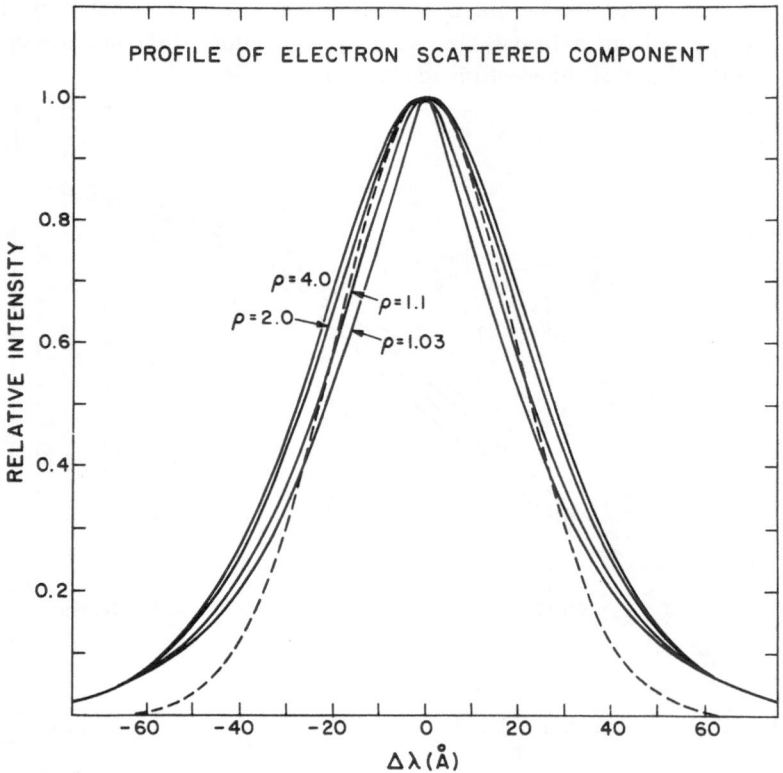

Fig. 6.   Profiles (solid lines) of the electron scattered component of Lα calculated for several values of ρ (units of $R_\odot$) using an isothermal coronal model with $T_e = 1.5 \times 10^6$ K. For comparison a pure Gaussian curve with a width corresponding to the thermal width is also plotted (dashed line).

Sun-center to the point where the observer's line-of-sight intersects the plane of the solar disk. For large values of ρ where the angular size of the solar disk is small (as viewed from the corona), these curves approach the value $\Delta\lambda_e = 2\Delta\lambda_0 \sin(\psi/2)$.

Figure 6 illustrates the effect of the above ψ dependence on the shape of the profile of the electron scattered Lα line. Theoretical profiles for an isothermal corona with $T_e = 1.5 \times 10^6$ K are given for several values of ρ. The profiles have been normalized to have the same central intensity. The dashed line is a Gaussian curve with a width corresponding to the thermal width $\Delta\lambda = 27.3$ Å $= (\lambda_0/c)\sqrt{2kT_e/m_e}$. The change in shape of the profile with ρ is caused by the decrease (with increasing ρ) of the angular size of the solar disk (as viewed from the coronal plasma elements scattering the radiation) as well as by the decreasing density gradient which determines the distance along the line of sight over which the coronal plasma makes a significant contribution to the total scattered radiation. Note that for $\rho \gtrsim 1.1$, the theoretical profiles are broader (FWHM) than the Gaussian curve. For large values of ρ ($\gtrsim 5R_\odot$) the width of the scattered line is a factor of $\sqrt{2}$ larger than the width of a Gaussian profile with a thermal width $\Delta\lambda_0 = (\lambda_0/c)\sqrt{2kT_e/m_e}$. The L$\alpha$ profiles calculated by Hughes (1965) appear to be too narrow by a factor of approximately 1.4, perhaps due to the neglect of this factor of $\sqrt{2}$.

Since the intensity of the scattered radiation measured at a given position is an integral over the contributions of plasma elements distributed along the line of sight, one cannot simply fit a Gaussian curve to a measured profile and determine the electron temperature from the width. One must account for the geometry of the features in the line of sight (see Section 4). This can be accomplished by modeling techniques. For typical observations the uncertainty in T_e resulting from uncertainties in the geometry should be less than 25%. This source of uncertainty occurs for any electron scattered spectral feature whose shape is used to determine electron temperatures.

6. Determination of T_i from Spectral Line Profiles of Ions

Although much information can be derived from the H I Lα line, a much more complete description is obtained when other UV and EUV spectral lines are also observed. There are a number of UV and EUV ($\lambda > 500$ Å) spectral lines (e.g. from C IV, N V, O VI, Ne VIII, Mg X, Si XII, Fe XII) that should be sufficiently intense to be measurable at distances of $r = 3$ to $5R_\odot$ and perhaps further from Sun-center with a suitably designed coronagraphic instrument (Kohl et al., 1981; Kohl and Withbroe, 1982). Also of interest are strong XUV ($150 < \lambda < 500$ Å) lines, particularly He II $\lambda 304$ (see Ahmad, 1977) which provides information on alpha particles and lines from numerous ionization stages of iron, Fe IX to Fe XVI, which could be used to monitor the ionization states of iron in the solar wind near the Sun and probe the atmosphere using ions with different charges, but the same mass.

For most of these EUV and XUV lines the emergent intensity is a combination of collisionally excited and resonantly scattered components. The collisionally excited component has an intensity (cf. Withbroe, 1970):

$$I_c(\lambda) = 0.86\,\lambda \int_{-\infty}^{\infty} A_{el}R_i N_e^2 C_{12}\phi_\lambda \, dx, \tag{10}$$

where $A_{el} = N_{el}/N_p$, N_{el} is the number of ions cm^{-3} summed over all stages of ionization of the atomic species producing the ion i, C_{12} is the collisional excitation rate coefficient, $R_i = N_i/N_{el}$ is the ionization balance term, N_i is the number of ions cm^{-3} of the ion species producing the line and ϕ_λ is the profile function (e.g. Gaussian for a Maxwellian particle velocity distribution function). The intensity of the resonantly scattered component is given by an equation identical to that for H I Lα (Equation (6)) except for the inclusion of the abundance A_{el} and use of R_i appropriate to the ion species involved. There may also be differences in the coefficients in the angular scattering function (see Equation (4)). For most ions collisional excitation dominates in the low corona, while at higher levels both radiative and collisional excitation are important (e.g. Figure 8a). The electron scattered component of these lines will be weak and blend into the 'continuum' produced by electron scattered disk radiation from the numerous weak lines in the EUV and XUV, the H I Lα continuum, He I continuum, etc.

The profile of the resonantly scattered component will be slightly narrower than that of the collisionally excited component due to the angular dependence of the scattering process (see Section 4), but the difference is very small and easily modeled (cf. Kohl and Withbroe, 1982). The relative contributions to the total intensity of the resonantly scattered and collisionally excited components can be evaluated using Equation (10) and the corresponding equation for the resonantly scattered component. Alternatively, for many lines one can use an empirical technique for separating the contributions of the two components as will be discussed in Section 7. The widths of EUV coronal lines are typically of the order of 0.05 to 0.1 Å for $T_i \approx 10^6$ K.

Spectral line profiles from ions can provide measurements of kinetic temperatures T_i of atomic species heavier than hydrogen. Measured in combination with the resonantly and electron scattered components of H I Lα, the spectral lines of heavier ions can provide critical data on mass-dependent or charge-to-mass dependent processes in the solar wind acceleration region.

7. Outflow Velocities

The intensity of a coronal resonance line depends on the number of particles in the line of sight capable of scattering radiation in the line and the intensity of the incoming radiation from the solar disk and lower levels of the corona. The number of scatterings is a function of the outflow velocity of the solar plasma. In a static atmosphere, the central wavelength of the coronal scattering profile is identical to that of the disk profile. However, in a region with solar wind flow the scattering profile is Doppler-shifted with respect to the disk profile, hence there is less efficient scattering resulting in a reduction in intensity of the scattered radiation. This effect is known as Doppler dimming (see

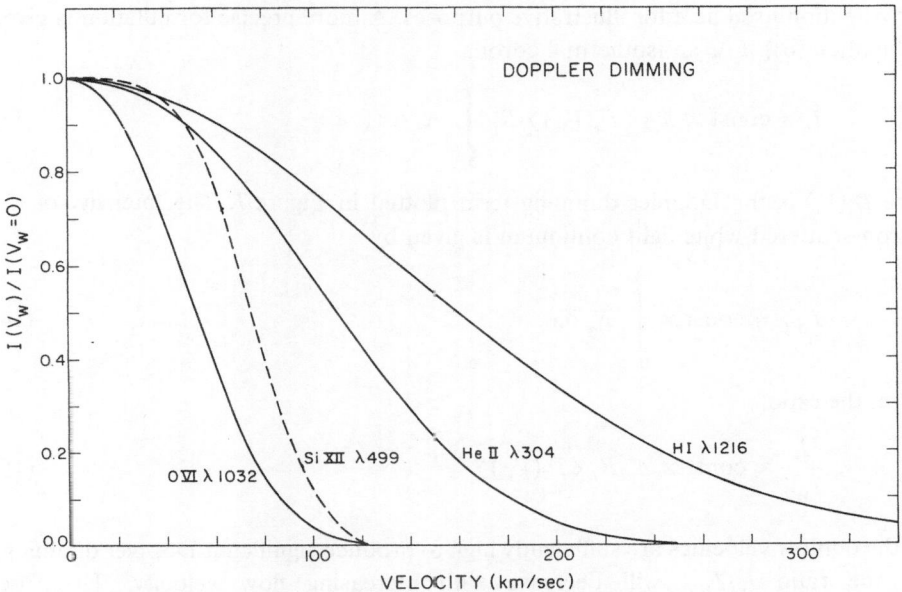

Fig. 7. Doppler dimming calculated for an isothermal corona with $T = 1.5 \times 10^6$ K (see text).

Hyder and Lites, 1970; Beckers and Chapman, 1973). Examples of the effect of Doppler dimming for several lines are illustrated in Figure 7. We see that the hydrogen Lα line is sensitive to flow velocities greater than about 100 km s^{-1}, while the other lines are sensitive to velocities above 30 to 60 km s^{-1} depending on the line.

How does one determine the amount of Doppler dimming? There are a variety of techniques for doing this (e.g. Kohl and Withbroe, 1982). These methods depend on comparing the intensity of a spectral feature that is affected by Doppler-dimming with the intensity of a spectral feature that is not. For example, a method suggested by G. Noci (personal communication) for hydrogen Lα makes use of the ratio of the intensity of the resonantly scattered component of a spectral line I_r and the intensity of the electron-scattered white light continuum. The intensity of the resonantly scattered component is given by (see Equation (6))

$$I_r = \text{const} \times \int_0^\infty A_{el} R_i N_e \bar{J} \, dx \tag{11}$$

where

$$\bar{J} = \int_{-\infty}^\infty J_\lambda \phi(\lambda - \lambda_w) \, d\lambda, \tag{12}$$

J_λ is the intensity of the disk radiation at wavelength λ, ϕ is the normalized absorption profile and λ_w is the Doppler shift introduced by the solar wind. (Equation (12) is an approximation used here for illustrative purposes. A more precise formulation is given by Equation (6)). For an isothermal corona

$$I_r = \text{const} \times A_{el} \langle D_i(V_w) \rangle R_i \int_0^\infty N_e \, dx \,, \tag{13}$$

where $D_i(V_i)$ is the Doppler dimming term plotted in Figure 7. The intensity of the electron-scattered white light continuum is given by

$$I_{WL} = \text{const} \times \int_0^\infty N_e \, dx \,. \tag{14}$$

Hence, the ratio

$$\frac{I_r}{I_{WL}} \approx \text{const} \times A_{el} R_i \langle D_i(V_w) \rangle \,. \tag{15}$$

If the outflow velocities are sufficiently high to produce significant Doppler dimming, then the ratio I_r/I_{WL} will decrease with increasing flow velocity. Therefore, measurements of the ratio I_r/I_{WL} as a function of radius can be used to determine the amount of Doppler-dimming and, hence, the bulk outflow velocity of the observed ion. A first order estimate of the outflow velocity can be obtained by assuming (1) that the ionization balance term R_i is constant (which occurs if the corona is isothermal or the ionization balance is frozen in) and (2) that the elemental abundance A_{el} is constant (for hydrogen $A_{el} = 1$). In this case application of Equation (15) to measurements at several heights will provide relative outflow velocities. A more definitive analysis requires a self-consistent model of the observed region which makes use of temperatures, densities and geometry derived from the observations and calculated values of R_i. The requirement that the outward particle flux be conserved provides an additional constraint.

It is highly desirable to use redundant checks on outflow velocities. For example, in place I_{WL} one can use the intensity of the electron scattered component of Lα which also depends on $\int N_e \, ds$. This is particularly useful in the case of Doppler dimming determined from Lα observations, since the ratio of the intensities of the resonantly and electron scattered components of Lα is independent of the absolute calibration of the instrument used to make the measurements. Another check is provided by the ratio of the intensities of the resonantly scattered (I_r) and collisionally excited (I_c) components of spectral lines from ions. This ratio is sensitive to Doppler dimming. One can easily show that

$$\frac{I_r}{I_c} \approx \text{const} \times \langle D_i(V_w) \rangle \, \frac{\displaystyle\int_0^\infty N_e \, dx}{\displaystyle\int_0^\infty N_e^2 \, dx} \tag{16}$$

which is independent of A_{el} and R_i. The ratio I_r/I_c can be determined empirically for spectral lines in the lithium (e.g. N v, O vi, Ne viii, Na ix, Mg x, Al xi, and Si xii) and sodium (e.g. Fe xvi) isoelectronic sequences, which account for the majority of the strong coronal resonance lines in the UV and EUV. The ions in these isoelectronic sequences produce a pair of resonance lines whose intensity ratio $I_{\lambda 1}/I_{\lambda 2}$ is a function of I_r/I_c (see Mariska, 1977; Kohl and Withbroe, 1982). (The resonantly scattered components of the two lines (e.g. Mg x $\lambda 610$, $\lambda 625$) have a $4:1$ ratio, while the collisionally excited components have a $2:1$ ratio.)

The Doppler-dimming techniques described above provide a means of obtaining information on outward flow velocities as a function of height above the solar limb. Doppler shifts of spectral lines provide information about flows directed toward or away from the observer. Flows that are symmetrical about the plane of the solar disk (e.g. spherically symmetric flows) can introduce line broadening in optically thin spectral lines measured above the limb, while unsymmetrical flows (e.g. plasma flowing outward in a bright streamer not in the plane of the disk) can produce Doppler shifts of spectral line profiles. The use of Doppler shifts for measuring flow velocities is particularly useful for observations of the low corona made using UV and EUV emission lines which can be observed on the disk. Such measurements can be used to map flow patterns in the transition region and lowest levels of the corona (cf. Cushman and Rense, 1976; Brueckner, 1980, Rottman *et al.*, 1981, 1982).

8. Charge States and Chemical Abundances

As discussed in Section 2, the ionization state of the solar wind is 'frozen in' within a few solar radii of the solar surface due to the rapid decline of density in the corona. Consequently, *in situ* measurements of charge states in the solar wind plasma at large distances from the Sun can provide information about the ionization balance in the solar wind acceleration region. However, due to the uncertainties in the physical conditions (temperatures, densities, flow velocities and geometry) in the acceleration region, the height at which 'freezing in' occurs cannot be determined reliably from these *in situ* solar wind measurements. An additional difficulty is that some charge states e.g. O viii, may be affected significantly by nonthermal tails in the electron velocity distribution function (e.g. Owocki, 1982; Owocki and Scudder, 1982). Optical observations can provide critical data on physical conditions in the solar wind acceleration region that are needed for interpreting charge state measurements; the combination of the two types of measurements (spectroscopic and charge state) provide complementary data that can be used to probe the physics of this important region of the heliosphere.

Determinations of the chemical abundances in the solar wind plasma at 1 AU indicate that the abundances of elements heavier than hydrogen are extremely variable, by more than a factor of 4 (see review by Hirshberg, 1975). These large variations appear to be associated with dynamical phenomena in the solar corona and the types of coronal structures where the wind originates. Thus, measurements of abundances using spectroscopic information acquired from optical measurements of radiation from the solar wind acceleration region are highly desirable in order to provide information on the magnitude and location of coronal abundance variations and thereby place constraints on the mechanisms producing these variations.

Measurements of the intensities of spectral lines can be used to obtain the desired information on chemical abundances and charge states. For example, for an isothermal corona Equation (6) yields for the ratio of the intensity of the resonantly scattered component of a spectral line at wavelength λ to that of hydrogen Lα:

$$\frac{I_r(\lambda)}{I_r(\lambda\,1216)} = \text{const} \times \frac{A_{el}R_i D_i(V_w)}{R_{\text{H I}}D_{\text{H I}}(V'_w)} \,. \tag{17}$$

The Doppler-dimming terms can be evaluated as described above in Section 7. The electron temperature can be measured using the profile of the electron-scattered component of H I Lα, while the ionization balance term R_i can be calculated from empirically derived knowledge of T_e, N_e and the outflow velocity. Thus, one can obtain a determination of A_{el}. In some cases one may be able to measure several ions from the same parent element and thereby check the calculated ionization balance.

Clearly, it is highly desirable to acquire spectroscopic measurements in the solar wind plasma that later is sampled *in situ* by a spacecraft. For example, coronagraphic instruments in Earth orbit could observe radiation from the coronal plasma in a polar region whose plasma outflow is being studied by the International Solar Polar Mission spacecraft.

9. Observational Considerations

Figure 8a gives predicted intensities as a function of radius for H I Lα and Mg x λ610. These intensities were calculated using Saito's (1970) density model. For simplicity, the atmosphere was assumed to be isothermal with a temperature $T = T_e = T_p = T_i = 2 \times 10^6$ K. Intensities were calculated for a static model and one with solar wind outflow where the particle flux $nVr^2 = $ constant with the constant corresponding to a proton flux at 1 AU equal to 3.8×10^8 cm^{-2} s^{-1} (cf. Feldman *et al.*, 1977). Intensities in streamers could be a factor of 5 to 50 times higher than those derived from Saito's model because of the higher densities in these features. In coronal holes the intensities could be a factor of 2 or more lower depending upon the spectral line, density in the hole and the outflow velocities (e.g. Kohl and Withbroe, 1982). We will discuss Lα further in Section 10 when we consider applications to existing observations.

With a UV/EUV coronagraph such as the one defined for Spacelab (Kohl *et al.*, 1981) an integrated intensity of 5×10^6 photons cm^{-2} s^{-1} str^{-1} yields a count rate of a 1 Hz or more (for a spatial resolution of $0.15' \times 4'$ and spectral resolution 0.1 Å) which is adequate for observing many spectral lines out to 3 to $5R_\odot$ and even further for the strongest lines, particularly Lα (Kohl and Withbroe, 1982). For measurements of the electron scattered H I Lα an integrated intensity of 10^6 photons cm^{-2} s^{-1} sterad^{-1} yields a count rate of 1 Hz (for $4' \times 4'$ spatial resolution, 10 Å spectral resolution). This is adequate for line profile measurements out to $4R_\odot$ in equatorial regions.

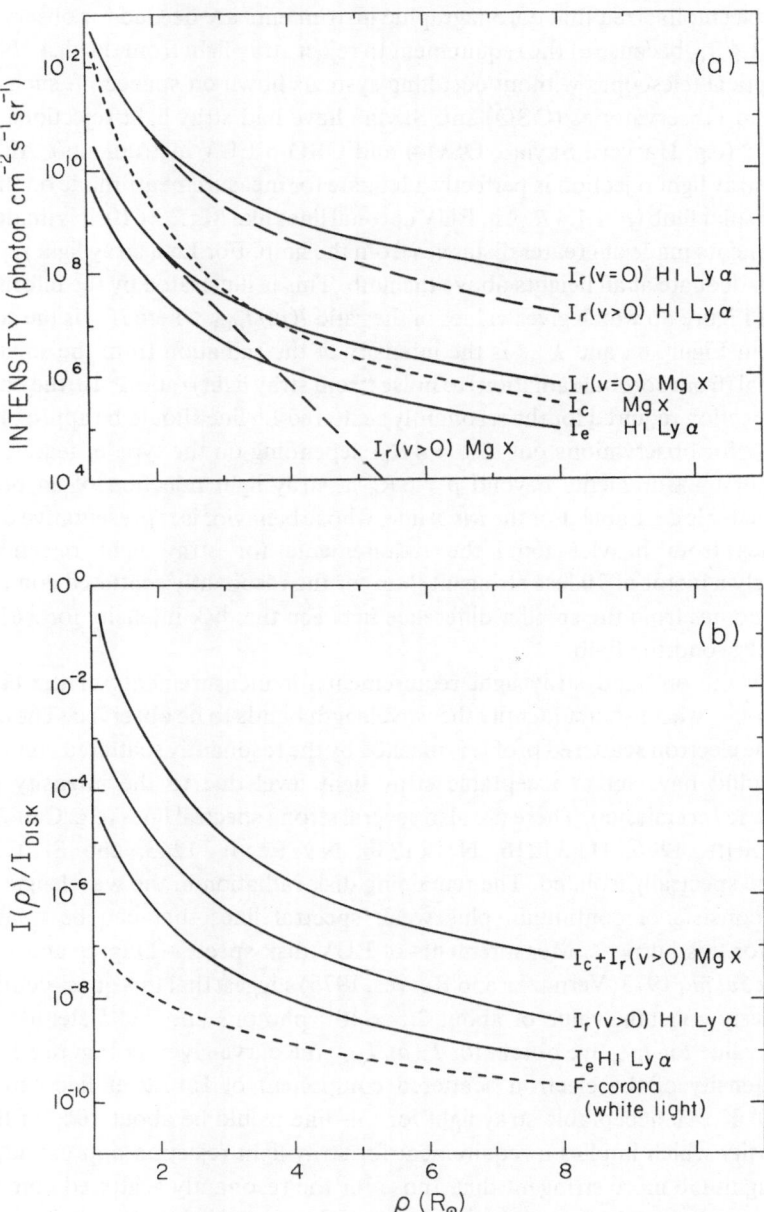

Fig. 8. (a) Intensity as a function of distance ρ from Sun center calculated using a model for typical equatorial region. Curves are given for the resonantly scattered I_r and electron scattered I_e components of H I Lα and the resonantly scattered and collisionally excited I_c components of Mg $\lambda 610$. Intensities $I_r (V = 0)$ were calculated for a static model and $I_r (V > 0)$ were calculated for a model with solar wind flow (see text). (b) Intensity in units of the disk intensity as a function of ρ for a model of typical equatorial region with solar wind flow. The curve for electron scattered H I Lα is described in the text. For comparison the intensity of the white light F-corona is also given. In the UV and EUV the F-corona may be less intense (in units of the disk brightness) by a factor of 2 to 3 (see text).

It should be emphasized that coronagraphic instruments are needed for observations beyond $\rho \gtrsim 1.4 R_\odot$ because of the requirement to reject stray light from the disk. Normal incidence optical telescopes without occulting systems flown on spacecraft such as the Orbiting Solar Observatories (OSO) and Skylab have had stray light rejections of the order of 10^{-3} (e.g. Harvard Skylab, OSO-4 and OSO-6 EUV instruments). Although this level of stray light rejection is perfectly adequate for measurements made on the disk and near the solar limb ($\rho \lesssim 1.4 R_\odot$) in EUV coronal lines like Mg x $\lambda 610$, it is inadequate for measurements made at greater distances from the limb. For Lα a stray light rejection of 10^{-3} is inadequate at all heights above the limb. This is illustrated by the information presented in Figure 8b which gives values of the ratio $I(\rho)/I_{\mathrm{disk}}$ where $I(\rho)$ is the coronal intensity from Figure 8a and I_{disk} is the intensity of the radiation from the solar disk.

For a signal (from coronal radiation) to noise (from stray light) ratio $\gtrsim 10$, the on-band stray light rejection required for the resonantly scattered Lα line should be approximately 10^{-7} to 10^{-8} for observations out to $\rho \approx 4 R_\odot$ depending on the type of feature being observed. For measurements beyond $\rho \approx 6 R_\odot$ a stray light rejection of an order of magnitude better is desirable. For the Mg x line, whose behavior is representative of EUV spectral lines from heavier ions, the requirements for stray light rejection are approximately a factor of 10 less stringent than for the resonantly scattered component of Lα. This comes from the smaller difference between the disk intensity for $\lambda 610$ and its intensity beyond the limb.

To specify the on-band stray light requirement for measurements of the electron scattered profile, we must first identify the wavelength bands to be observed. The central 4 to 5 Å of the electron scattered profile is masked by the resonantly scattered component and also would have an unacceptable stray light level due to the intensity of the chromospheric Lα emission. There are also several strong spectral lines (e.g. C III $\lambda 1175$, N I $\lambda 1200$, Si III $\lambda 1206$, H I $\lambda 1216$, N v $\lambda 1238$, N v/Fe xII $\lambda 1243$, and Si III $\lambda 1264$) which can be spectrally isolated. The remaining disk radiation in the wavelength range of interest consists of continuum plus weak spectral lines that can be treated as continuum for this purpose. Measurements of EUV disk spectra (Dupree and Reeves, 1971; Dupree et al., 1973; Vernazza and Reeves, 1978) suggest that this quasi-continuum has a relatively constant value of about 2.5×10^{12} photons cm^{-2} s^{-1} sterad^{-1} Å^{-1}. Taking this value for I_{disk} we obtain for $I_e(\rho)/I_{\mathrm{disk}}$ the curve given in Figure 8b, where I_e is the intensity of the electron scattered component of H I Lα at line center for $T_e \approx 2 \times 10^6$ K. An acceptable stray light for this line would be about 10% of the line center intensity which implies a requirement for stray light rejection approximately an order of magnitude more stringent than those for the resonantly scattered component (but an order of magnitude less stringent than for white light observations of the K-corona). If the instrumental stray light contribution is known so that it can be subtracted from the observations, then a higher level of stray light can be tolerated. One possible technique for doing this is through measurements of the intensity at the wavelengths of strong lines such as Si III $\lambda 1206$ and C II $\lambda 1335$ whose scattered coronal radiation is expected to be much lower than that of EUV coronal lines and H I Lα due to the small abundance at coronal temperatures of the scattering ions. One can also make broad-band measurements at wavelengths outside the spectral region where the coronal lines contribute significantly to the intensity.

A capability for suppression of off-band photospheric radiation from the visible and near UV is also required. This is extremely critical because of the large intensity of the photospheric radiation at these wavelengths. For observations at $\rho \lesssim 3.5 R_\odot$ the necessary suppression of on-band and off-band radiation has been demonstrated by a rocket UV coronagraph which uses an externally and internally occulted telescope together with a scanning spectrometer and solar blind photoelectric detector (Kohl *et al.*, 1978, 1980).

Interpretation of measurements of Lα coronal radiation must also take into account the effects of geocoronal absorption (scattering) and emission ($I \approx 10^9$ photon cm^{-2} s^{-1} sterad^{-1} near solar maximum). Neutral hydrogen in the geocorona scatters photons out of the central portion of the coronal Lα profile, while Lα emission from the geocorona contributes photons near line center. Close to the Sun, where the coronal

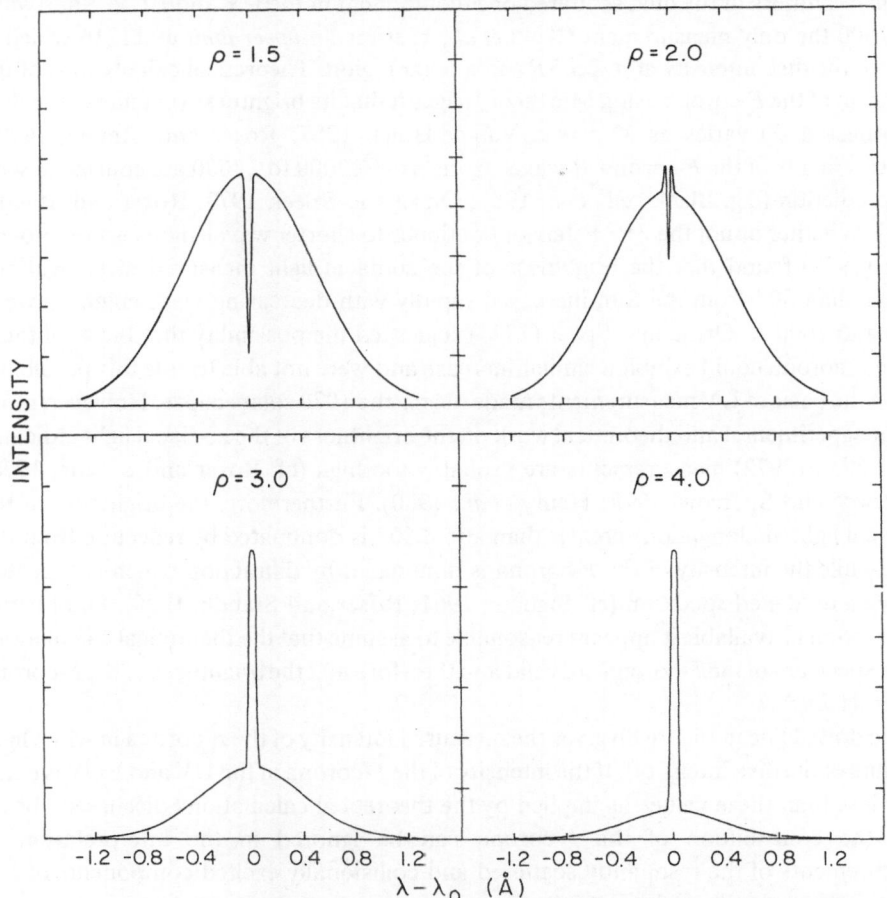

Fig. 9. Calculated geocoronal absorption/emission contribution to profiles of H I Lα from a typical equatorial region.

intensities are large, the effects of the geocoronal absorption are more important than those due to the geocoronal emission; at distances beyond about $2R_\odot$ the effects of the geocoronal emission are more important. This is illustrated in Figure 9 for a set of profiles with intensities typical of equatorial regions. Because the geocoronal absorption/emission feature is much is much narrower than the profile of the solar line, the effects of the geocorona can be accounted for through use of measurements made with sufficiently high spectral resolution (≈ 0.1 Å), a value readily attainable. There is another narrow emission component of much lower intensity ($I \approx 4 \times 10^7$ photon cm^{-2} s^{-1} sterad^{-1}), the interplanetary Lα emission component whose contribution is negligible for measurements made at $\rho < 3.5 R_\odot$. At greater distances where its contribution may be non-negligible, its effects can be eliminated through use of measurements made with good spectral resolution (≈ 0.1 Å).

Scattering by dust in the solar corona and interplanetary space also contributes to the coronal emission spectrum. At the present time there is very little empirical information on the brightness of the dust scattered or F-component in the UV and EUV. Shortward of $\lambda 1500$ the only measurement (Kohl et al., 1980) is an upper limit at $\lambda 1216$ which is 10^{-7} of the disk intensity at $\rho = 3.5 R_\odot$ in a polar region. Theoretical calculations of the spectrum of the F-corona using Mie theory suggest that its brightness (in units of the disk brightness at λ) varies as $\lambda^{1/2}$ (e.g. Van de Hulst, 1957; Roser and Staude, 1978). Measurements of the F-corona at wavelengths from $\lambda 2000$ to $\lambda 8000$ are consistent with this prediction (e.g. Blackwell et al., 1967; Orrall and Speer, 1973; Roser and Staude, 1978). Whether or not the $\lambda^{1/2}$ behavior continues to shorter wavelengths is not known. Lillie (1972) found that the brightness of the zodiacal light measured at elongations greater than 50° from the Sun increased rapidly with decreasing wavelength between 2000 and 1600 Å. Orrall and Speer (1973) suggested the possibility that the brightness of the F-corona could exhibit a similar increase and were not able to rule this possibility out on the basis of UV measurements made during the 1970 solar eclipse. However, more recent experimental and theoretical work on the brightness of the zodiacal light indicates that Lillie's (1972) measurements are probably too high (cf. Roser and Staude, 1978; Weinberg and Sparrow, 1978; Henry et al., 1980). Furthermore, the brightness of the zodiacal light at elongations greater than about 50° is dominated by reflection from the dust, while the intensity of the F-corona is dominated by diffraction which is expected to give a reddened spectrum (cf. Ingham, 1961; Roser and Staude, 1978). Until better information is available it appears reasonable to assume that the theoretical calculations of the spectrum of the F-corona are valid and therefore that the brightness of the F-corona varies as $\lambda^{1/2}$.

The dotted line in Figure 8b gives the measured intensity of the F-corona in white light (in units of the disk intensity). If the intensity of the F-corona in the UV and EUV is equal to or less than these values as implied by the theoretical calculations discussed above, then the contribution of the F-corona can be ignored in the interpretation of measurements of the resonantly scattered and collisionally excited components of UV and EUV spectral lines for $\rho \lesssim 8 R_\odot$.

There are two components of the F-coronal radiation that need to be considered in the interpretation of observations of the electron scattered Lα radiation. The first of these is the dust scattered Lα disk radiation. This component, which has a FWHM ≈ 0.5 Å, can be easily separated from the electron scattered radiation due to the two order of magnitude difference in the line widths. (The same argument applies to separation of the resonantly scattered radiation, which as indicated above, is expected to have an intensity much larger than that of the F-corona.) The on-band contribution from the F-corona poses more of a problem. This radiation is due to the dust scattering of the UV 'continuum' near Lα. If the intensity of the F-corona at wavelength λ (in units of the disk intensity at λ) varies as $\lambda^{1/2}$, then at wavelengths near $\lambda 1216$ this component will have an intensity approximately a factor of 2 smaller than the values given by the dotted curve in Figure 8b. If this is the case, then the contribution of the F-corona can be ignored in the interpretation of observations of the electron scattered in the low corona at heights corresponding to $\rho < 2R_\odot$. For observations made at greater distances the contribution of the F-corona should be taken into account. In equatorial regions the intensity of the F-corona is expected to be a factor of 5 to 10 smaller than the total intensity of the electron scattered component of Lα at $\rho = 2.5R_\odot$ and not reach equal intensity until $\rho \gtrsim 8R_\odot$. This is marked contrast to the situation in the visible where the intensity of the white light F-corona equals that of the electron scattered or K-corona in equatorial regions for $\rho \approx 2.5R_\odot$ and is several times larger by $\rho \approx 5R_\odot$.

Because of the lack of experimental verification that the brightness of the F-corona in the UV is as predicted by theory, improved measurements are needed. This is particularly important for achieving reliable measurements of electron scattered Lα radiation from the corona. This can be accomplished by making measurements at strong lines such as Si III $\lambda 1206$ as suggested earlier for measuring instrumental stray light. The latter contribution can be determined in the laboratory. Whether the undesired scattered disk radiation is due to stray light in the instrument or dust near the Sun is not significant for the purpose of measuring T_e, what is important is a capability of separating the signal from the dust and instrumental stray light from the signal due to scattering from coronal electrons. An alternative possiblity for separating the contributions of the electron scattered and F-components is to make use of the polarization of the electron scattered radiation as is commonly done in the visible.

10. Results of Initial Observations

At the present time there exists only a very limited amount of UV and EUV spectroscopic data from the solar corona beyond $r = 1.5R_\odot$. These data were acquired with a UV coronagraph employing reflecting optics (cf. Kohl et al., 1978). This instrument and a companion white light coronagraph were carried above the UV absorbing layers of the terrestrial atmosphere by a sounding rocket. There have been two flights, both near the time of solar maximum, one in 1979 and the other in 1980 (Kohl et al., 1980; Wieser et al., 1981).

The primary objective of the 1980 flight was to study a coronal hole located at the southern solar pole. Coronal holes are thought to be a major source of solar wind,

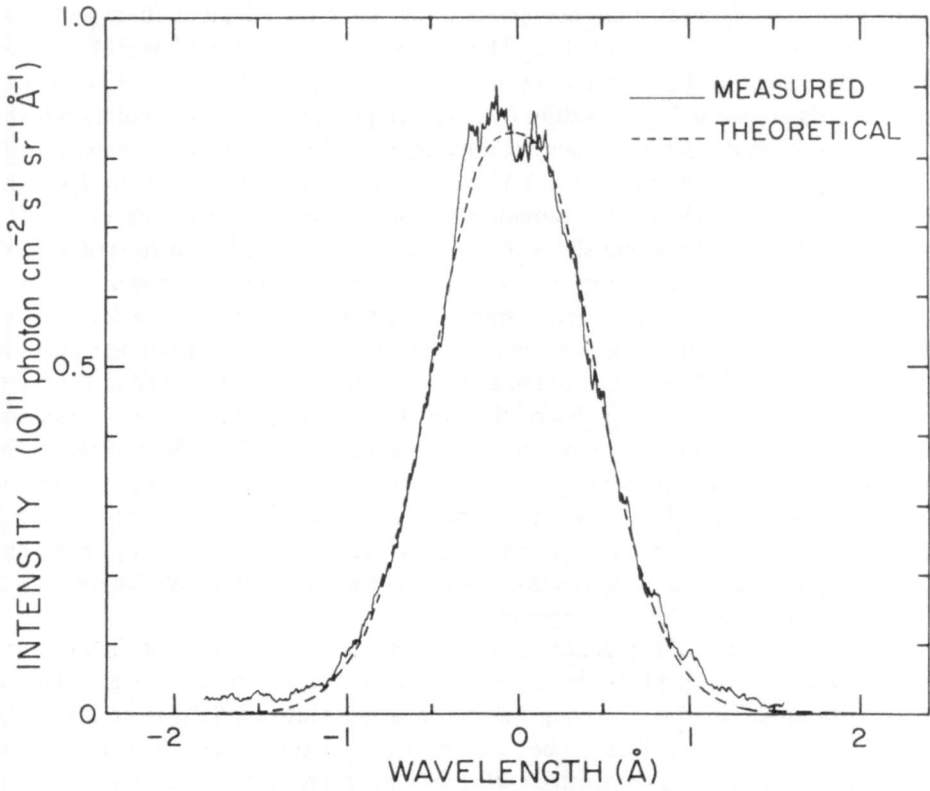

Fig. 10.   A comparison of an empirical (solid line) and theoretical (dashed line) profiles of the resonantly
scattered component of Lα. The empirical profile was measured in a polar region near the edge of a coronal
hole. The theoretical profile has been convolved with the instrumental profile which has a
FWHM = 0.35 Å.

particularly high speed solar wind streams. Consequently, coronagraphic determinations
of plasma parameters of these features are of considerable importance. An example of
the quality of the spectroscopic data that can be obtained with a UV coronagraphic
instrument is shown in Figure 10 which presents an empirical profile (solid line) of the
resonantly scattered component of hydrogen Lα measured at $\rho = 1.8 R_\odot$ in the polar
region observed in the 1980 rocket flight. The parameter ρ is the distance measured in
solar radii from Sun-center to the point where the line of sight intersects the plane of the
disk. The empirical profile has been fit with a profile calculated for an isothermal corona.
This illustrates how well this particular observation is represented by a Maxwellian
velocity distribution function for the hydrogen atoms along the line of sight. The
theoretical profile has lower intensities in the line wings; however, until a more detailed
analysis of the data is completed, we will not know whether or not this is significant.

As discussed in Sections 3 and 4 measurements of spectral line profiles provide
information on kinetic temperatures in the region where the spectral line radiation

originated. By measuring profiles at several heights in the corona one can obtain information on the temperature gradient in the observed region. Figure 11 shows hydrogen kinetic temperatures determined from Lα profiles measured at several positions ρ on a radius vector directed along the axis of the coronal hole that was centered on the south solar pole during February 1980. The profiles measured at $\rho = 1.5, 2.5,$ and $3.0\,R_\odot$ had nearly identical widths implying that the hydrogen kinetic temperature was nearly constant ($T_{HI} \approx 10^6$) over the height range where the line profiles were formed $r = 1.5\text{–}4\,R_\odot$. Measured temperatures are plotted at the radii ($r = 1.7, 3.1,$ and $3.8\,R_\odot$) corresponding to the mean heights where the radiation observed at $\rho = 1.5, 2.5,$ and $3.0\,R_\odot$ originated (see Withbroe *et al.*, 1982). The magnitude of the error estimates given depend upon the uncertainties in fitting the profiles with a Gaussian curve and the uncertainties in the correction for the effects of geocoronal emission/absorption near line center (cf. Withbroe *et al.*, 1982). The solid line is the inferred run of temperature. It should be emphasized that this is a kinetic temperature and includes the effects of both thermal and non-thermal motions broadening the Lα line profiles.

For comparison we have plotted temperatures predicted by a simple two-fluid model with no plasma heating above the base of the corona (cf. Hartle and Sturrock, 1968;

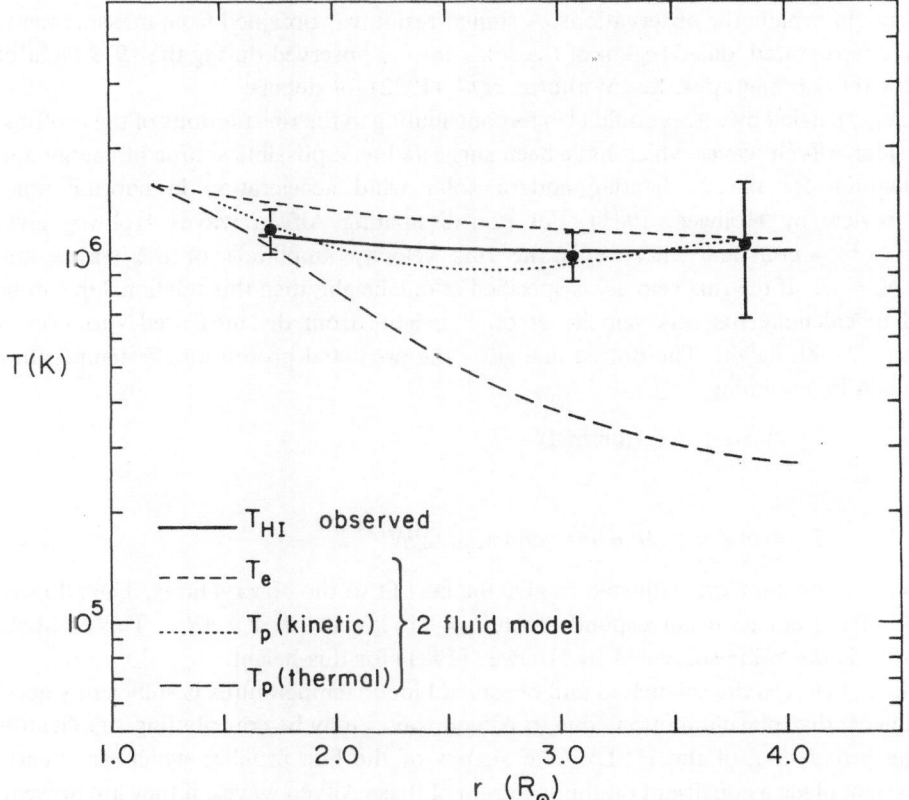

Fig. 11. A comparison of empirical and theoretical temperatures in a polar coronal hole (see text).

Nerney and Barnes, 1977; Hollweg, 1978). The short dash line gives the run of electron temperature with radius, while the long dash line gives the predicted proton thermal temperatures. The electron and proton temperatures diverge with height due to the rapid decrease in electron-proton coupling with decreasing density. The electron temperature has a shallow temperature gradient due to the high thermal conductivity of electrons, while the proton temperature falls off nearly aiabatically. We obtained an estimate of the electron temperature from the ratio of the Lα and white light intensities. This ratio depends on the ionization balance of hydrogen which is primarily a function of the electron temperature (see Gabriel, 1971). The value obtained, 1.5×10^6 K for $r = 1.7 R_\odot$, provides a boundary condition for the assumed two fluid model.

Because of the strong coupling between coronal hydrogen atoms and protons in the observed height range, one expects the hydrogen and proton kinetic temperatures to be equal. The difference between the observed coronal hydrogen kinetic temperatures and the calculated proton thermal temperatures indicates that the assumed model is inadequate. One way of bringing the calculated and observed temperatures into agreement is to increase the rms velocity of the protons. There are several ways of accomplishing this. One way is through extended proton heating in the region 1.5 to $4 R_\odot$. Addition of thermal energy by a mechanism with an energy dissipation length of about $4 R_\odot$ could explain the observations. A similar result was obtained from measurements in an unstructured 'quiet' region of the solar corona observed during the 1979 flight of the rocket coronagraphs. See Withbroe et al. (1982) for details.

Energy carried by waves could also be contributing to the rms motions of the protons. Consider Alfvén waves which have been suggested as a possible source of energy and momentum for plasma heating and/or solar wind acceleration in coronal holes (see review by Hollweg, 1981). For non-dissipating Alfvén waves Hollweg gives $N^{1/2} \langle v^2 \rangle = $ constant where v is the rms velocity amplitude of the waves and $N = N_e = N_p$. If the rms velocity is specified at one height, then this relationship can be used to calculate the rms velocity at other heights from the measured variation of density N with height. The dotted line gives the predicted proton kinetic temperature obtained by assuming

$$T_p(\text{kinetic}) = T_p(\text{thermal}) + T_A$$

with

$$T_A = m \langle v^2 \rangle / 2k = m \times \text{constant}/2kN^{1/2} ,$$

where the constant was adjusted to give the best fit to the observations. The adopted value of the constant corresponds to $v_{\text{rms}} = 115$ km s^{-1} at $r = 4R_\odot$. This is about two-thirds the value suggested by Hollweg (1981) for this height.

The fit between the calculated and observed kinetic temperatures is sufficiently good to suggest that plasma motions due to Alfvén waves may be contributing significantly to the broadening of the H I Lα. The shapes of the line profiles, which are nearly Gaussian, place a constraint on the spectrum of these Alfvén waves, if they are present. It is important to note that there are other explanations for the nearly constant width

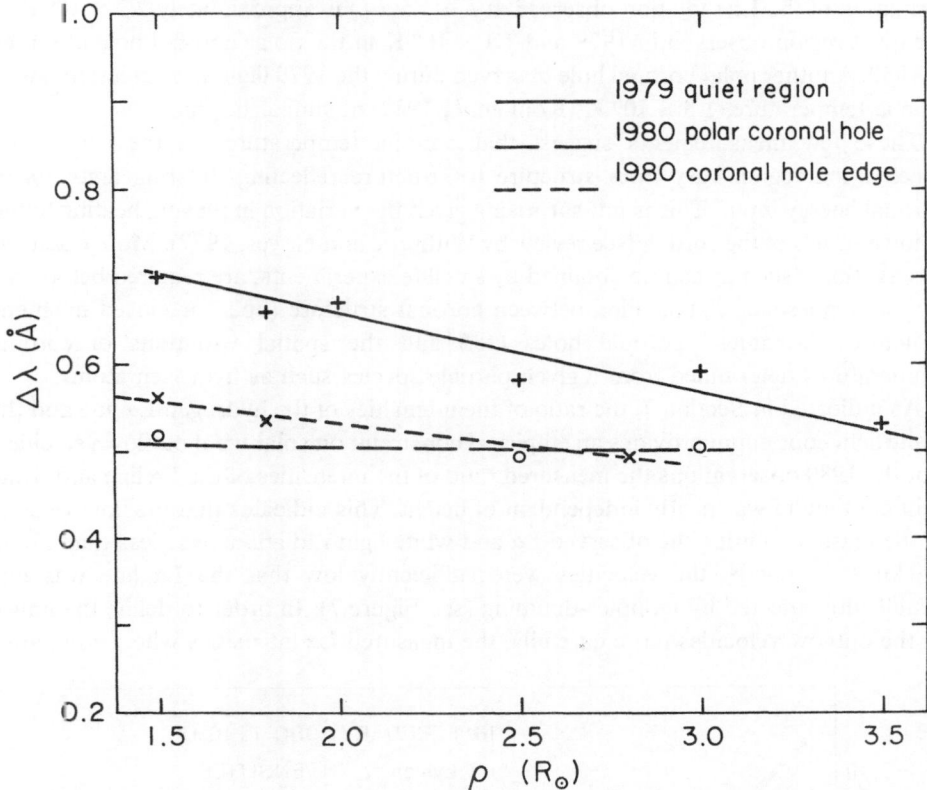

Fig. 12. Widths ($1/e$ half width $\approx 0.6 \times$ FWHM) of resonantly scattered H I Lα measured in 3 coronal regions.

of the Lα line, such as the above mentioned extended proton heating. In order to distinguish between thermal and non-thermal line broadening mechanisms, additional empirical constraints are needed, such as measurements of spectral lines from ions with different masses.

In order to illustrate the differences that can occur between different regions of the corona we have plotted in Figure 12 hydrogen Lα widths ($1/e$ half width $\approx 0.6 \times$ FWHM) for the three best observed regions in the 1979 and 1980 flights of the Rocket Lyman Alpha Coronagraph. For these three sets of data the spectral line widths are either constant or decreasing with increasing radius for $1.5 \leq \rho \leq 3.5\,R_\odot$. This indicates that the hydrogen kinetic temperatures were nearly constant or decreasing with increasing radius in the observed regions. Because of the sensitivity of the hydrogen kinetic temperature to energy deposition mechanisms that heat protons, this places tight upper limits on the amount of direct proton heating between $r = 1.5$ and $4\,R_\odot$ (cf. Withbroe *et al.*, 1982). The differences in the magnitude of the line widths indicates that significant differences in proton kinetic temperatures between different regions are possible. For example, the kinetic temperature at $r = 1.7\,R_\odot$ (the mean height of formation of the Lα radiation observed at $\rho = 1.5\,R_\odot$) is approximately 2.2×10^6 K in the quiet region observed in 1979 and 1.1×10^6 K in the polar coronal hole observed in 1980. Another polar coronal hole observed during the 1979 flight had an intermediate kinetic temperature, 1.8×10^6 K (Kohl *et al.*, 1980) at similar heights.

These few measurements suggest that coronal temperatures in the solar wind acceleration region vary from structure to structure reflecting differing amounts of coronal energy input. This is not surprising given the variation in plasma heating found at lower levels of the corona (see review by Withbroe and Noyes, 1977). More extensive observations, such as can be obtained by satellite experiments, are required before one can seek possible relationships between coronal structure (open or closed magnetic structures, streamers, coronal holes etc.) and the spatial variations of coronal temperatures determined from a given particle species such as hydrogen atoms.

As indicated in Section 7, the ratio of the intensities of the hydrogen Lα line and the white light continuum provides an empirical constraint on solar wind outflow velocities. For the 1980 observations the measured ratio of the intensities of the Lα line and white light continuum was nearly independent of height. This indicates that the flow velocity of the plasma emitting the observed Lα and white light radiations was less than about 100 km s^{-1}, that is, the velocities were sufficiently low that the Lα line was not significantly affected by Doppler-dimming (see Figure 7). In order to define the limits on the outflow velocities more carefully, the measured Lα intensities where compared

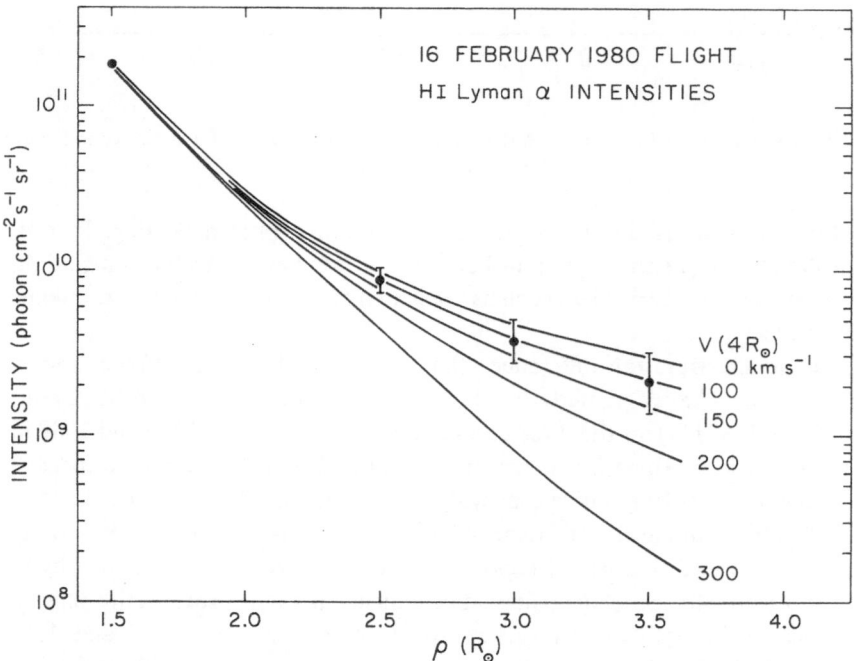

Fig. 13. A comparison of measured (points) and calculated (curves) H I Lα intensities as a function of distance from Sun center. The curves give values calculated for models with different solar wind fluxes parameterized here by the wind velocity at $4R_\odot$ (see text).

with those calculated from a series of coronal models (see Figure 13). The intensity of the scattered Lα radiation depends on the shape of the coronal scattering profile, the number of neutral hydrogen atoms in the line of sight (which can be calculated from the electron density and the electron temperature which determines $N_{H I}/N_e$) and the solar wind velocity. The width of the Lα scattering profile was measured and the electron densities were determined from measurements of the polarization and brightness of the white light corona. The electron temperature at $r = 1.7 R_\odot$ was determined from the observations as discussed above and for other heights was assumed to vary as $r^{-2/7}$ (e.g. the same as in Figure 11). Use of a model with an isothermal electron temperature yields similar results due to the insensitivity of the Lα intensity to variations in the electron temperature (e.g. Withbroe et al., 1982). For the radial variation of the solar wind velocity V was assumed a constant outward particle flux NVr^2.

The upper curve in Figure 13 is for a static atmosphere. The other curves show the predicted Lα intensities for models with different outward particle fluxes parameterized by the velocity at $r = 4 R_\odot$. At low heights where the density is high and solar wind velocity is low there is little Doppler dimming. However, due to the steady increase in flow velocity with increasing height, the amount of Doppler dimming increases with height causing the intensity to diverge from that calculated for the static model. A comparison of the calculated intensities with those measured confirms that the amount of Doppler dimming over the observed range of heights is small, corresponding to flow velocities at $r = 4 R_\odot$ of less than about 150 km s^{-1}. Given that the sound speed for a corona with $T_e = 1$ to 1.5×10^6 K is 130 to 160 km s^{-1}, the observations suggest that the solar wind flow in the observed plasma was subsonic for $r \lesssim 4 R_\odot$ and thus that the critical point was at $r \gtrsim 4 R_\odot$. Lα measurements in an unstructured 'quiet' region of the corona observed in April 1979 also showed little or no Doppler-dimming consistent with subsonic flows for $r \leq 4 R_\odot$ (Withbroe et al., 1982). A more detailed inhomogeneous model for the polar region observed in 1980 is being developed from ground-based synoptic coronal data, 1980 eclipse measurements and data from the rocket coronagraphs (Munro et al., 1982). This model should yield tighter constraints on the range of possible flow velocities in this region.

Analyses of observations acquired in the 1979 and 1980 flights of the CfA/HAO rocket coronagraphs yields the following empirical constraints on theoretical models for the solar wind acceleration region:
 — nearly Gaussian H I Lα profiles;
 — nearly constant or decreasing hydrogen kinetic temperatures or $1.5 < r < 4 R_\odot$;
 — subsonic flow for $r < 4 R_\odot$ (critical point at $r \gtrsim 4 R_\odot$);
 — an upper limit of 140 km s^{-1} for the rms velocity of waves capable of broadening the Lα line for $r \approx 4.0 R_\odot$;
 — some evidence for extended proton heating or a non-thermal contribution to the motions of H I atoms in the observed regions.

For a more detailed discussion of results of the 1979 and 1980 flights of the rocket coronagraphs see Kohl et al. (1980, 1982), Munro et al. (1982); and Withbroe et al. (1982).

11. Future Observational Programs

In the above sections we have discussed some of the possibilities for probing the physical conditions in the solar wind acceleration region through use of UV and EUV spectroscopy. We have also briefly summarized results of preliminary steps in implementing UV spectroscopy of this critical region of the solar atmosphere made possible by brief rocket flights of coronagraphic instruments on sounding rockets. Future sounding rocket or Detached Shuttle Payload flights are expected to provide additional information about the physical conditions in the solar wind acceleration region and thereby provide additional empirical constraints on solar/stellar wind theory. For example, the existing Rocket Lyman Alpha Coronagraph is being modified to measure the integrated intensities of the O VI resonance lines at $\lambda 1032$ and $\lambda 1037$. These lines are much more sensitive to low speed flows ($30 < V < 100$ km s^{-1}) than the Lα line (see Figure 7). Measurements with the O VI line will provide tighter constraints on the magnitude of the solar wind flow velocities within a few solar radii where subsonic flows have been indicated in the first two flights of the Lyman Alpha Coronagraph. A subsequent modification under study is the addition of an array detector that will permit measurement of the electron scattered component of the hydrogen Lα line.

Sounding rocket and Detached Shuttle Payload flights are extremely useful for proof testing the spectroscopic diagnostic techniques. However, to achieve the full capabilities of UV/EUV coronagraphic instruments, experiments in long term orbital flights are required to provide the observing time needed (1) to measure the weak coronal emission over a wide range of heights in a variety of coronal structures, (2) to obtain synoptic observations for probing the three dimensional structure of the corona and its temporal variations, and (3) observing coronal transient phenomena. The availability of long observing times is particularly important for measuring profiles of spectral lines other than resonantly scattered H I Lα at $\rho \gtrsim 2R_\odot$ and to undertake measurements of other coronal parameters such as magnetic fields (Bommier and Sahal-Brechot, 1982). A detailed definition study of a UV/white light coronagraphic instrument package suitable for flight on Spacelab or Space Platform has been conducted (cf. Kohl et al., 1981). NASA science working groups have proposed several missions in which UV/white light coronagraphs play an important role, the Solar Coronal Explorer (SCE), the Solar Terrestrial Observatory (STO) and the Pinhole/Occulter Facility (POF). The SCE has an instrument complement consisting of a white light coronagraph, UV/EUV coronagraph, soft X-ray imaging telescope and XUV spectrometer designed to study the corona over a period of several years from a spacecraft similar to SMM or a Space Platform (Orrall et al., 1981). The STO (Canfield et al., 1981) is an instrument package with two groups of experiments mounted on Shuttle-serviced orbiting platform. One group of instruments studies the solar radiative output and the corona as a source of solar wind, while the other group studies the response of the terrestrial atmosphere to variations in the radiative and particle flux from the Sun. The POF (Hudson et al., 1981) employs an occulter on a 50 m boom (see Figure 14) which casts a large shadow permitting the use of large aperture (~ 1 m) coronagraphic instruments. The occulter also

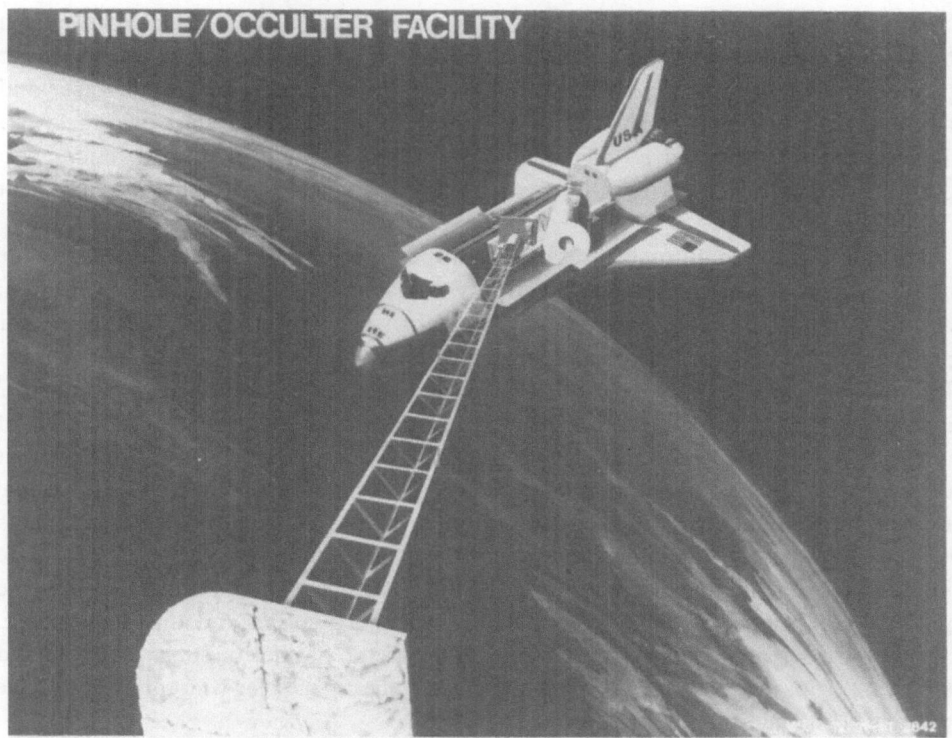

Fig. 14.   Illustration of the Pinhole/Occulter Facility (POF) with its 50 m boom deployed from the Orbiter bay. POF would provide a distant external occulter capable of shielding large aperture coronagraphic optics from bright solar disk radiation and also provide a pinhole array for high resolution imaging in hard X-ray radiation.

serves as a multi-pinhole array for high resolution ($\lesssim 1$ arc sec) hard X-ray imaging. Because of the large photon collecting capability of the POF coronagraphs, they can provide measurements with high spatial resolution (~ 1 arc sec, an order of magnitude or more better than more conventional instruments) and good time resolution for studying transient phenomena, as well as providing a capability for measuring weak coronal lines at large distances beyond the limb of the Sun.

12. Summary

Through use of the plasma diagnostic information contained in UV and EUV emission lines it is possible to greatly expand our knowledge about the physical conditions in the solar wind acceleration region $r \gtrsim 1.3 R_\odot$. Coronagraphic measurements of the intensities and profiles of the resonantly scattered and electron scattered components of hydrogen Lα coupled with measurements of the intensity and polarization of the electron scattered white light continuum can provide information on T_e, T_p, N_e, $N_{H\,I}$, and outflow velocities greater than 100 km s^{-1}. A more complete description of conditions in the upper corona is obtained when measurements of spectral lines from heavier ions are simultaneously measured with Lα to yield information on ion kinetic temperatures, non-thermal particle velocities, ion densities, chemical abundances and flow velocities in the range $30-100$ km s^{-1}, plus additional constraints on T_e and N_e. The information provided by combined UV/EUV and white light observations is a considerable improvement over having only N_e obtained from traditional white light coronagraphs.

Plasma diagnostic information derived from spectroscopic measurements can provide critical empirical constraints on mechanisms for plasma heating, solar wind acceleration, and transport of mass, momentum and energy, as well as mechanisms responsible for producing differences in chemical composition in the solar wind acceleration region. Information on temperatures and flow velocities is particularly important for investigating plasma heating and solar wind acceleration mechanisms and separation of mechanisms that depend upon driving the solar wind thermally and those driving it through wave-particle interactions. In addition, empirical data obtained from optical measurements of the solar wind acceleration region can be related to *in situ* measurements of electron and ion temperatures, flow velocities, abundances, charge states, etc. made in the solar wind at large distances from the Sun. The overall goal is understanding the physical processes and mechanisms operating in the solar corona and inner heliosphere. This knowledge can advance our understanding of the physics of stellar coronae and stellar mass loss, as well as improving the understanding of solar wind physics and the physics of the interplanetary medium.

Acknowledgements

This work was supported by NASA under grants NSG 5128 to the Harvard College and NAGW–249 to the Smithsonian Institution and under Order No. W–13 998 to the High Altitude Observatory and by the Fluid Research Fund and Langley–Abbott Program of the Smithsonian Institution.

References

Ahmad, I. A.: 1977, *Solar Phys.* **53**, 409.
Allen, C. W.: 1963, *Astrophysical Quantities*, 2nd ed., University of London, Athlone Press.
Bame, S. J., Ashbridge, J. R., Feldman, W. C., and Kearney, P. D.: *Solar Phys.* **35**, 137.
Basri, G. S., Linsky, J. L., Bartoe, J. D. F., Brueckner, G. E., and Van Hoozier, M. E.: 1979, *Astrophys. J.* **230**, 924.
Beckers, J. M. and Chipman, E.: 1974, *Solar Phys.* **34**, 151.
Blackwell, D. E., DeWhirst, D. W., and Ingham, M. E.: 1967, *Astron. Astrophys.* **5**, 1.
Bommier, V. and Sahal-Brechot, S.: 1982, *Astron. Astrophys.*, in press.
Brueckner, G. E.: 1980, *Highlights of Astronomy* **5**, 557.
Canfield, R. C., Chappell, C. R., Eddy, J. A., Farmer, C. B., Fisk, L. A., Geller, M. A., Gosling, J. T., MacQueen, R. M., Nagy, A. F., Neugebauer, M. M., Paulikas, G. A., Russell, C. T., Russell, P. B., Tandberg-Hanssen, E. A., and Taylor, W. W. L.: 1981, *Solar Terrestrial Observatory*, Final Report of the Science Study Group, NASA Marshall Space Flight Center.
Cram, L. E.: 1976, *Solar Phys.* **48**, 3.

Cushman, G. W. and Rense, W. A.: 1976, *Astrophys. J.* **207**, L61.

Dupree, A. K. and Reeves, E. M.: 1971, *Astrophys. J.* **165**, 599.

Dupree, A. K., Huber, M. C. E., Noyes, R. W., Parkinson, W. H., Reeves, E. M., and Withbroe, G. L.: 1973, *Astrophys. J.* **182**, 321.

Feldman, U. and Behring, W. E.: 1974, *Astrophys. J.* **189**, L45.

Feldman, W. C., Asbridge, J. R., Bame, S. J., and Gosling, J. T.: 1977, in O. R. White (ed.), *The Solar Output and Its Variations*, Colorado Assoc. Univ. Press, p. 351.

Fite, W. L., Smith, A. C. H., and Stebbings, R. F.: 1962, *Proc. Roy. Soc. Ser. A* **268**, 527.

Gabriel, A. H.: 1971, *Solar Phys.* **21**, 392.

Gabriel, A. H., Garton, W. R. S., Goldberg, L., Jones, T. J. L., Jordan, C., Morgan, F. J., Nicholls, R. W., Parkinson, W. H., Paxton, H. J. B., Reeves, E. M., Shenton, D. B., Speer, R. J., and Wilson, R.: 1971, *Astrophys. J.* **169**, 595.

Gouttebroze, P., Lemaire, P., Vial, J. C., and Artzner, G.: 1978, *Astrophys. J.* **225**, 655.

Hartle, R. E. and Sturrock, P. A.: 1968, *Astrophys. J.* **151**, 1155.

Henry, R. C., Anderson, R. C., and Fastie, W. G.: 1980, in I. Halliday and B. A. McIntosh (eds.), *Solid Particles in the Solar System*, D. Reidel Publ. Co., Dordrecht, Holland, p. 41.

Hirshberg, J.: 1975, *Rev. Geophys. Space Phys.* **13**, 1059.

Hollweg, J. V.: 1978, *Rev. Geophys. Space Phys.* **16**, 689.

Hollweg, J. V.: 1981, in S. Jordan (ed.), *The Sun as a Star*, NASA SP-450, NASA, Washington, D.C., p. 355.

Holzer, C. J. van: 1950, *Bull. Astron. Inst. Neth.* **11**, 160.

Hudson, H. S., Kohl, J. L., Lin, R. P., MacQueen, R. M., Tandberg-Hansen, E., and Dabbs, J. R.: 1981, *The Pinhole/Occulter Facility*, NASA Technical Memorandum NASA TM-82413, NASA, Marshall Space Flight Center, Alabama.

Hughes, C. J.: 1965, *Astrophys. J.* **142**, 321.

Hundhausen, A. J.: 1972, *Coronal Expansion and Solar Wind*, Springer-Verlag, New York.

Hyder, C. L. and Lites, B. W.: 1970, *Solar Phys.* **14**, 147.

Ingham, M. F.: 1961, *Monthly Notices Roy. Astron. Soc.* **122**, 157.

Joselyn, J. A. and Holzer, T. E.: 1978, *J. Geophys. Res.* **83**, 1019.

Kohl, J. L. and Withbroe, G. L.: 1982, *Astrophys. J.* **256**, 263.

Kohl, J. L., Reeves, E. M., and Kirkham, B.: 1978, in K. A. van der Hucht and G. Vaiana (eds.), *New Instrumentation for Space Astronomy*, Pergamon, New York, p. 91.

Kohl, J. L., Weiser, H., Withbroe, G. L., Noyes, R. W., Parkinson, W. H., Reeves, E. M., Munro, R. H., and MacQueen, R. M.: 1980, *Astrophys. J.* **241**, L117.

Kohl, J. L., Withbroe, G. L., Weiser, H., MacQueen, R. M., and Munro, R. H.: 1981, *Space Sci. Rev.* **29**, 419.

Kohl, J. L., Weiser, H., Parkinson, W. H., Withbroe, G. L., and Munro, R. H.: 1982, (in preparation).

Lillie, C. F.: 1972, in *The Scientific Results from the Orbiting Astronomical Observatory OAO-2*, NASA SP-310, p. 95.

Linert, C., Richter, I., Pitz, E., and Planck, B.: 1981, *Astron. Astrophys.* **103**, 177.

Mariska, J. T.: 1977, Ph.D. Thesis, Harvard University.

Moe, O. K. and Nicolas, K. R.: 1977, *Astrophys. J.* **211**, 579.

Munro, R. H. and Jackson, B. V.: 1977, *Astrophys. J.* **213**, 874.

Munro, R. H. and Mariska, J. T.: 1977, *Bull. Am. Astron. Soc.* **9**, 370.

Munro, R. H., Kohl, J. L., Weiser, H., and Withbroe, G. L.: 1982, (in preparation).

Nerney, S. and Barnes, S. A.: 1977, *J. Geophys. Res.* **62**, 3213.

Newkirk, G., Jr.: 1967, *Ann. Rev. Astron. Astrophys.* **5**, 213.

Newkirk, G., Dupree, R. G., and Schmahl, E. J.: 1970, *Solar Phys.* **15**, 15.

Orrall, F. Q. and Speer, R. J.: 1973, *Solar Phys.* **29**, 41.

Orrall, F. Q., Barnes, A., Burlaga, L. F., Kahler, S. W., Munro, R. H., Pneuman, G. W., Sheeley, N. R., Walker, A. B. C., and Withbroe, G. W.: 1981, *Solar Coronal Explorer*, Science Working Group Report, NASA Goddard Space Flight Center.

Owocki, S. P.: 1982, Ph.D. Thesis, U. Colorado.

Owocki, S. P. and Scudder, J.: 1982, submitted to *Astrophys. J.*

Percival, I. C.: 1966, *Nucl. Fusion* **6**, 182.

Perry, R. M. and Altschuler, M. D.: 1972, *Solar Phys.* **28**, 435.

Prinz, D. K.: 1974, *Astrophys. J.* **187**, 369.

Roser, S. and Staude, H. J.: 1978, *Astron. Astrophys.* **67**, 381.

Rottmann, G. J., Orrall, F. Q., and Klimchuk, J. A.: 1981, *Astrophys. J.* **247**, L135.
Rottmann, G. J., Orrall, F. Q., and Klimchuk, J. A.: 1982, submitted to *Astrophys. J.*
Saito, K.: 1970, *Ann. Tokyo Astron. Obs. Ser. 2* **12**, 53.
Spitzer, L., Jr.: 1962, *Physics of Fully Ionized Gases,* Interscience, New York.
Vaiana, G. S. and Rosner, R.: 1978, *Ann. Rev. Astron. Astrophys.* **16**, 393.
van de Hulst, H. C.: 1947, *Astrophys. J.* **105**, 471.
van de Hulst, H. C.: 1950, *Bull. Astron. Inst. Neth.* **11**, 135.
Vernazza, J. E. and Reeves, E. M.: 1978, *Astrophys. J. Suppl.* **37**, 485.
Weinberg, J. L. and Sparrow, J. G.: 1978, in McDonnell (ed.), *Cosmic Dust,* Wiley, New York, p. 75.
Weiser, H., Kohl, J. L., Parkinson, W. H., and Withbroe, G. L.: 1981, *Bull. Am. Astron Soc.* **12**, 917.
Wilson, D. C.: 1977, *The Three Dimensional Solar Corona – A Coronal Streamer, NCAR/CT-40,* Univ. of Colo.
    and Nat. Center Atmos. Res.
Withbroe, G. L.: 1970, *Solar Phys.* **11**, 42.
Withbroe, G. L. and Noyes, R. W.: 1977, *Ann. Rev. Astron. Astrophys.* **15**, 363.
Withbroe, G. L., Kohl, J. L., Weiser, H., Noci, G., and Munro, R. H.: 1982, *Astrophys. J.* **254**, 361.
Zirker, J. B.: 1981, in S. Jordan (ed.), *The Sun as a Star,* NASA SP-450, NASA, Washington, D.C.

X- AND XUV-RADIATION OF THE SOLAR CORONA*

G. ELWERT

Lehrstuhl für Theoretische Astrophysik der Universität Tübingen

Abstract. First a survey of the ionization states and emission lines of the ions existing in the corona is given. Then instruments for taking pictures of the Sun in the X- and the XUV-region as well as for measuring spectra emitted in interesting locations on the Sun are presented. Methods of plasma diagnostics, in particular for the determination of the mean temperature and the differential emission measure are described.

In the following review of observations, which are related to the topic of the workshop, types of coronal structures especially coronal holes, active regions and large scale structures are described. Their relations to the photospheric magnetic fields are dealt with; methods to calculate coronal magnetic fields are briefly discussed. As for temporal variations results of the analysis of expanding X-ray arches and of structures becoming visible in the outer corona in white light are mentioned. Finally, plasma diagnostics by means of high-resolution spectra are dealt with, in particular methods for the determination of the particle density by lines of He-like ions and of the local temperature by Li-like satellites lines. Thus non-thermal random velocities and outward moving plasma can be inferred during flares.

1. Introduction

The investigation of the X- and XUV radiation of the Sun has developed to an extensive field of research during its three decades existence. Therefore it is impossible to give a complete survey over this field within the scope of a review paper. The author has to confine himself to giving an introduction to the observation methods and the theoretical methods applied for the interpretation of the observations; then he will select some special aspects which are related to the topic of the workshop.

2. Ionization State of the Coronal Plasma

In the optical spectral region the solar corona can be observed only during total eclipses or by means of the coronograph, since its brightness is about a million times weaker than the photosphere. Therefore, the inner part of the corona is visible only at the solar limb but not on the solar disk. This is different in the X-ray and XUV region, since the photosphere is black in these wavelength bands due to the relatively low photospheric temperature.

The first information about the physical state of the corona could be obtained a long time ago from optical observations, in particular from the green and red coronal lines; they are hyperfine structure lines of the highly ionized atoms Fe XIV and Fe X. Fe XIV is an iron ion, which has lost 13 electrons of the 26 electrons of the neutral Fe, i.e., it has in addition to the 2 K and 8 L electrons 3 electrons in the M shell.

The ionization energy of Fe XIII is 350 eV. Since the ionization is caused mainly by electron collisions due to the low energy density of X-rays in the corona, the electrons

---

* Paper presented at the IX-th Lindau Workshop 'The Source Region of the Solar Wind'.

Space Science Reviews **33** (1982) 53–82.   0038–6308/82/0331–0053$04.50.

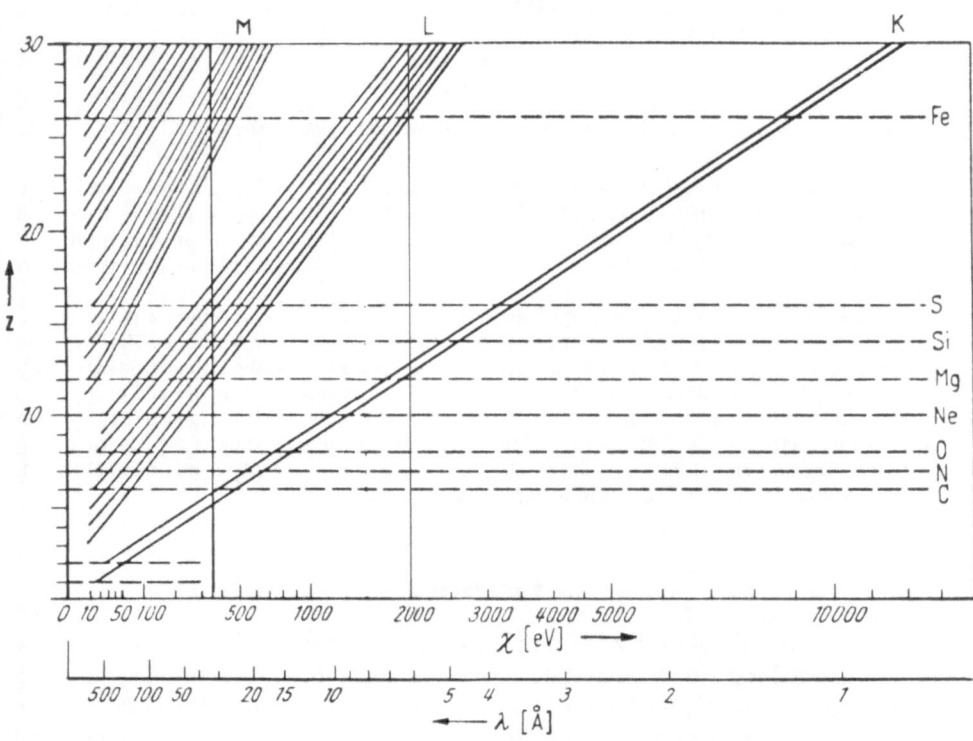

Fig. 1.   Ionisation energies χ for isoelectronic ions as a function of the nuclear charge z. The second scale
on the abscissa indicates the wavelengths of the series-limit continua corresponding to χ.

must have sufficient kinetic energy. Using a Maxwellian energy distribution a temperature of about 10^6 K could be inferred. A survey on the ionization stages of the coronal
atoms which follow from this temperature can be seen in Figure 1. On the abscissa the
ionization energy is given and on the ordinate the nuclear charge. The various curves
correspond to different electron systems, beginning with one-, two-, three-electron
systems, corresponding to H-, He-, Li-like ions, respectively. For a given ionization
energy the ionization state of a considered electron system can be read off. The
ionization energy of 350 eV is indicated by a vertical line. One can verify that Fe has
only 3 electrons in the M shell. Now one can read off from the figure that under these
conditions the ions of S and Si still contain some electrons in the M shell. The ions of
Mg, Ne, and O, however, have only two K electrons, i.e., they are helium-like ions. In
the usual nomenclature of the ionization stages they are denoted as Mg XI, Ne IX, and
O VII.

These statements obtained from a look to the figure of course give only a survey of
the ionization stages attained. A detailed calculation with the help of the ionization
formula shows that not only the ionization stages mentioned, but also a few neighbouring
ones occur. The resonance lines of these ions have wavelengths of about 10 to 25 Å.
Figure 2 shows that in fact lines of Mg XI, O VII, O VIII, and various Fe ions have been

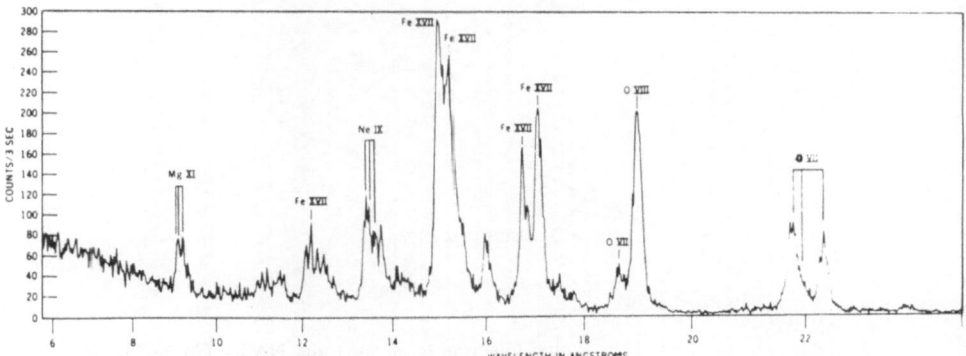

Fig. 2.   Solar spectrum between 6 and 25 Å recorded under undistrubed conditions by OSO-5 (after Neupert, 1971).

observed in spectra of the X-radiation from the whole Sun. On the other hand, the corona can also be observed in the light of Fe lines produced by higher shell transitions which have wavelengths of several hundreds Å, by Fe lines in the XUV region as well as by XUV lines of other ions.

These statements refer to the undistrubed corona. Under disturbed conditions, especially during flares, ionization stages with ionization energies in the keV range can be obtained. They are indicated in Figure 1 by an additional vertical line at 2 keV which corresponds to a temperature of about 10^7 K. One can see that under these conditions O, N, and C are fully ionized and that Si, S, and the heavier atoms exist as two-electron systems, i.e., helium-like ions. The same is true for iron in a broad temperature range up to several 10^7 K.

3. Instruments for the Observation of X- and XUV-Radiation

The observation of the X-radiation of the whole Sun, of course, provides information only on an average state of the total emitting atmosphere. The real purpose of the observation is the measurement of spectra at interesting locations of the corona or the picturing of the corona in the light of X-rays in wavelength bands as narrow as possible. This requires the construction of a spectroheliograph.

A very powerful instrument is the Wolter-Giacconi telescope (Wolter, 1952; Giacconi et al., 1965). In order to get reflection of X-radiation on mirrors one has to work with grazing incidence. Aberrations for non-axial rays can be avoided to a large extent by double reflection of the X-rays at a paraboloid and a confocal hyperboloid. The path of rays in the mostly used type of a Wolter–Giacconi telescope is shown in Figure 3.

The X-ray telescope flown by the American Science and Engineering, Inc. (AS & E), Cambridge, U.S.A. on Skylab (Vaiana et al., 1974, 1977) had a resolution of 2″ near

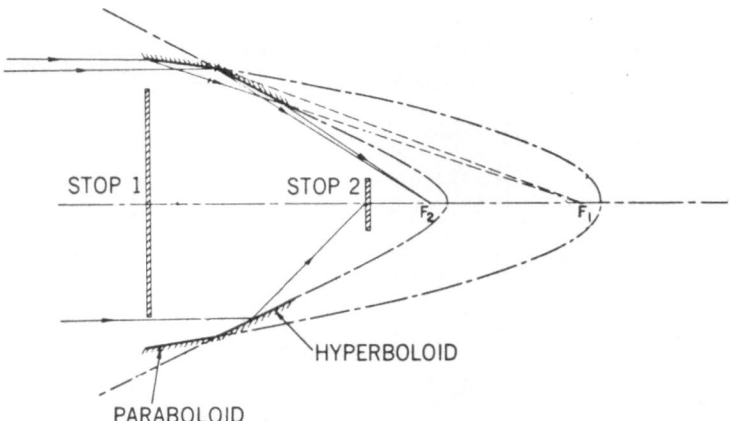

Fig. 3.   Schematic diagram of the Wolter-Giacconi telescope (after NASA TM X–73369).

the optical axis. The narrow central peak of the point spread function is, however, surrounded by broad wings originating from scattering of the X-radiation. About half of the energy of a point source image is outside a radius of about 50″. An exact quantitative analysis therefore requires a deconvolution of the pictures.

Besides this X-ray telescope there was another one from the Aerospace Corporation on Skylab with similar imaging qualities (Burke *et al.*, 1974; Henze *et al.*, 1976). In both telescopes X-ray filters with various wavelength pass bands between 3 and 55 Å were used, which are rather broad bands. Therefore, the picturing is by no means the really wanted monochromatic one.

This is different for 3 other picturing devices which belong to the XUV or UV range, viz.

(1) The XUV spectroheliograph of the NRL, also flown on Skylab (Tousey, 1976; Tousey *et al.*, 1977). It is based on the fact that a nearly perpendicular total reflection on mirrors is possible in the wavelength region of the much softer XUV-radiation above about 100 Å. Using a concave grating one obtains pictures of the Sun from different directions which for a given grating constant are determined by the wavelengths of strong XUV lines (cf. Figure 4). If these wavelengths are very close together, the pictures may partly overlap. The instrument was built for the 170 to 600 Å wavelength region. In this wavelength band besides the chromospheric lines, in particular the Ly-α line of He II, there are some coronal lines, in particular of Fe XV at 284 Å and of Fe XVI at 263 Å. The spatial resolution of the spectroheliograph is 2″.

The overlaping of spectrally pure pictures can be avoided, if they are produced by two-dimensional scanning instead of direct imaging. This procedure was applied to several instruments.

(2) The XUV spectroheliograph of the Harvard Observatory. Here a surface element of the Sun is imaged by a mirror system on the entrace aperture of a concave grating.

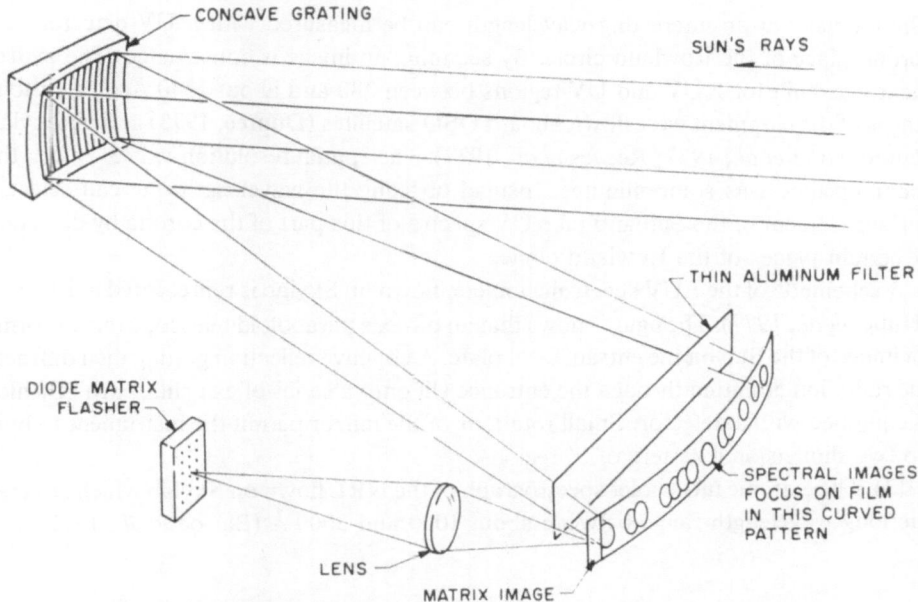

Fig. 4.   Optical layout of the NRL slitless spectrograph (after Widing, 1975).

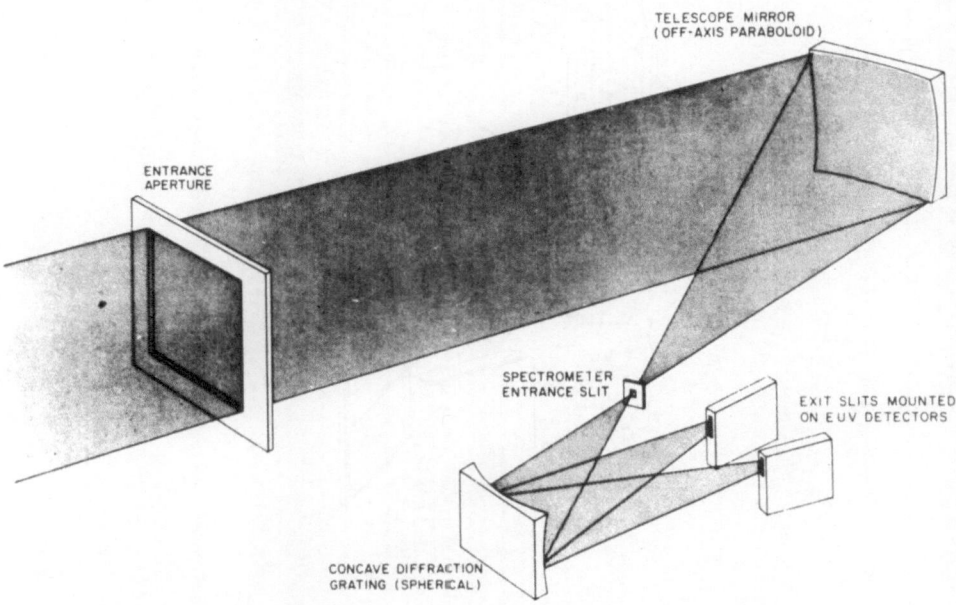

Fig. 5.   Path of rays in the EUV-spectroheliograph of the Harvard College Observatory on OSO-7.

The intensity at an interesting wavelength can be measured with a UV detector at a certain place of the Rowland circle. By scanning an image is constructed. The instrument was built for XUV and UV regions between 280 and about 1340 Å. The various stages of development were flown aboard OSO satellites (Dupree, 1973) and on Skylab (Underwood *et al.*, 1977; Reeves *et al.*, 1977). The spatial resolution was $5'' \times 5''$, the scanning time was some minutes. Instead of fixing the wavelength one can select a surface element of the Sun and take UV spectra of this part of the corona by detectors at certain places of the Rowland circle.

A schematic of the EUV spectroheliometer flown on Skylab is represented in Figure 5 (Huber *et al.*, 1977). The figure shows that an off-axis paraboloid telescope mirror forms an image of the Sun on the entrance-slit plate. A concave reflection grating then diffracts the radiation admitted through the entrance slit onto a series of exit slits, each of which is equipped with a detector. Small rotations of the mirror permit the instrument to built up two-dimensional rasters of 5' regions.

(3) The extreme ultraviolet spectrograph of the NRL flown on Skylab which covered the long wavelength range between about 1000 and 3000 Å (Bartoe *et al.*, 1977).

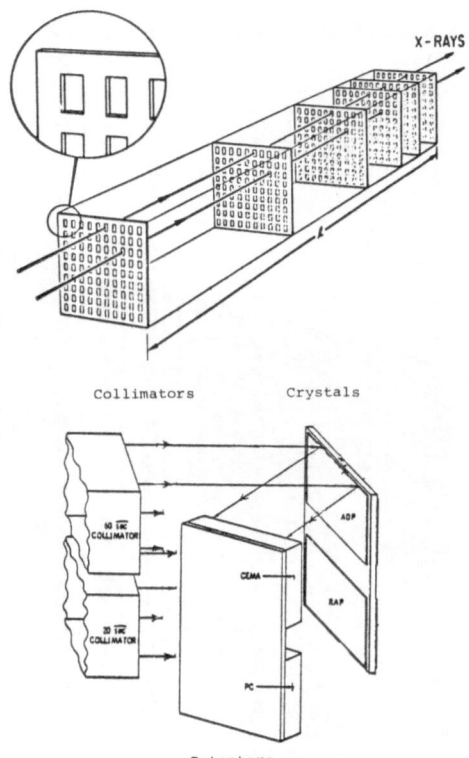

Fig. 6. Principle of a multiple X-ray collimeter and layout of the SOLEX collimeters crystals and detectors (after Landecker *et al.*, 1981).

(4) Instead of imaging the Sun with the help of a mirror system one can use multi-grid collimators to mask out small solid angles. This method of image production is the basis of two new instruments. One of them is a spectroheliograph denoted SOLEX of the Space Science Laboratory of the Aerospace Corporation which is flown on the Air Force P78–1 satellite (McKenzie et al., 1976; Landecker et al., 1981). The collimator masks out only one solid angle and the images are again produced by scanning (Figure 6). An other instrument is used in the HXIS experiment of the Space Research Laboratory in Utrecht which was flown on the SMM satellite. At high energies where the HXIS experiment observes the hard X-radiation during flares, imaging by collimators is the only possibility to produce pictures, since the reflectivity of mirrors for X-rays with wavelengths below 3 Å drops very rapidly towards zero, even for grazing incidence. A description of the HXIS experiment is given in the accompanying paper of E. Haug.

On the P78–1 satellite two multi-grid collimators with resolutions of 20″ and 60″ were employed. Behind them Bragg spectrometers of flat organic crystals with large lattice constants are attached, where the X-radiation between 3 and 25 Å is reflected according to the Bragg law. The scanning pattern is 5′ × 5′ in case of the high spatial resolution of 20″ which is used for a detailed study of an interesting region. Applying the lower resolution a map of the whole Sun can be scanned. The scanning time required is of the order of 1 and 8 min, respectively.

The instrument can also be used in the spectrometer mode. Then it is directed to an interesting location of the Sun and spectra of the X-radiation between 3 and 25 Å can be recorded by scanning.

Aboard the same satellite is also the SOLFLEX (= Solar flare) X-ray experiment built by NRL (Doschek et al., 1979; 1980; Landecker et al., 1981). It employs 4 Bragg crystal spectrometers designated to record extremely high-resolution spectra of important X-ray emission lines in narrow wavelength bands between about 2 and 8 Å.

4. Plasma Diagnostics: Determination of Temperature and Differential Emission Measure

The theory of radiation of a plasma in the X-ray region allows one to calculate the line and continuum emission E in erg cm^{-3} s^{-1} $\Delta\lambda$ for given element abundances as a function of the wavelength and the temperature of the plasma (Elwert, 1954; Tucker and Koran, 1971; Kato, 1976). It is based on the assumption that the ionization state is determined by a stationary equilibrium of ionization and recombination proceses (Elwert, 1952; Burgess, 1965; Jordan, 1969, 1970). The ionization occurs by electronic impact, the recombination under direct emission of photons and by dielectronic recombination. The lines are excited by electron impact. To a good approximation the emission is proportional to the square of the electron density

$$E(\lambda, T) = n_e^2 P(\lambda, T). \tag{1}$$

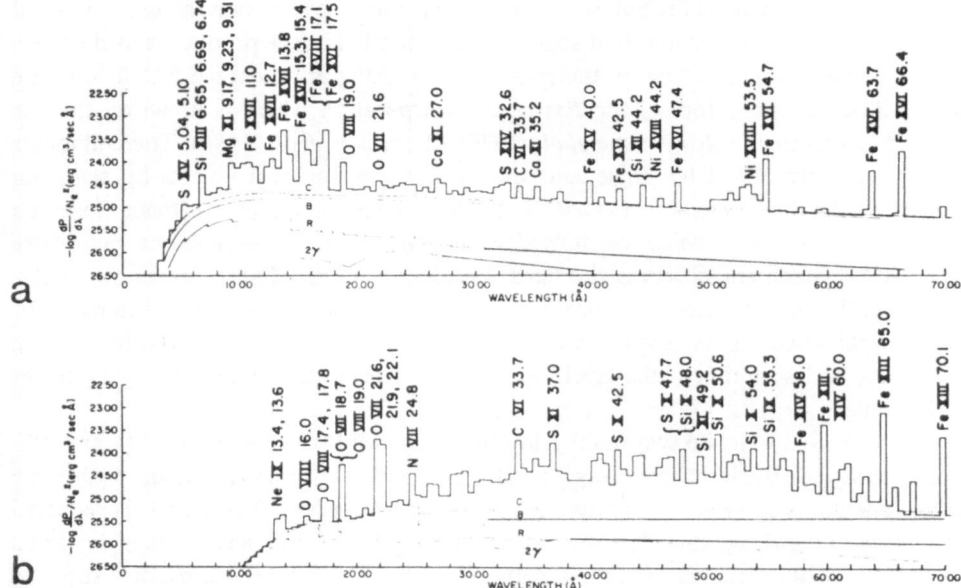

Fig. 7. Spectrum of the coronal plasma for (a) $T = 5 \times 10^6$ K and (b) 1.6×10^6 K, if a resolution of 0.5 Å is assumed (after Tucker and Koran, 1971).

The reason for the proportionality to n_e^2 is that the number of direct line excitations is proportional to the electron density n_e and to the ion density, which for a given temperature is also proportional to n_e. Examples of the spectral functions $P(\lambda, T)$ are given in Figure 7 for temperatures $T = 1.6 \times 10^6$ and 5×10^6 K.

Since the effects of absorption are of no importance to X-radiation, the line intensity at Earth distance originating from the emitting volume V is

$$I_\lambda = C \int n_e^2 P(\lambda, T) \, \mathrm{d}V . \tag{2}$$

with a known factor C including the distance between Sun and Earth.

The intensity I originating from a surface element of the solar disk corresponding to the resolution of the imaging device can be measured. The volume integral then becomes an integral on the line of sight l. If a monochromatic X-ray picture taken in the light of a spectral line L can be reproduced, the line intensity is given by an integral along the line of sight l.

Assuming the temperature along the line of sight to be constant, $P_L(T)$ can be put before the integral

$$I_L = C' P_L(T) \int n_e^2 \, \mathrm{d}l = C' P_L(T) S , \tag{4}$$

where S is temperature independent. Therefore, the temperature T which has the meaning of an average temperature, can be determined from the intensity ratio of two lines: $I_{L_1}/I_{L_2} = P_{L_1}(T)/P_{L_2}(T)$. If the temperature T is known, the quantity $S = \int n_e^2 \, dl$, the so-called differential emission measure, follows for any line of sight. Thus, a two-dimensional distribution of the temperature and the differential emission measure on the solar disk can be derived. Assuming a reasonable thickness of the radiating plasma, for instance, equal to its lateral extension, average values of the electron density can be estimated.

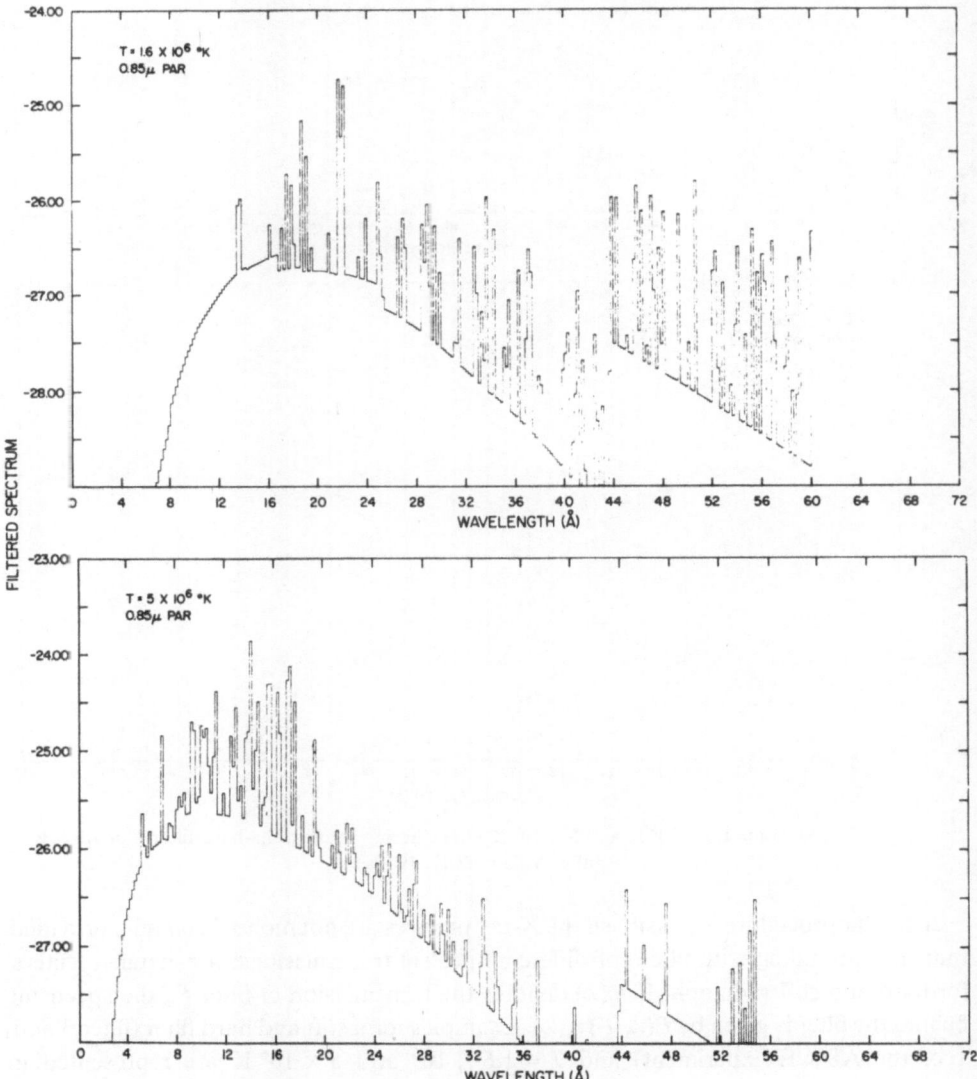

Fig. 8. Spectra for 1.6×10^6 K and 5×10^6 K after filtering through the soft filter 3 of AS & E (after Vaiana *et al.*, 1973a).

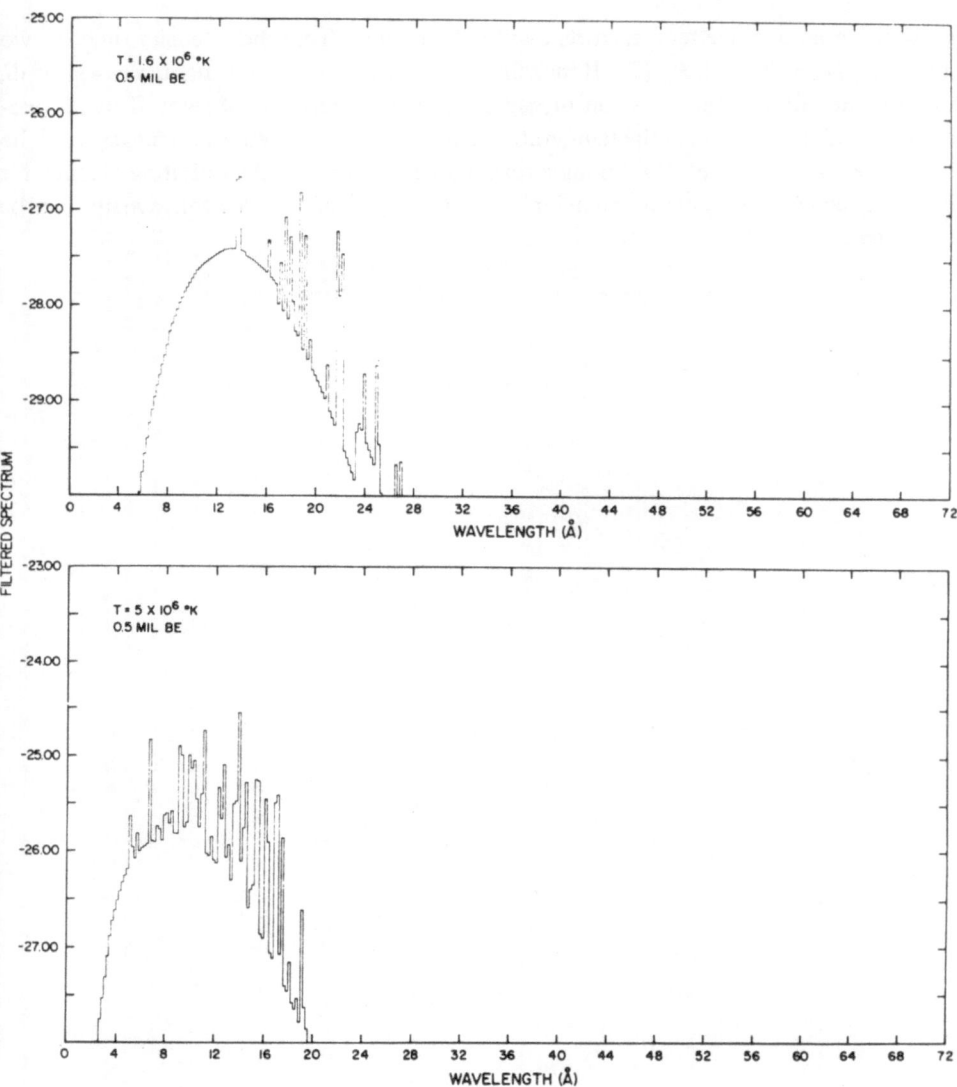

Fig. 9.   Spectra for 1.6×10^6 K and 5×10^6 K after filtering through the hard filter 1 of AS & E
(after Vaiana *et al.*, 1973a).

A similar procedure is possible if the X-ray pictures are not monochromatic, provided
that they are taken with filters of different spectral transmissions, for instance, filters
for hard and soft radiation. If $\eta_i(\lambda)$ denotes the transmission of filter F_i, the spectrum
behind the filter is given by $P(\lambda, T)\,\eta_i(\lambda)$. Examples for a soft and hard filter (filter 3 and
1 of the AS & E-Experiment) and $T = 1.6 \times 10^6$ and 5×10^6 K are represented in
Figures 8 and 9. The whole flux transmitted through filter F_i is

$$I_{F_i} = C'' \int n_e^2 \left(\int P(\lambda, T)\,\eta_i(\lambda)\,\mathrm{d}\lambda \right) \mathrm{d}l . \tag{5}$$

Assuming again that the temperature is constant along the line of sight, one obtains

$$I_{F_i} = C'' \int P(\lambda, T)\,\eta_i(\lambda)\,\mathrm{d}\lambda \times S. \tag{6}$$

The quotient of I_{F_i} and I_{F_k}, the so-called spectral hardness $R_{ik} = I_{F_i}/F_{F_k}$ is a function of the temperature. For filters 1 and 3 considered above R_{ik} is represented in Figure 10.

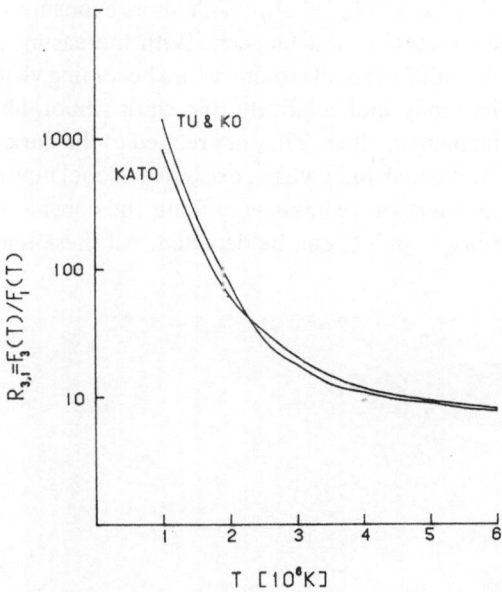

Fig. 10. Spectral Hardness Index R_{ik} for filters 3 and 1 of AS & E (after Maute *et al.*, 1981).

The energy of the X-ray quanta transmitted through the two filters F_i and F_k can be derived from the film densities. Thus the averaged temperature, and using the absolute intensity measured with one filter the differential emission measures can be determined; then the averaged electron density can be estimated. From these data values for the mean energy density $\varepsilon = 3n_e kT$ and the gas presure $\eta = 2n_e kT$ follow.

In certain temperature intervals it is possible to derive these parameters from a single X-ray picture (Kahler, 1976). Introducing the energy density into the above formula, one obtains

$$I_F = C'' \int \frac{\varepsilon^2 \int P(\lambda, T)\,\eta_F(\lambda)\,\mathrm{d}\lambda}{(3kT)^2} \tag{7}$$

In certain temperature intervals between 10^6 and 10^7 K the integral over λ is approximately proportional to T^2: $\int P(\lambda, T)\,\eta_F(\lambda)\,\mathrm{d}\lambda \propto T^2$. Then $I_F = C''' \int \varepsilon\,\mathrm{d}l$ where C''' is a new constant dependent on instrumental parameters. Hence the energy density ε and the pressure can directly be determined from the radiation flux measured by one picture only.

5. Types of Coronal Structures and Their Relations to the Photospheric Magnetic Field

Figure 11 shows a picture of the X-ray Sun taken with the As & E-telescope and a mean exposure time of 16 s (Vaiana *et al.*, 1973b). With short exposure times only the bright central parts of the active regions can be seen. With increasing exposure time inter-connecting loops and extended coronal structures are becoming visible. With still further increasing exposure time long and relatively thin dark ribbon-like structures can be noticed, the so-called filament cavities. They are related to the dark filaments which can be seen in Hα-pictures and which indicate the existence of cool matter. Extended regions appear where the X-ray emission remains very faint, the coronal holes. In addition, a great number of small bright points can be detected. All these features are to be seen in Figure 11.

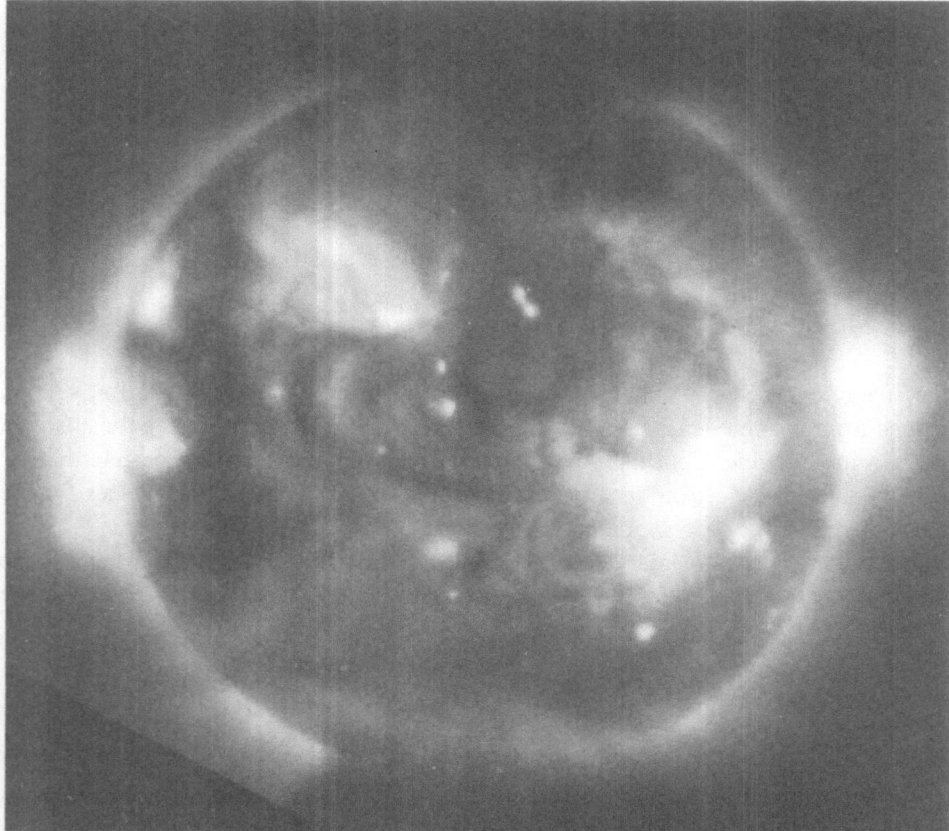

Fig. 11.   X-ray picture of the solar corona taken by the AS & E-telescope (after Vaiana *et al.*, 1973b).

Al these structures of the corona are closely correlated with the underlying photospheric magnetic field. This can clearly be seen when comparing X-ray or the XUV-pictures with the isolines of the photospheric magnetic field. Figure 12 shows isolines of the longitudinal magnetic field for September 7, 1973 according to measurements of the Mount Wilson Observatory. The pertinent XUV picture represented in Figure 13 is a spectroheliogram in the light of the line of Fe xv at 284 Å taken by the NRL at the same day. This picture is a negative. The correlations are particularly evident if one considers the cores of active regions which are brightest in the XUV light.

Furthermore there are strong correlations to the radio emission in the cm wavelength range. This is demonstrated in Figures 14 and 15. Figure 14 shows the Fe xv spectroheliogram for August 31, 1973, Figure 15 isolines of the radio emission observed at the same day with the Effelsberg radio dish at the wavelength of 2.8 cm. The centers of the radio intensity peaks coincide with the position of the X-ray sources as well. As described in the last section mean values of temperature, differential emission measure,

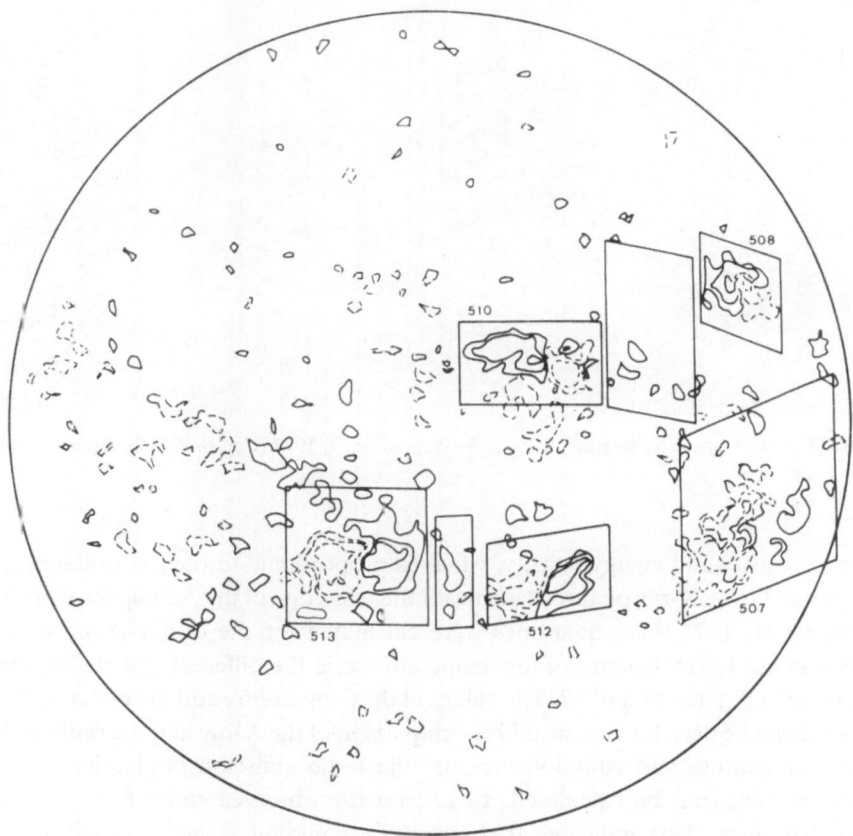

Fig. 12. Mount Wilson magnetogram of September 7, 1973. The parallelograms show projected tangential planes on which footpoints of field lines were placed (after Elwert *et al.*, 1982).

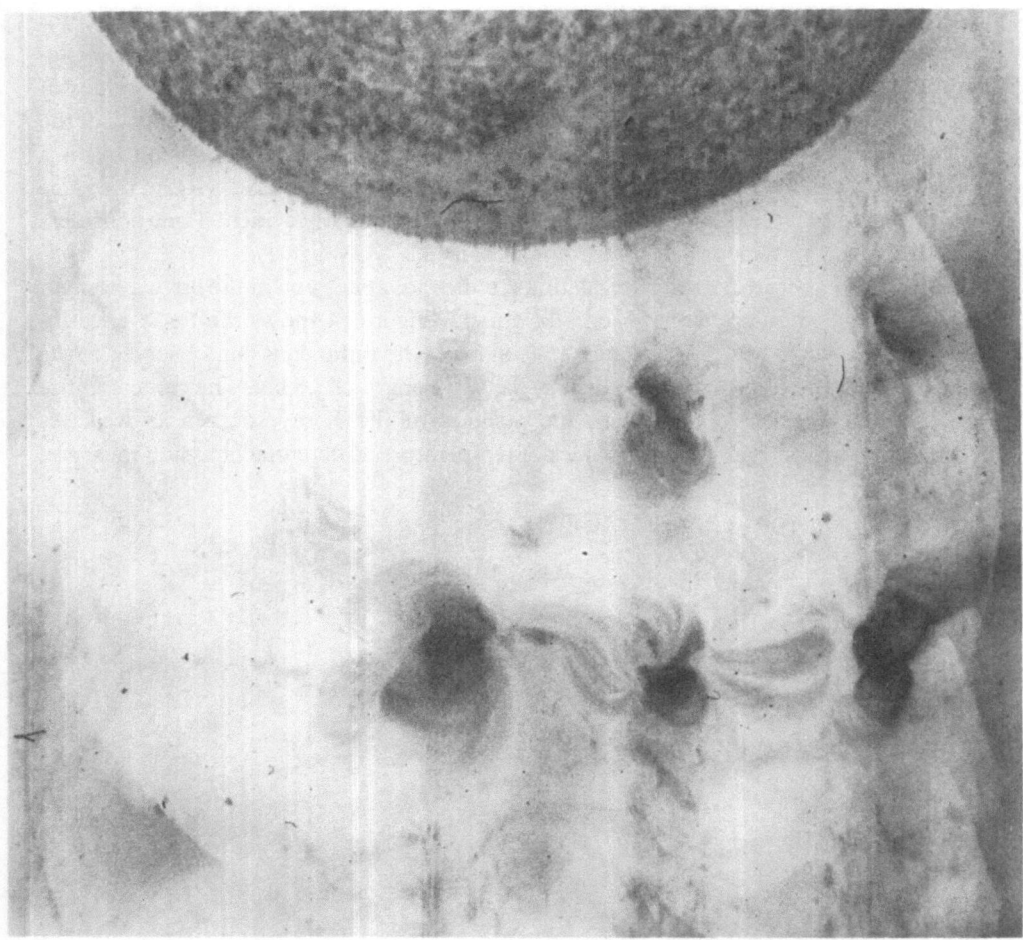

Fig. 13. Fe xv spectroheliogram, λ284 Å for September 7, 1973 (Courtesy G. E. Brückner NRL).

electron density and energy density, which is proportional to the gas pressure, can be calculated. Using X-ray pictures taken with the telescope of the Aerospace Corporation at August 31, 1973 these quantities were calculated for the center group of activity (Elwert *et al.*, 1981). Isolines of the temperature and the differential emission measure are drawn in Figures 16 and 17. The values of the temperature and the emission measure obtained can be used for a quantitative comparison of the X-ray and the radio emission. From temperature and emission measure the radio emission originating as thermal bremsstrahlung can be calculated. In general the observed radio fluxes exceed the calculated ones. This indicates that the radio emission is not typically collisional bremsstrahlung and that there is an additional contribution of magnetic bremsstrahlung. Some typical values obtained for temperature, electron density and energy density in

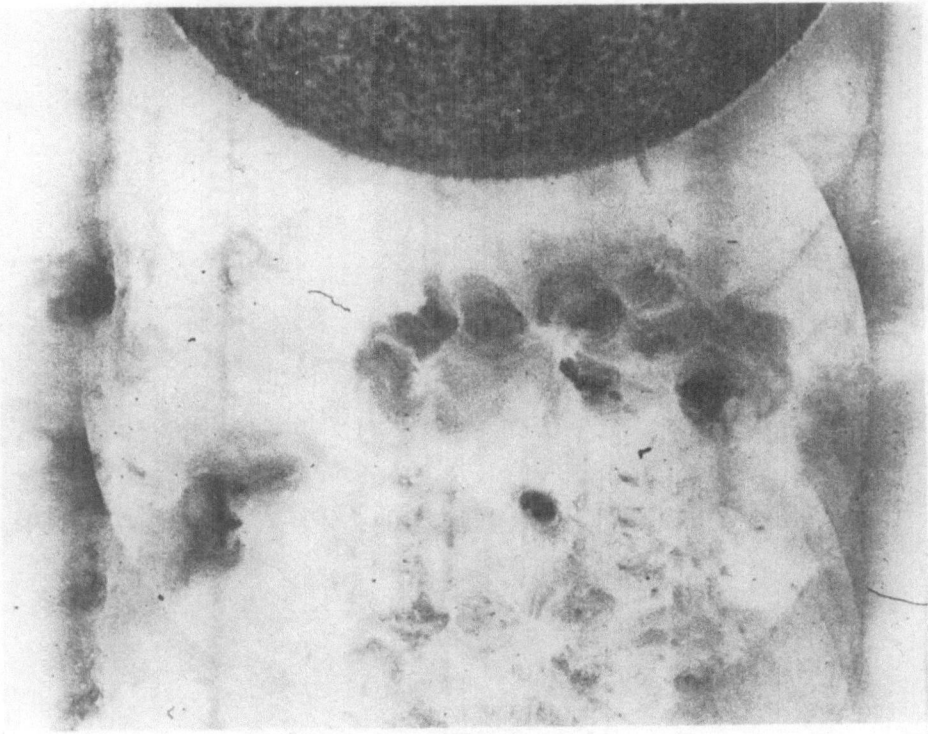

Fig. 14.   Fe XV spectroheliogram, $\lambda 284$ Å for August 31, 1973 (Courtesy G. E. Brückner NRL).

case of coronal holes, active regions, and large-scale structures are given in Table I (cf. Vaiana, 1976).

The value for the temperature of a coronal hole cannot be obtained directly, since no emission is seen in the harder filters. Only an estimate based on the emission of the softest filter can be given. This is possible by the measurements of the radial intensity decrease above coronal holes at the solar limb, if one assumes that the corona is in hydrostatic equilibrium and that it is isothermal. The electron density is given by the barometric formula $n_e = n_{e0}\, e^{-h/H}$ where the scale height $H = (kT\mu/mg_\odot)$ contains the

TABLE I

	$T\,(\mathrm{K})$	$n_e\,(\mathrm{cm}^{-3})$	$\varepsilon = 3n_e kT\,(\mathrm{erg\,cm}^{-3})$
Coronal hole	1.3×10^6	3×10^8	0, 1
Active region	2.5×10^6	7×10^9	7
Large scale structure	1.6×10^6	8×10^8	0, 5

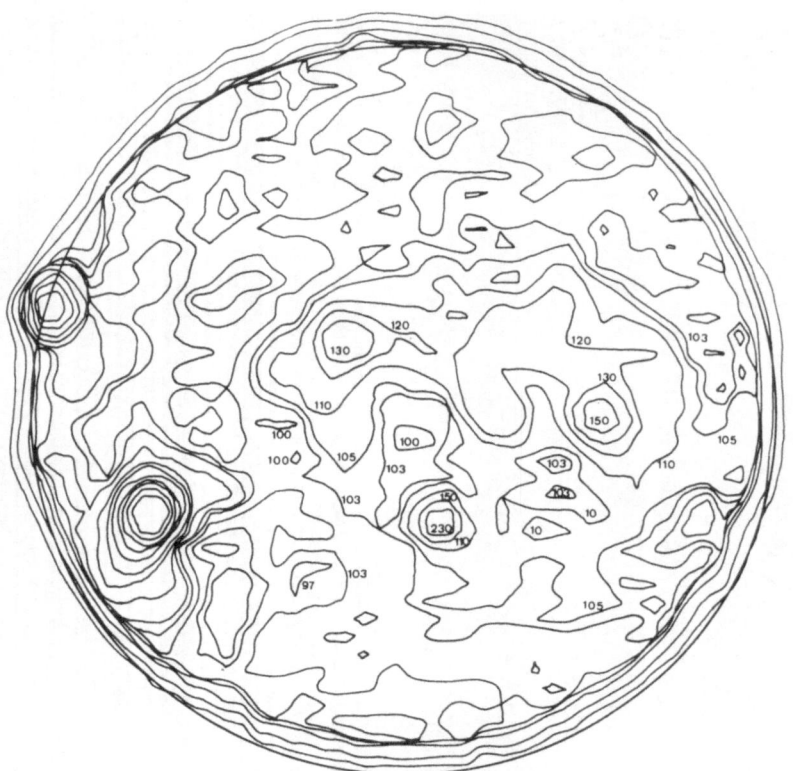

Fig. 15.   Radio maps scanned with the Effelsberg radio dish (λ2.8 cm) for August 31, 1973 (Courtesy E. Fürst, MPI für Radioastronomie, Bonn).

temperature T. The exponential density decrease results in a exponential law for the differential emission measure along the line of sight. By fitting this exponential slope of the intensity with the measurements one obtains a barometric temperature of 1.3×10^6 K (Vaiana *et al.*, 1973a; Krieger *et al.*, 1973).

For the understanding of the structure of the chromosphere and the transition region to the corona the XUV lines originating from this region are of great importance. They have in particular been observed with the Harvard spectroheliograph on OSO6 and aboard Skylab (Dupree *et al.*, 1973; Huber *et al.*, 1974; Vernazza *et al.*, 1978). These observations indicate that in holes the intensity of typical XUV lines as those of Mg x and Si xii which originate in layers of a temperature of about 10^6 K is drastically reduced compared to quiet regions; for lines emitted in regions of lower temperature the difference is smaller. The interpretation of the line intensities shows that the density and the temperature gradient in hole regions is smaller than in quiet regions. For a more accurate interpretation the construction of an atmospheric model of the transition region is necessary in which the observed line intensities are represented and the equations for the hydrostatic equilibrium as well as for the energy and the momentum balance are

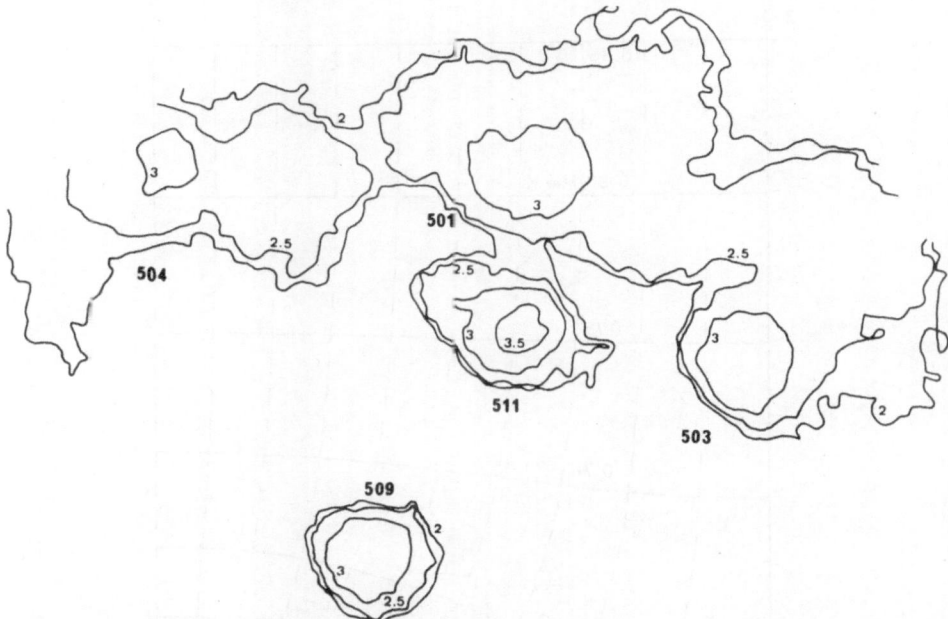

Fig. 16. Maps of calculated temperature distribution for August 31, 1973. The temperatures are given in units of 10^6 K (after Elwert *et al.*, 1981).

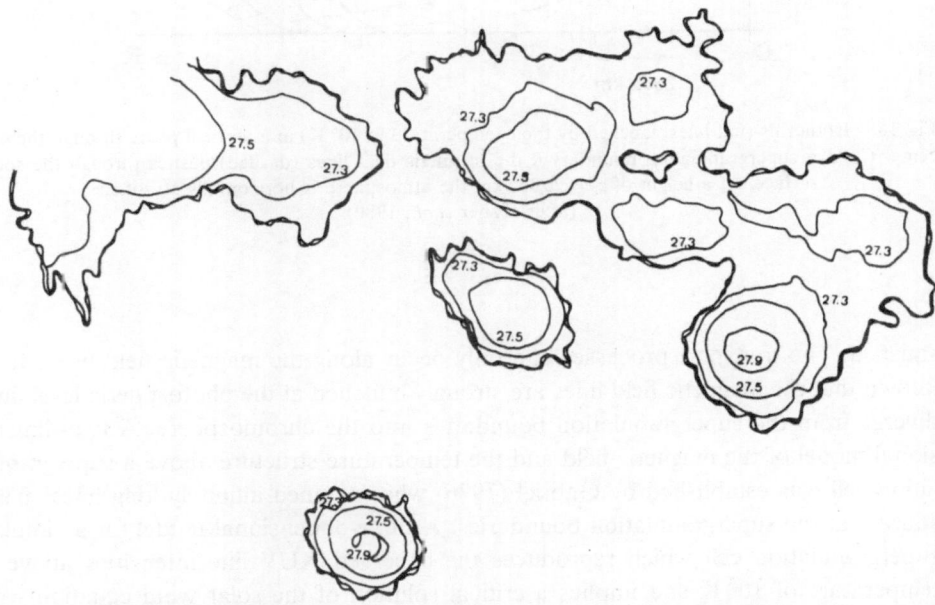

Fig. 17. Maps of the line-of-sight emission measure S for August 31, 1973. The given numbers are $\log_{10} S$ (after Elwert *et al.*, 1981).

G. ELWERT

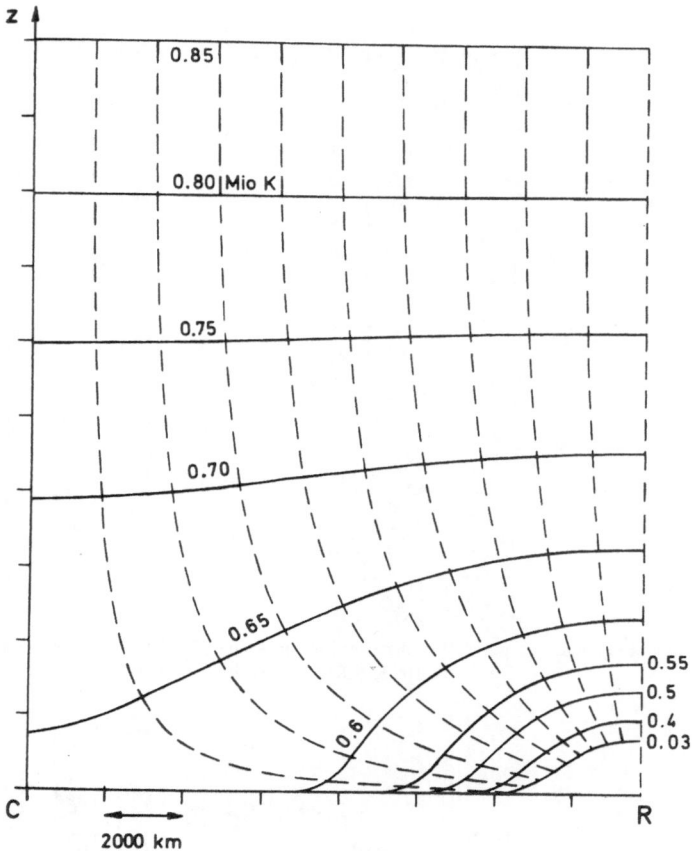

Fig. 18.   Isotherms (full lines, labelled by their temperature in 10^6 K) in a vertical plane through the cell center C. At a supergranulation boundary R the magnetic field lines (dashed) pushes through the solar surface. At a height of $z_c = 20\,000$ km the atmosphere is horizontally stratified (after Elzner *et al.*, 1980).

satisfied. The transport processes can only occur along the magnetic field lines. It is known that the magnetic field lines are strongly bunched at the photospheric level and diverge from the supergranulation boundaries into the chromosphere. A two-dimensional model of the magnetic field and the temperature structure above a supergranulation cell was established by Gabriel (1976) who assumed infinitely long linear field sources at the supergranulation boundaries. A three-dimensional model for a circular supergranulation cell which reproduces the observed XUV line intensities above a temperature of 10^5 K and implies a critical solution of the solar wind equation was derived by Elzner *et al.* (1980). The magnetic field lines and the temperature structure obtained is represented in Figure 18.

6. Calculations of Coronal Magnetic Fields

It is of great interest for the interpretation of the X-ray pictures to know the structure of the magnetic field lines in the corona which emerge from the photosphere. Since the coronal magnetic field cannot be observed by means of the Zeeman effect, methods for its computation from the observed longitudinal components of the photospheric magnetic field have been developed. Neglecting electric currents in the corona one has to find out a potential function φ. The derivative of φ in the line-of-sight direction must agree with the longitudinal components of the magnetic field observed in the photosphere. Assuming that the magnetic field does not change during a solar rotation, this mathematical problem was approximately solved by an expansion of the potential function in spherical harmonics (Altschuler et al., 1977). In this procedure, however, the photospheric field actually existing for the time considered is not used. During one solar rotation that photospheric field observed in the neighbourhood of the central meridian is taken. This method yields therefore satisfactory results only for large-scale magnetic-field structures which do not change within a solar rotation period. It results e.g. in the statement that the coronal holes are located in such parts of the solar atmosphere which are limited by magnetic fields of the same polarity so that the magnetic field is open and the field lines diverge outwards (Levine et al., 1977). Figure 19 shows an example.

On the other hand, there is also a method to calculate the instantaneous magnetic field actually existing for a day considered. By this method the closed magnetic fields in the inner corona can approximately be calculated. In the neighbourhood of active regions normal components of the field strength in the photosphere can be computed from the line-of-sight components; from these normal components one can calculate the magnetic field in the corona (Elwert et al., 1982).

This procedure is not only possible for a potential field but also for force-free fields, i.e., if there exist electric currents parallel to the magnetic field, and if their strength is proportional to the magnetic field strength. The comparison of the map of field lines (Figure 20) calculated by this method and the XUV picture (Figure 13) shows rather good agreement referring to the observed loops. This holds for loops which connect opposite polarities of different active regions as well as opposite polarities within the same region. It might be of interest for solar wind research that there are small locations of diverging field lines within active regions. They are coincident with the dark spots appearing as bright spots of the photographic negative of the XUV picture and which might be named coronal mini-holes.

7. Expansions of X-ray Arches into the Outer Corona

White light photographs established that expulsions of coronal gas occur very often. These events are mostly associated with eruptions of solar prominences or with large flares. It was, however, not clear whether the material comprising these coronal transients observed in the outer corona were to be identified with the prominence

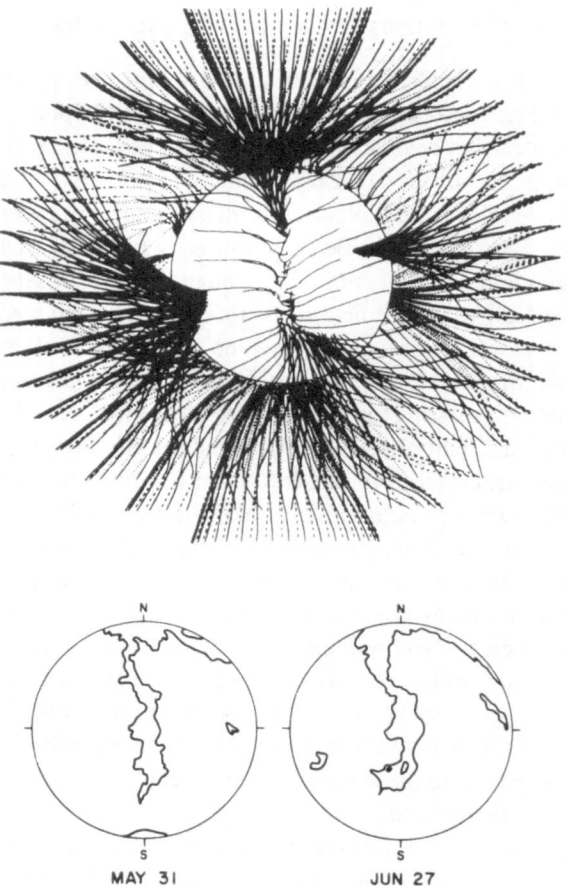

Fig. 19. Open-field lines calculated from surface magnetic data during rotation 1602/1603 compared with
the outline of coronal holes (after Levine, 1977).

material or the hotter gas in the lower corona. During some events this could be
investigated by comparison with Hα and X-ray pictures. During an event which took
place on August 13, 1973, enhanced coronal X-ray emission near the solar limb was
observed (Rust and Hildner, 1976). The upper picture in Figure 21 shows the X-ray loop
near the limb. The feature expanded within one hour by about 10^4 km into the outer
corona. This can be seen most clearly by representing the ratios of the X-ray intensities
at various times, e.g. the intensity at a later time related to the intensity at an earlier time.
The results are shown in the lower pictures in Figure 21. Bright spots indicate the
locations where the hot matter has moved in, dark spots indicate locations where it has
moved out.

The velocity of the outward moving structures visible in the pictures is about
20 km s^{-1}, half an hour later it is 30 km s^{-1}, i.e., an acceleration occurs. The structures

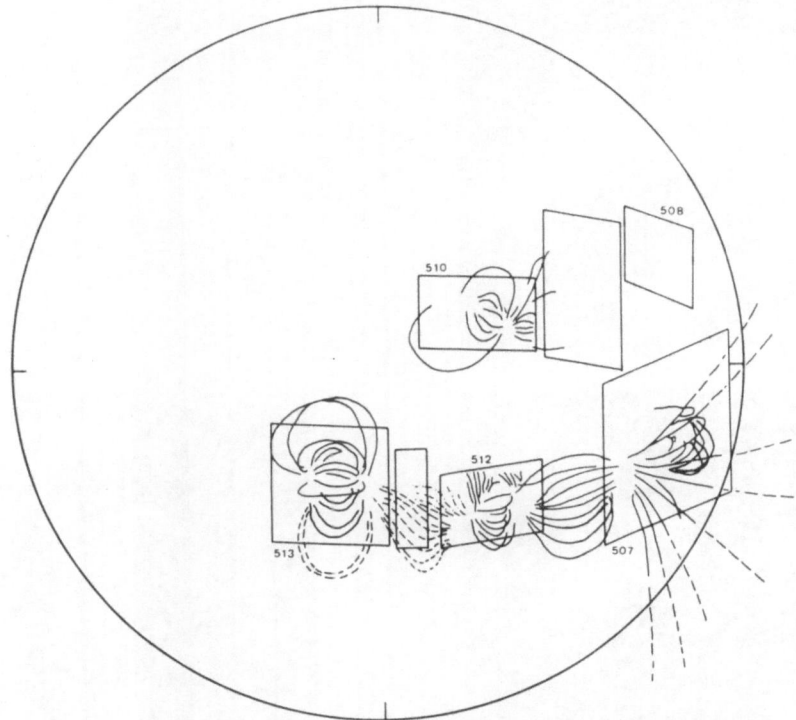

Fig. 20.   Line-of-sight projection of magnetic field lines calculated for September 7, 1973. The full lines are calculated for potential fields, the broken lines for force-free fields (after Elwert *et al.*, 1982).

visible in X-rays and in white light are drawn in Figure 22 for various times. The white light arch appears to originate from the material expelled from the X-ray arch.

In Figure 23 the height of the expanding arch versus time is plotted. The propagation of the expanding arch may be described approximately by a constant outward acceleration of about 12 m s^{-2}. From the flux of the white light the excess mass in the outer corona between two and six solar radii can be estimated as 2×10^{15} g.

By quantitative analysis of the X-ray picture one can deduce the temperature and the differential emission measure of the hot material observed near the limb. With a plausible length of the line of sight of about 10^{10} cm the electron density can be estimated. Allowing for the spatial extension of the X-ray structure the total number of electrons can be derived. It results in approximately 10^{39} electrons corresponding to a total mass of about 10^{15} g. From this agreement one can conclude that the arches of the unstable active region visible in X-rays expand and transport nearly the total mass. A similar event was observed by Skylab on January, 1974 (Webb *et al.*, 1981). In this case sequential and overlapping data were obtained in Hα as well as in soft X-rays in white light.

These data made it possible to follow the trajectory of the near-surface cool gas visible in Hα and the hot gas with a temperature of the order of 10^6 K from the limb to large

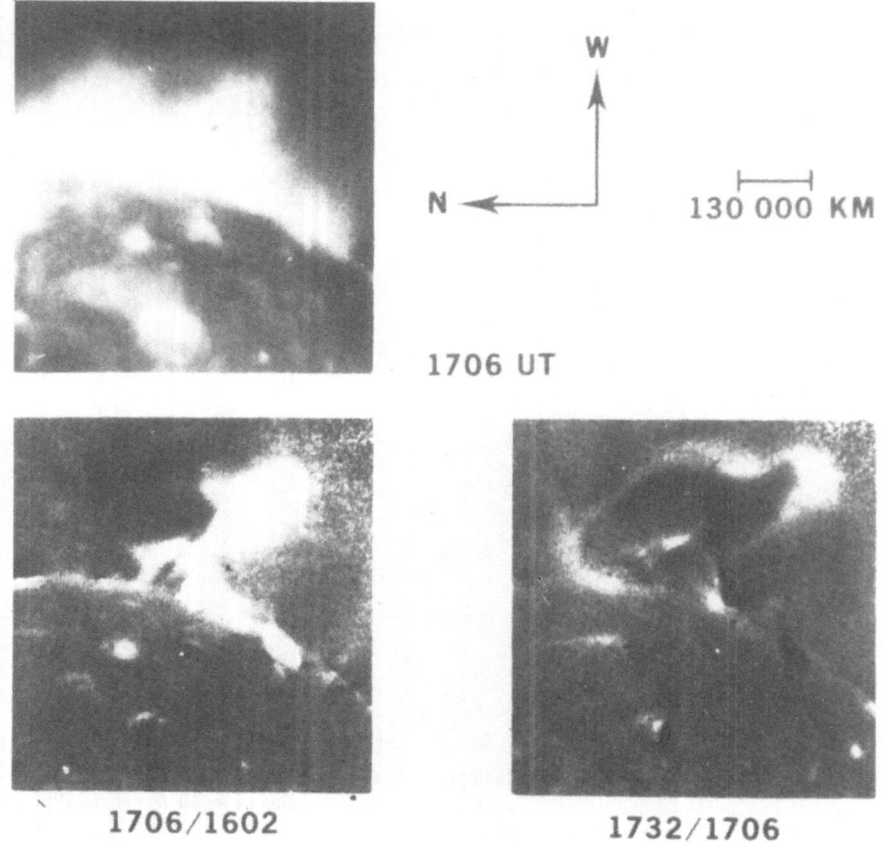

Fig. 21. Top: X-ray image of the corona over McMath 12472, August 13, 1973. Bottom left: Computed ratio of 'before and after' X-ray images shows depletion region (dark), moving material (bright) and hook-shaped enhancement associated with the prominence eruption. Bottom right: Ratio of images during arch expansion: bright = position at 17:32 UT; dark = position at 17:06 UT (after Rust *et al.*, 1976).

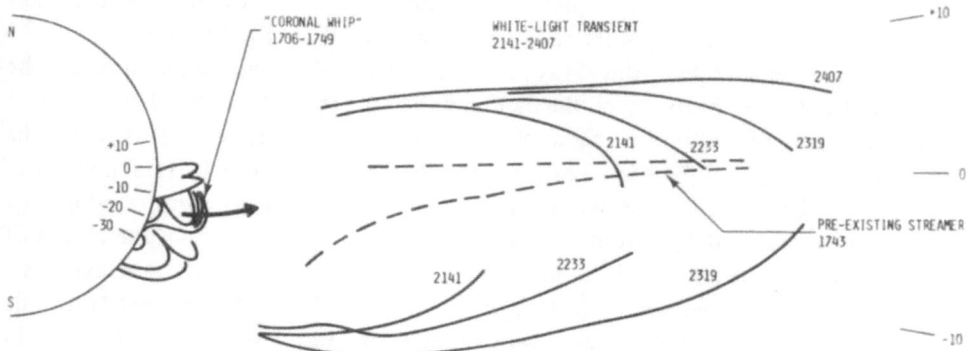

Fig. 22. Composite drawing of the mass ejection as deduced from the X-ray photographs and white light images. Heavy lines indicate the edges of the white light transient at the times indicated (after Rust *et al.*, 1976).

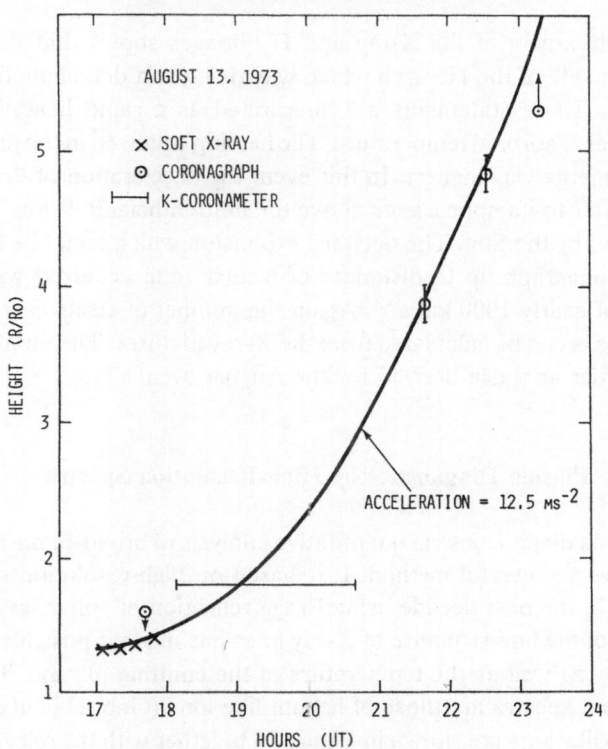

Fig. 23.   Height vs time plot showing progress of the expanding X-ray arch in the inner corona and of the leading edge of the white light bubble in the outer corona (after Rust *et al.*, 1976).

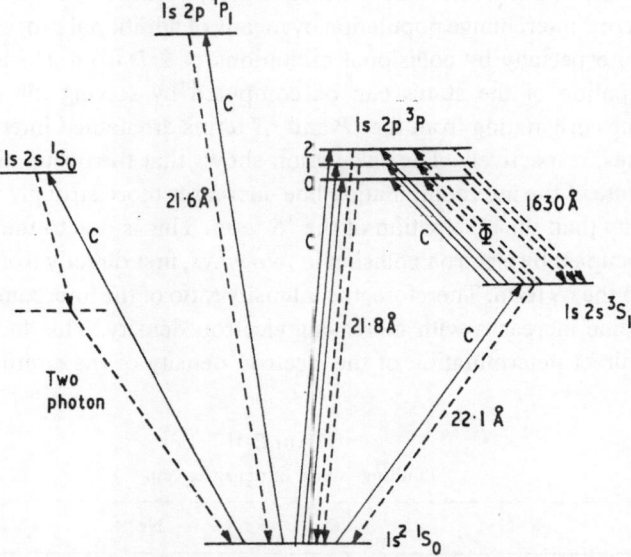

Fig. 24.   Schematic representation of the lower energy levels of a helium-like ion showing processes taken into account in the analysis. The wavelengths indicated relate to OVII (after Gabriel *et al.*, 1972).

distances. The comparison of the X-ray and Hα images shows that the X-ray arch coincided with that side of the Hα arch which was rising and detaching from the solar surface the fastest. These statements are interpreted as a rapid heating of the cool prominence material to coronal temperature. The heating occurred in the part of the arch where the kinetic energy was highest. In this event the acceleration of the prominence occurred mainly prior to its appearance above the limb in heights below 3.5×10^4 km which were occulted by the Sun. The outward expansion which could be followed with the white light coronagraph up to distances of 6 solar radii occurred with an almost constant velocity of nearly 1000 km s^{-1}. Again the number of electrons of the emitted gas and the total mass can be calculated from the X-ray pictures. The numbers obtained are of the same order as those derived for the August event.

8. Plasma Diagnostics by High-Resolution Spectra

In addition to plasma diagnostics via quantitative analysis of broad-band X-ray pictures there is a second very powerful method. It is based on high-resolution spectra which could be obtained in the past decade using Bragg reflection of soft X-rays at crystals. The measurement of the fine-structure of X-ray lines has made it possible to determine the electron density as well as the temperature of the emitting plasma. The prototype of density dependent spectra are those of helium-like ions (Gabriel *et al.*, 1969, 1972).

The terms of He-like ions are shown in Figure 24 together with the relevant excitation and decay mechanisms. Collision processes are drawn as full lines, radiation processes as dashed lines. The ^1P term is populated by electron impact which is balanced by radiative decay producing the resonance line. However, because of their long lifetimes the 2^3P and 2^3S terms interchange population by means of additional processes between the excited states, especially by collisional excitations of 2^3P from 2^3S and radiative decay. The occupation of the states can be computed by solving the coupled rate equations. The lines originating from the ^3P and ^3S terms are named intercombination and forbidden lines, respectively. The calculation shows that the occupation of the ^3P term, the initial state of the intercombination line, increases more strongly with increasing electron density than the occupation of the ^3S term. This is due to the fact that the ^3P term can be occupied by electron collision in two ways, first directly from the ground level, secondly via the ^3S term. Therefore, the intensity ratio of the intercombination line to the forbidden line increases with increasing electron density. This fact provides a possibility for a direct determination of the electron density of the emitting plasma.

TABLE II

Limiting values of densities (cm^{-3})

C v	N vi	O vii	Ne ix	Mg xi
1.4×10^8	1.2×10^9	7.2×10^9	1.5×10^{11}	1.8×10^{12}

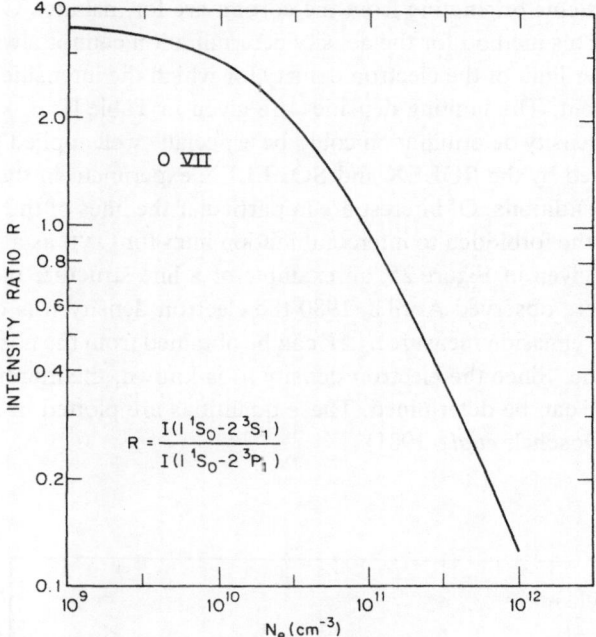

Fig. 25. Intensity ratio of the forbidden to intercombination line for O VII as a function of the electron density (after Doschek *et al.*, 1981).

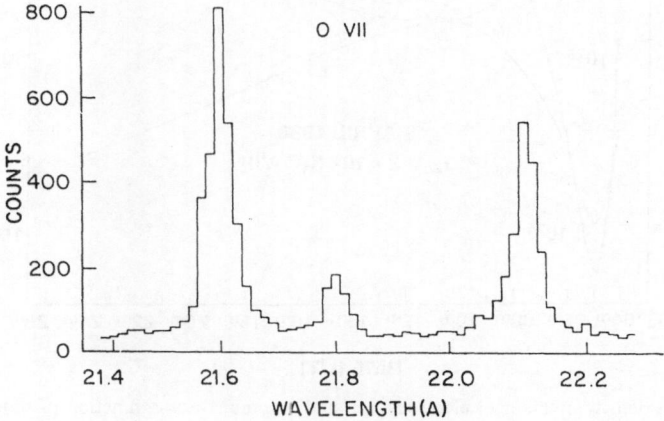

Fig. 26. Observed fine structure of the O VII line (after Acton *et al.*, 1972).

Two-electron systems originating from the corona are for instance C V, N VI, O VII, Ne IX, and Mg XI. This method for the density determination cannot always be applied since there is a lower limit of the electron density for which the intensities of both lines are distinctly different. The limiting densities are given in Table II.

This method of density determination could be especially well applied to high resolution spectra observed by the SOLEX and SOLFLEX experiment in the spectrometer mode under flare conditions. Of interest are in particular the lines of the ion O VII. The intensity ratio R of the forbidden to intercombination lines for O VII as a function of the electron density is given in Figure 25, an example of a line structure measurement in Figure 26. For a flare observed April 8, 1980 the electron density was derived. Additionally, the volume emission measure $n_e^2 \Delta V$ can be obtained from the absolute intensity of the resonance line. Since the electron density n_e is known, the number of particles in the volume $n_e \Delta V$ can be determined. These quantities are plotted in Figure 27 as a function of time (Doschek *et al.*, 1981).

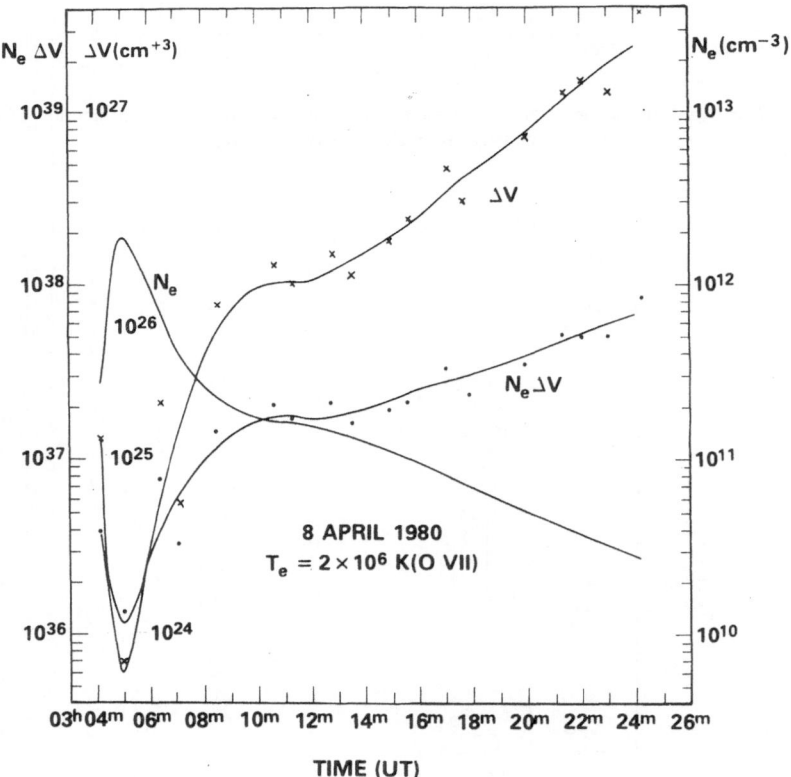

Fig. 27.   Electron density, number of electrons $n_e V$ and volume ΔV as a function of time for a flare on April 8, 1980. The volume ΔV refers to the plasma volume in which the O VII line was emitted (after Doschek *et al.*, 1981).

Further information about the physical state of the plasma can be obtained by using spectra with even higher resolution measured by the SOLFLEX experiment. In this experiment spectra can be scanned in four diagnostically important short wavelength bands between about 2 and 8 Å. They contain, for instance, lines of the He-like ions Fe XXV and Ca XIX and the lithium-like ions Fe XXIV and Ca XVIII. The lithium-like satellite lines are produced by dielectronic recombination to He-like ions. In this process an electron colliding with an ion is captured in a highly excited state of the ion. The surplus energy of the impact electron instead of being directly radiated is used to excite another bound electron of the recombining ion. The latter having been doubly excited can undergo a transition to the ground level by photon emission. In this way dielectronically excited satellite lines are produced.

The ratios of these dielectronic lithium-like satellite lines to the He-like resonance lines can be used to determine the electron temperature, as theory shows that these ratios are temperature dependent (cf. Gabriel, 1972; Vainstein *et al.*, 1973; Bhalla *et al.*, 1975; Safronova *et al.*, 1978). Examples are the satellite lines of Fe XIV and Ca XVIII and the

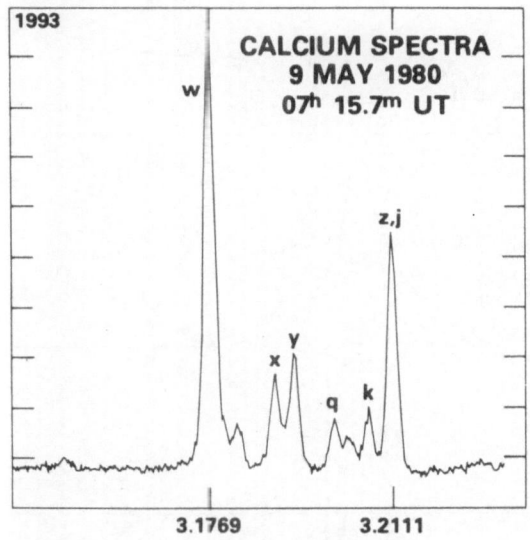

Fig. 28. Ca SOLFLEX-spectrum for a flare on May 9, 1980. The temperature of the regions emitting the lines of Ca XIX can be derived from the intensity ratio of the dielectronically excited line *k* to the resonance line *w* (after Doschek *et al.*, 1981).

resonance lines of Fe XXV and Ca XIX (Figure 28). This temperature determination is independent of whether or not the plasma is in ionization equilibrium. During the rise phase of flares temperatures of about 2×10^7 K are obtained (Feldman *et al.*, 1980a).

From these locally derived temperatures further interesting conclusions can be drawn. Using these temperature values the thermal width of the lines can be derived. As the spectral resolution is very high, i.e., of the order 10^{-3} Å at 1 Å, the observed width can be measured and compared with the theoretical one. Most of the line profiles can be represented by a Gaussian distribution, but the observed widths are broadened in excess of the predicted thermal widths. Thus the excess widths can be interpreted as a non-thermal random velocity (Doschek *et al.*, 1979, 1980). In the rise phase of a flare the motions are largest and attain a velocity of 150 km s^{-1}.

In some flares (class M in 1–8 Å X-rays), however, a blue-shifted component was observed during the rise phase. It appears as an asymmetrical extension of the blue wing of the resonance line, for instance in the spectrum of Ca xix (Feldman *et al.*, 1980b).

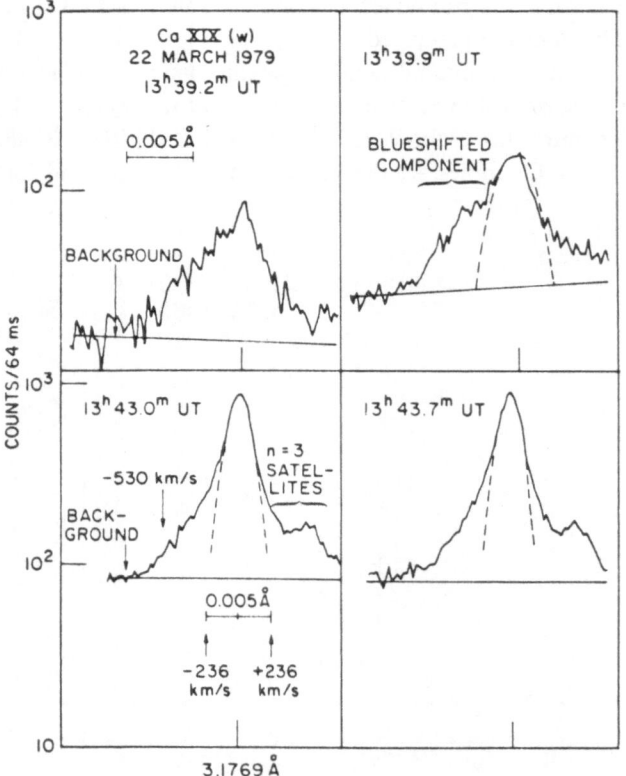

Fig. 29. Spectra of Ca xix showing a blueshift component during the rise phase and the maximum flux for a flare on March 22, 1979 (after Feldman *et al.*, 1980).

Figure 29 shows line profiles for different times during the rise phase of the flare and the time of maximum flux. The blue-shifted component is indicated, the dashed lines are Gaussian fits to the unshifted line components. From this asymmetrical line profile it can be concluded that the hot plasma with a temperature of about 2×10^7 K is moving outwards from the Sun with a velocity along the line of sight of about 400 to 600 km s^{-1}, i.e., velocities which are typical of low temperature ejecta earlier seen in Hα.

References

Acton, L. W., Catura, R. C., Meyerott, A. J., Wolfson, C. J., and Culhane, J. L.: 1972, *Solar Phys.* **26**, 183.

Altschuler, M. D., Levine, R. L., Stix, M., and Harvey, J.: 1977, *Solar Phys.* **51**, 345.

Bartoe, J.-D. F., Brueckner, G. E., Purcell, J. D., and Tousey, R.: 1977, *Appl. Opt.* **16**, 879.

Bhalla, C. P., Gabriel, A. H., and Presnyakow, L. P.: 1975, *Monthly Notices Roy. Astron. Soc.* **172**, 359.

Burgess, A.: 1965, *Astrophys. J.* **141**, 1588.

Burke, H. O. and Davis, A. J.: 1974, *Proc. Soc. Photo-Optical Instr. Engin.* **44**, 775.

Doschek, G. A., Kreplin, R. W., and Feldman, U.: 1979, *Astrophys. J.* **233**, L137.

Doschek, G. A., Feldman, U., and Kreplin, R. W.: 1980, *Astrophys. J.* **239**, 725.

Doschek, G. A., Feldman, U., Landecker, P. B., and McKenzie, D. L.: 1981, *Astrophys. J.* **249**, 372.

Dupree, A. K., Huber, M. C. E., Noyes, R. W., Parkinson, W. H., Reeves, E. M., and Withbroe, G. L.: 1973, *Astrophys. J.* **182**, 321.

Elwert, G.: 1952, *Z. Naturforschung* **7a**, 432.

Elwert, G.: 1954, *Z. Naturforschung* **53**, 637.

Elwert, G., Villing, W., Vorpahl, J., and Broussard, R. M.: 1981, *Astron. Astrophys.* **101**, 150.

Elwert, G., Müller, K., Thür, L., and Balz, P.: 1982, *Solar Phys.* **75**, 205.

Elzner, L. R. and Elwert, G.: 1980, *Astron. Astrophys.* **86**, 188.

Feldman, U., Doschek, G. A. and Kreplin, R. W.: 1980a, *Astrophys. J.* **238**, 365.

Feldman, U., Doschek, G. A. and Kreplin, R. W.: 1980b, *Astrophys. J.* **254**, 1175.

Gabriel, A. H.: 1972, *Monthly Notices Roy. Astron. Soc.* **160**, 99.

Gabriel, A. H.: 1976, *Phil. Trans. Roy. Soc. London* **A281**, 239.

Gabriel, A. H. and Jordan, C.: 1969, *Monthly Notices Roy. Astron. Soc.* **145**, 241.

Gabriel, A. H. and Jordan, C.: 1972, in E. W. McDaniel and M. R. C. McDowell (eds.), *Case Studies in Atomic Collision Physics II*, Chapter 4.

Giacconi, R., Reidy, W. P., Zehnpfennig, T., Lindsay, J. C., and Muney, W. S.: 1965, *Astrophys. J.* **142**, 1274.

Henze, W., Jr. (ed.): 1976, NASA TM X-73369.

Huber, M. C. E. and Timothy, J. G.: 1977, *Space Sci. Instr.* **3**, 389.

Huber, M. C. E., Foukal, P. V., Noyes, R. W., Reeves, E. M., Schmahl, E. J., Timothy, J. G., Vernazza, J. E., and Withbroe, G. L.: 1974, *Astrophys. J.* **194**, L115.

Jordan, C.: 1969, *Monthly Notices Roy. Astron. Soc.* **142**, 499.

Jordan, C.: 1970, *Monthly Notices Roy. Astron. Soc.* **149**, 1.

Kahler, S.: 1976, *Solar Phys.* **48**, 255.

Kato, T.: 1976, *Astrophys. J. Suppl. Ser.* **30**, 397.

Krieger, A. S., Timothy, A. F., and Roelof, E. C.: 1973, *Solar Phys.* **29**, 505.

Levine, R. H., Altschuler, M. D., and Harvey, J. W.: 1977, *Astrophys. J.* **215**, 636.

Landecker, B. P. and McKenzie, D. L.: 1981, Space Sci. Lab. Rep. No. SSL-81(6960-01)-6.

McKenzie, D. L., Landecker, P. B., and Underwood, J. H.: 1976, *Space Sci. Instr.* **2**, 125.

Maute, K. and Elwert, G.: 1981, *Solar Phys.* **70**, 273.

Neupert, W. N.: 1971, in Macris, C. J. (ed.), *Physics of the Solar Corona*, p. 237.

Reeves, E. M., Timothy, J. G., Huber, M. C. E., and Withbroe, G. L.: 1977, *Appl. Opt.* **16**, 849.

Rust, D. M. and Hildner, E.: 1976, *Solar Phys.* **48**, 381.

Safronova, U. I., Urnov, A. M., and Vainstein, L. A.: 1978, P. N. Lebedev Physical Institute, Preprint No. 212.

Tousey, R.: 1976, *Phil. Trans. Roy. Soc. London* **A281**, 359.

Tousey, R., Bartoe, J.-D. F., Brückner, G. E., and Purcell, J. D.: 1977, *Appl. Opt.* **16**, 870.

Tucker, W. H. and Koran, M.: 1971, *Astrophys. J.* **168**, 283.

Underwood, J. H., Milligan, J. E., de Loach, A. C., and Hoover, R. B.: 1977, *Appl. Opt.* **46**, 858.

Vaiana, G. S.: 1976, *Phil. Trans. Roy. Soc. London* **A281**, 365.

Vaiana, G. S., Krieger, A. S., and Timothy, A. F.: 1973a, *Solar Phys.* **32**, 81.

Vaiana, G. S., Davis, J. M., Giacconi, R., Krieger, A. S., Silk, J. K., Timothy, A. F., and Zombeck, M.: 1973b, *Astrophys. J.* **185**, L47.

Vaiana, G. S., Krieger, A. S., Petrasso, R., Silk, J. K., and Timothy, A. F.: 1974, *Proc. Soc. Photo-Opt. Instr. Engin.* **44**, 185.

Vaiana, G. S., van Speybroeck, L., Zombeck, M. W., Krieger, A. S., Silk, J. K., and Timothy, A. F.: 1977, *Space Sci. Instr.* **3**, 19.

Vernazza, J. E. and Reeves, M. E.: 1978, *Astropys. J. Suppl.* **37**, 485.

Webb, D. F. and Jackson, B. V.: 1981, *Solar Phys.* **73**, 341.

Widing, K. G.: 1975, in R. S. Kane (ed.), 'Solar Gamma-, X-, and EUV Radiation', *IAU Symp.* **68**, 153.

Wolter, H.: 1952, *Ann. Phys.* **10**, 94.

HARD X-RAYS FROM THE SUN*

E. HAUG

Lehrstuhl für Theoretische Astrophysik der Universität Tübingen, F.R.G.

Abstract. Temporal and spectral characteristics of solar hard X-ray bursts are briefly reviewed. The merits of non-thermal and thermal flare models are discussed. The validity of these models may be checked by future measurements of X-ray polarization. Finally, some important results of recent satellite experiments are described providing information on the spatial distribution of hard X-ray sources: the multi-spacecraft observation of X-ray bursts and the imaging of X-ray sources by means of the HXIS instrument.

1. Introduction

Less than 24 yr ago, in March, 1958, the first hard X-rays originating from the solar atmosphere were detected during a flare (Peterson and Winckler, 1959). Since that time a great number of measurements from balloons, rockets, and satellites have supplied a bulk of data showing the significance of solar hard X-radiation for understanding the physical state of the disturbed solar atmosphere, in particular the corona. Nevertheless, the available information is far from being sufficient to distinguish between various models suggested to explain the phenomena taking place during hard X-ray bursts.

Solar hard X-radiation with photon energies above some 15 keV is produced in observable intensity only during flares. A typical X-ray flare can be roughly divided into three phases (Kane, 1974): precursor, impulsive (flash) phase and gradual phase, which are schematically illustrated in Figure 1. The onset of the X-ray flare is indicated by a slow enhancement of soft X-rays announcing a possible occurrence of the gradual and/or impulsive phase within the following 10 min or so. The impulsive phase, which has a duration of a few tens of seconds to a few minutes, is characterized by a rapid increase and decrease in the radiation flux. The flash phase is most pronounced in hard X-rays above 20 keV, and the typical rise and decay times are shorter the higher the photon energy is (Kane and Anderson, 1970). With high time resolution the radiation flux may be decomposed into several short-lived spikes, the so-called 'elementary flare bursts' (van Beek *et al.*, 1974; De Jager and De Jonge, 1978), which may have a quasiperiodic structure (Figure 2). Then the duration of the impulsive phase may be essentially longer than in simple cases.

In the gradual phase the soft X-ray flux has a smooth time profile, and the intensity reaches a maximum *after* the impulsive phase peak, and then decreases slowly to the pre-flare level. The total duration of the gradual phase is ≳ 10 min. It is important to stress that the three phases are not necessarily present in all flares: the impulsive hard component is missing in many soft X-ray bursts (Kane, 1969). However, an impulsive burst is always associated with a gradual phase in the soft X-region.

According to these characteristics I will primarily deal with the impulsive phase of X-ray bursts comprising the main part of hard X-radiation. The impulsive phase can

---

* Paper presented at the IX-th Lindau Workshop 'The Source Region of the Solar Wind'.

Space Science Reviews **33** (1982) 83–97. 0038–6308/82/0331–0083$02.25.

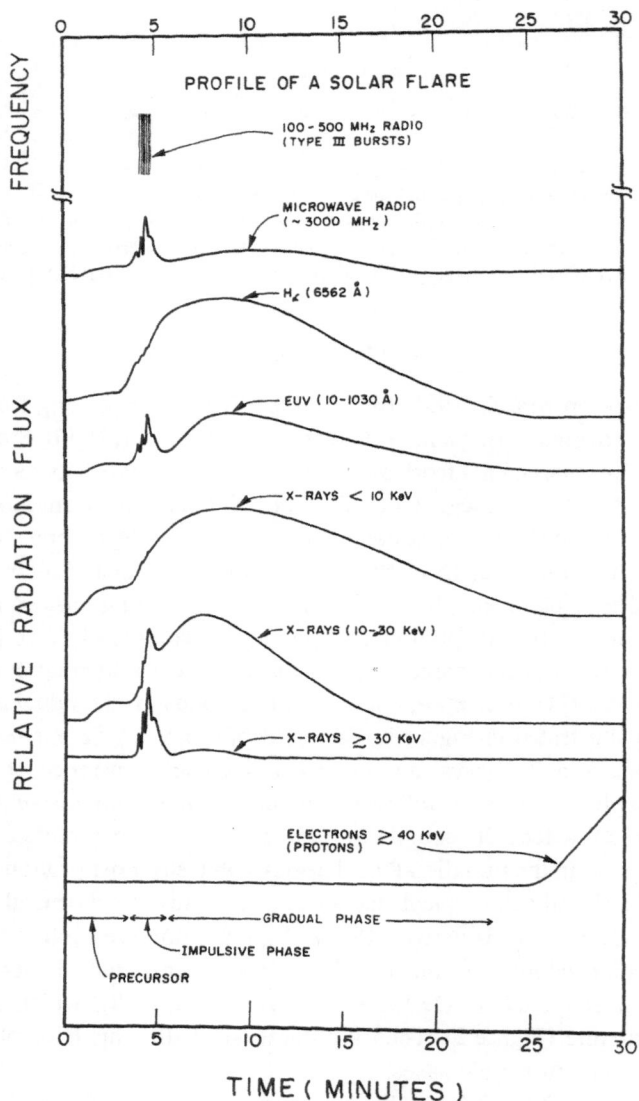

Fig. 1. Schematic representation of the different phases of a solar flare as observed in electromagnetic and particle radiation. (After Kane, 1974.)

also be observed in EUV (10–1030 Å) (Kane and Donnelly, 1971; Donnelly and Kane, 1978) and at radio frequencies (Kundu, 1961; Kane, 1981b; Kosugi, 1981). The radio bursts most closely correlated to impulsive X-ray bursts are the microwave bursts ($\lambda \lesssim 30$ cm, $\nu \gtrsim 1000$ MHz) and the type III radio bursts at decimeter and meter wavelengths. The simultaneity of impulsive hard X-ray, EUV and microwave bursts and their often remarkable correspondence even in temporal fine structures indicate their common origin from an electron population which is accelerated to high energies. This,

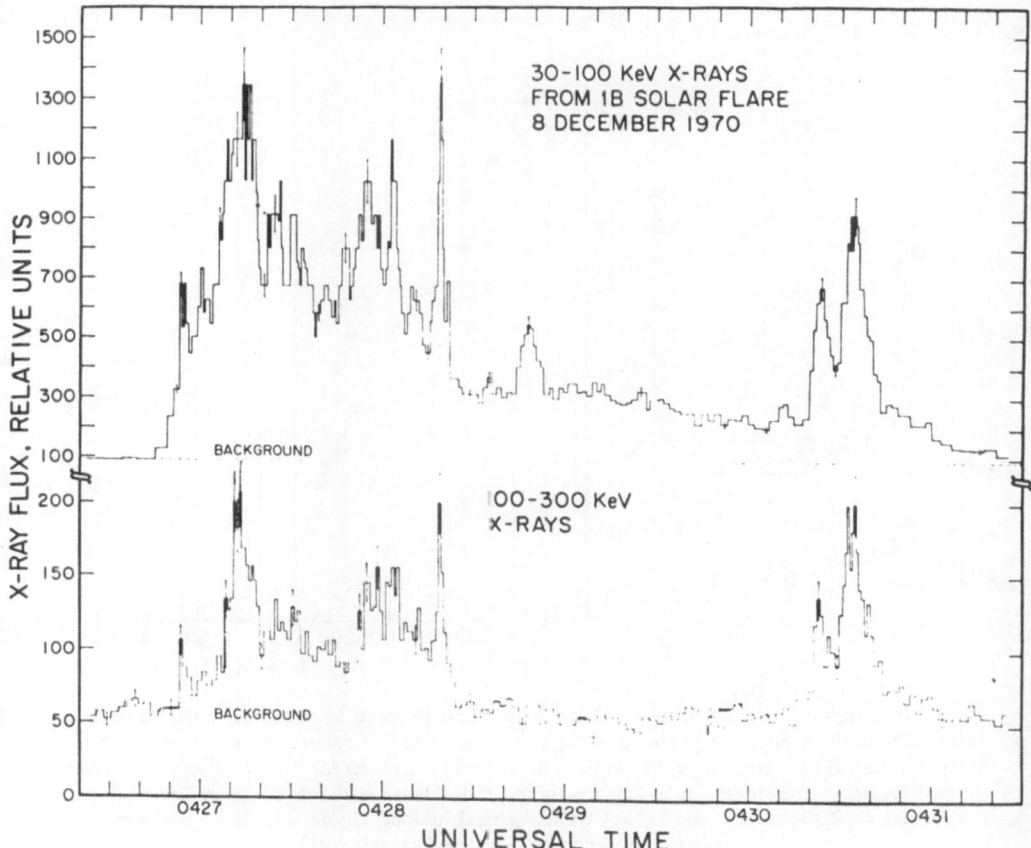

Fig. 2. Complex hard X-ray burst during a 1B solar flare on 8 December 1970 consisting of many elementary flare bursts. (After Anderson and Mahoney, 1974.)

however, does not necessarily mean that the various radiations are emitted from the same spatial region (cf. Section 4). The rapid time variations in the hard X-ray, microwave and type III radio bursts indicate an electron acceleration process which is continuous in time or more probably a continuous series of elementary flare bursts with short time constants which may be produced subsequently at different locations in the active region (Cheng *et al.*, 1981). In a few X-ray flares the occurrence of a gradual *hard* X-ray component gives evidence of a second stage of particle acceleration producing higher energy electrons and protons (Frost and Dennis, 1971; Hudson, 1978). Figure 3 shows an impulsive X-ray burst followed by a slower long lasting (40 min) burst which is most conspicuous at high photon energies above 225 keV. That is, the X-ray spectra during the slow burst are considerably harder than those during the impulsive flare.

The second stage of acceleration is highly correlated with type II radio bursts which are produced by shock waves propagating in the solar corona. The slower rise of the

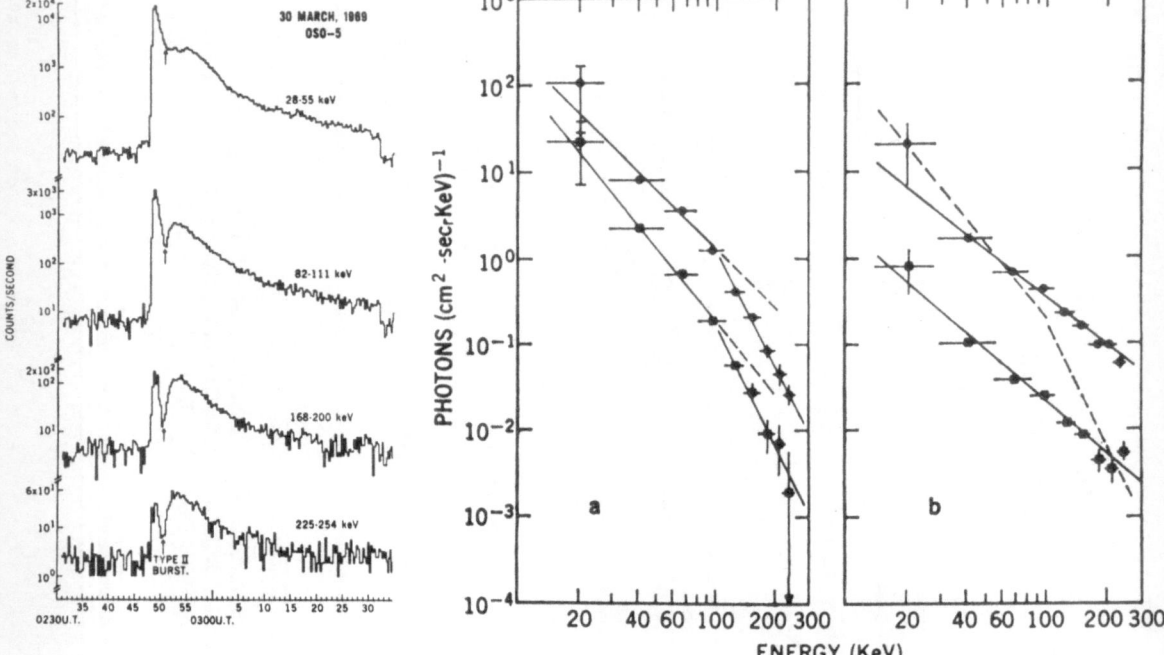

Fig. 3. Intensity-time profiles of an impulsive X-ray burst followed by a gradual burst which is most pronounced at the highest energy channels. (a) X-ray spectra at two different times during the impulsive burst; the dashed lines emphasize the break of the slope of the power laws near 100 keV. (b) Spectra at the peak (circles) and during the decay (squares) of the gradual burst; both spectra can be fitted with a power law of index $\gamma \approx 2.0$ between 15 and 250 keV without a break in slope. The dashed curve represents the lower spectrum of (a). (After Frost and Dennis, 1971.)

second burst, its harder spectrum, the larger number of energetic photons and the good correlation with type II bursts suggest that the acceleration mechanism of the second stage is different from the mechanism of the first stage (Frost and Dennis, 1971).

The knowledge about the mechanism which accelerates the electrons to energies up to several hundreds of keV is quite poor in spite of all the observational and theoretical efforts during the past 20 yr. In any case the magnetic fields and in particular the reconnection of antiparallel field lines will play a decisive part (Švestka, 1976). Also, the energy stored in the magnetic field around sunspots is by far sufficient to supply the power required for the acceleration of a large number of particles.

2. X-Ray Spectra and Flare Models

The X-ray spectrum at photon energies $h\nu$ above some 10 keV is in most cases consistent with a power law of the differential photon flux

$$I(h\nu) = K(h\nu)^{-\gamma} \tag{1}$$

(Kane and Anderson, 1970). Here K is a constant and the spectral index γ takes values between 2 (hard spectrum) and 8 (soft spectrum). Most frequent are spectral indices around $\gamma = 4$ to 5 whereas very hard radiation with $\gamma < 3$ is extremely rare. Above some critical energy $h\nu_0$ the spectrum steepens rapidly, i.e., the spectral index γ increases (Kane and Anderson, 1970). This may be expressed by an additional cut-off factor $e^{-\nu/\nu_0}$ in Equation (1). Apart from some large events where the power law extends to several hundreds of keV (Suri et al., 1975) the value of $h\nu_0$ lies in the range of 60–100 keV.

This hard X-radiation is produced as bremsstrahlung in collisions between energetic electrons and the ions of the solar atmosphere; electron-electron breemsstrahlung can be neglected (Haug, 1975). The X-ray spectrum is related to the energy distribution of the fast electrons in the X-ray source via the bremsstrahlung cross section. For example, the power-law spectrum (1) implies a power-law electron spectrum (Brown, 1971).

However, a power-law spectrum is by no means a unique fit to the observational data. The proportional and scintillation counters used as detectors have a rather poor spectral resolution. Because of the low photon fluxes at higher energies spectral information is generally limited to a few channels. So it is not surprising that any rapidly decreasing function can be fitted quite well to the data. This function is usually taken to be the power law (1). In some cases a better fit to the data is provided by a thermal spectrum of the form

$$I(h\nu) = K \, \frac{g(h\nu, kT)}{h\nu\sqrt{kT}} \, e^{-h\nu/kT} \tag{2}$$

(Crannell et al., 1978; Mätzler et al., 1978; Elcan, 1978) with $kT = 10$ to 100 keV; $g(h\nu, kT)$ is the Gaunt factor. Such a spectrum would be produced in an isothermal plasma of temperature T. Besides, a power-law X-ray spectrum may also be produced by a multi-temperature thermal distribution of electrons with temperatures between 10^8 and 10^9 K (Chubb, 1972; Brown, 1974). The controversy about the nature of the hard X-ray spectrum – thermal or non-thermal – is closely related to the validity of different models of X-ray bursts (Brown and Smith, 1980; Kane, 1981a). It will therefore be highly desirable to have more accurate measurements of the hard X-ray spectrum with high time resolution.

Assuming the power law (1) for the hard X-ray spectra, the energy distribution of the source electrons has also the form of a power law, i.e., the electron flux is given by

$$\Phi(E) = AE^{-\delta} \tag{3}$$

where the spectral index δ is dependent on the spectral index γ of the photon spectrum and A is a function of K, γ, and the particle density of the source (Lin and Hudson, 1971; Brown, 1971). Thus the measurement of the X-ray spectrum provides information on the population of energetic electrons.

A small fraction of less than 1% of the electrons accelerated during the flare escapes from the solar corona and can be detected in the interplanetary space by satellite-borne instruments (Lin, 1974). As is to be expected, these electrons have a power-law spectrum of the form (3). They could be detected up to energies of 50 MeV (Datlowe, 1971); their spectral index δ ranges from 2 to 5 with the great majority of the events falling between $\delta = 3$ and 4.5.

After having derived the energy distribution of the accelerated electrons from the X-ray spectra, some characteristics of the process can be determined, e.g. the total number and the kinetic energy of the electrons (above some lower limit E_0) and the efficiency of acceleration (Lin and Hudson, 1971). As a result the source electrons appear to carry more energy than any other particle component in the flare, and possibly a substantial fraction of the total flare energy. This leads to the difficulty that an improbably high acceleration efficiency is demanded and that the number of accelerated electrons is extremely high, comprising virtually all the electrons in the flare volume.

These are the reasons for the efforts to explain the hard X-radiation by thermal models where the source electrons are part of a distribution of all the electrons present (with mean energies $\gtrsim 10$ keV or $T \gtrsim 10^8$ K) rather than comprising a fast component in a 'cool' background (Brown, Melrose and Spicer, 1979; Smith and Lilliequist, 1979; Smith and Auer, 1980; Emslie, 1981). The essential point of these thermal models is that in a confined thermal plasma the only electron energy losses are by bremsstrahlung and the collisional loss to ions if these have a lower temperature than the electrons, whereas in a non-thermal source the electrons lose the predominant fraction of their energy by collisions. Thus nearly 100% of the available electron energy goes into radiation compared to less than 0.01% for a non-thermal component.

The most serious objections raised against the application of thermal models to hard X-ray bursts are (Kahler, 1971a, b; 1975): firstly, at the high temperatures needed, the collisional mean free paths and relaxation times exceed, respectively, feasible source sizes and burst durations for acceptable plasma densities. Therefore the distribution cannot be collisionally relaxed, i.e., the particles cannot have a Maxwellian energy distribution and a well defined temperature at any point of the X-ray source. Secondly, the limit of efficiency of an unconfined hot plasma is not given by the collisional energy losses but rather by the rapid conductive and convective cooling resulting from the high temperature and pressure gradients present.

These problems may be overcome by invoking relaxation of the electron energy distribution by noncollisional, i.e., collective, processes and anomalous conductivity effects (Smith and Lilliequist, 1978; Smith and Auer, 1980).

In an alternative form of thermal models which obviates the two problems mentioned above, the source is heated and cooled by successive adiabatic compression and expansion of a plasma kernel by magnetic-field variations (Mätzler et al., 1978).

In conclusion, thermal hard X-ray burst models seem capable of reducing the electron number and energy requirements by at least an order of magnitude from non-thermal model values and permit this energy to appear as a bulk electron energization with random motions rather than as directed electron beams.

3. Polarization of X-Radiation

A decision on the validity of thermal or non-thermal burst models could be possible by measurements of the linear polarization of hard X-rays. Up to now I have only considered the temporal variation and the spectra of the X-radiation. Additional information concerning the nature of the source electrons and their origin can be obtained by observing the X-ray polarization. In particular the anisotropy of the source electrons can be inferred from such measurements. Unfortunately the results reported so far are inconclusive due to the extreme difficulty of the experiments.

The linear polarization is characterized by the degree of polarization, P, and the orientation angle. A nonvanishing polarization can only be expected if there is some anisotropy in the source. This is *not* the case for instance in an isotropic thermal source, apart from a weak polarization caused by Compton backscattering (Bai and Ramaty, 1978). On the other hand, the non-thermal bremsstrahlung produced in solar flares should be polarized assuming that the accelerated electrons move predominantly into one direction, i.e., if they have an anisotropic velocity distribution (Elwert and Haug, 1970; Haug, 1972; Brown, 1972). This assumption is fulfilled in the non-thermal models of impulsive X-ray sources where the electrons spiral around magnetic field lines with a certain pitch-angle distribution. For a given electron spectrum and pitch-angle distribution the degree of polarization P can be computed as a function of the photon energy and the angle ϑ between the magnetic field and the line of sight. As an example, Figure 4 shows the calculated polarization curves for an electron power-law with exponential cut-off at high energies and for pitch-angle distributions of the form $\sin^2 \alpha$ and $\cos^2 \alpha$ (Haug, 1972).

The resulting values of P are maximum near $\vartheta = 90°$. According to the sign of P, the orientation of the polarization vector is perpendicular or parallel to the projected direction of the magnetic field, respectively.

The measurement of solar X-ray polarization is very difficult due to the low intensities and the rapid time variations. The polarimeters are based on the dependence of the Compton scattering cross section on the polarization state of the incident X-rays (Tindo *et al.*, 1970; Novick, 1975). All the solar X-ray polarimeters employed up to now failed to meet the ambitious requirements to determine the temporal and spectral features of solar X-ray polarization. Limited effective area, long integration periods, poor spectral resolution and inherently non-symmetric construction all contributed to limit severely the usefulness of these instruments. So far there exist only a few results on the X-ray polarization during flares. These have mainly been obtained by the Russian team of Tindo and co-workers with the aid of polarimeters aboard Intercosmos satellites (Tindo *et al.*, 1970, 1972a, b, 1973, 1976). The degree of polarization in a broad X-ray band centered at 15 keV was between 0 and 20% with large uncertainties. The orientation angle of the polarization showed large fluctuations in time. Hence the results of these measurements cannot answer the question which X-ray burst model would be valid.

Recently, the construction of a solar flare X-ray polarimeter has been completed at the Columbia Astrophysics Laboratory of the Columbia University, New York (Novick,

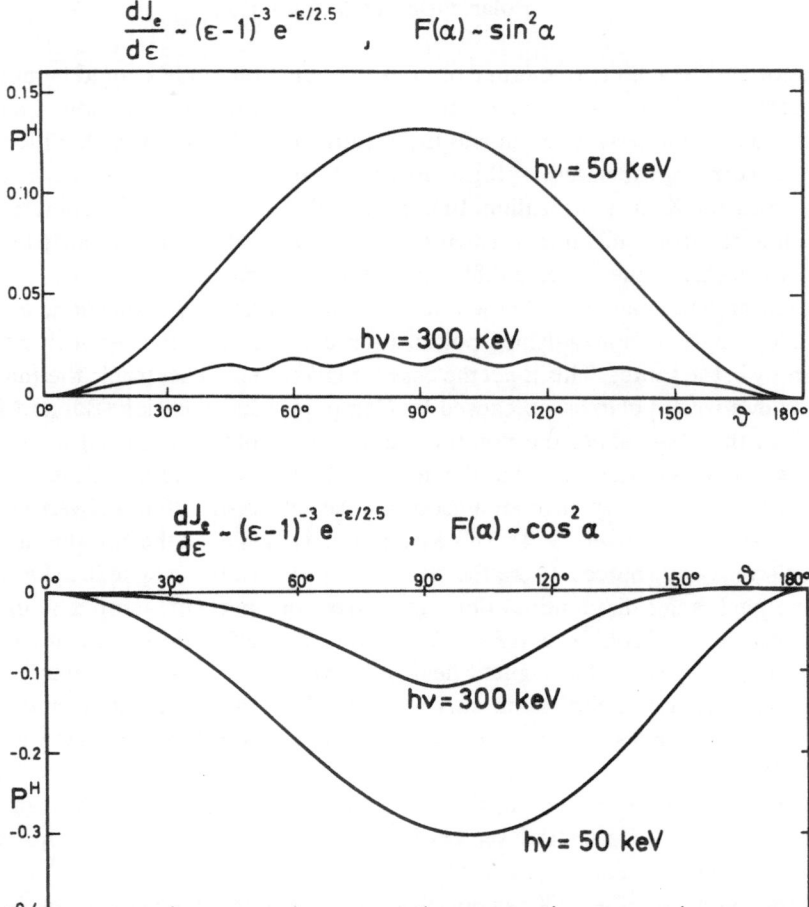

Fig. 4.   Calculated polarization of bremsstrahlung from electrons with the spectrum $\Phi(E) \sim E^{-3} e^{-E/E_0}$ (cut-off energy $E_0 \approx 767$ keV) as a function of the angle ϑ between the magnetic-field direction and line-of-sight for photon energies 50 and 300 keV. (a) Pitch-angle distribution of the electrons around the magnetic field $F(\alpha) \sim \sin^2 \alpha$. (b) Forward-cone pitch-angle distribution $F(\alpha) \sim \cos^2 \alpha$. (After Haug, 1972.)

1981). It is scheduled to fly on an early Space Shuttle mission and later on will hopefully be placed on a long-lived solar observatory. This instrument which is sensitive over the energy band from 5 to 30 keV is expected to be superior to all the previous ones. Extensive tests showed that it has the required sensitivity. Computer estimates proved that useful sensitivities can be achieved in energy bands between 5 and 25 keV in observing times as short as 10 s. It is hoped that this polarimeter will soon provide important new data on the X-radiation originating during the impulsive phases of X-ray bursts.

4. Size and Location of Hard X-Ray Sources

One of the important parameters largely unknown at present is the spatial distribution of the hard X-ray source. Until recently X-ray imaging has been performed only by grazing-incidence telescopes with upper energy threshold at a few keV (Vaiana *et al.*, 1973). Direct measurements of the location and size of the hard X-ray sources are difficult. Nearly all the instruments detecting solar hard X-rays were uncollimated and thus observed the full Sun. Up to the year 1980 only a single one-dimensional observation of an X-ray burst has been achieved by means of a satellite-borne modulation collimator (Takakura *et al.*, 1971). Therefore, information about the altitude of hard X-ray sources could be obtained mainly by the observation of behind-the-limb flares (Catalano and Van Allen, 1973; Roy and Datlowe, 1975; McKenzie, 1975). As a result hard X-ray sources are generally located at altitudes below 10 000 to 20 000 km above the photosphere, even though hard X-ray emission has also been observed from heights of $\approx 70\,000$ km (Hudson, 1978).

Another possibility to determine the vertical structure of the X-ray source are observations with two or more spacecraft separated in heliographic longitude or latitude (Catalano and Van Allen, 1973). Such multi-spacecraft observations can also make important contributions to the investigation of X-ray anisotropy; this problem is closely associated with the X-ray polarization (Elwert and Haug, 1970, 1971). Sixteen solar X-ray bursts have so far been observed simultaneously by detectors aboard the International Sun−Earth Explorer 3 (ISEE-3), orbiting the Earth, the Pioneer−Venus Orbiter (PVO), and HELIOS−B (Kane, 1981a). As an example for the results obtained from such observations I shall discuss the impulsive X-ray burst of 5 October, 1978 which is attributed to a relatively strong flare located about 15° behind the East limb of the Sun (Kane *et al.*, 1979). The ISEE-3 and PVO observations and the view angles of the two spacecraft are shown in Figure 5. Whereas the PVO detector could see the flare region down to an altitude of ≈ 600 km, ISEE-3 could only observe the part of the X-ray source located at altitudes $\gtrsim 25\,000$ km, the lower part being occulted. This occultation accounts evidently for the fact that the flux of energetic X-rays $\gtrsim 50$ keV at the ISEE-3 location was smaller by a factor of ≈ 600 compared to the PVO observations. Interestingly, the impulsive X-ray emission as seen by ISEE-3, which is consistent with a power-law spectrum (Figure 5b), can be clearly identified down to the lowest photon energies (≈ 5 keV) observable with the ISEE-3 spectrometer. Usually, i.e., in disk flares, this soft part of the spectrum is wholly dominated by the gradual emission with its thermal spectrum so that there is no indication of the flash phase in this energy region.

These observations show (1) that the hard X-ray source is extended from the chromosphere/transition region up to the corona but with the most intense part of the impulsive hard X-ray source located at altitudes $< 25\,000$ km, and (2) that the soft X-ray source responsible for the gradual phase of the burst decreases with increasing altitude much more rapidly than the brightness of the impulsive hard X-ray source and is located at altitudes well below 25 000 km above the photosphere. The fact that the spectrum of the impulsive coronal X-rays can be represented by a power law down to energies of ≈ 5 keV suggests a non-thermal emission process at least in altitudes above 25 000 km.

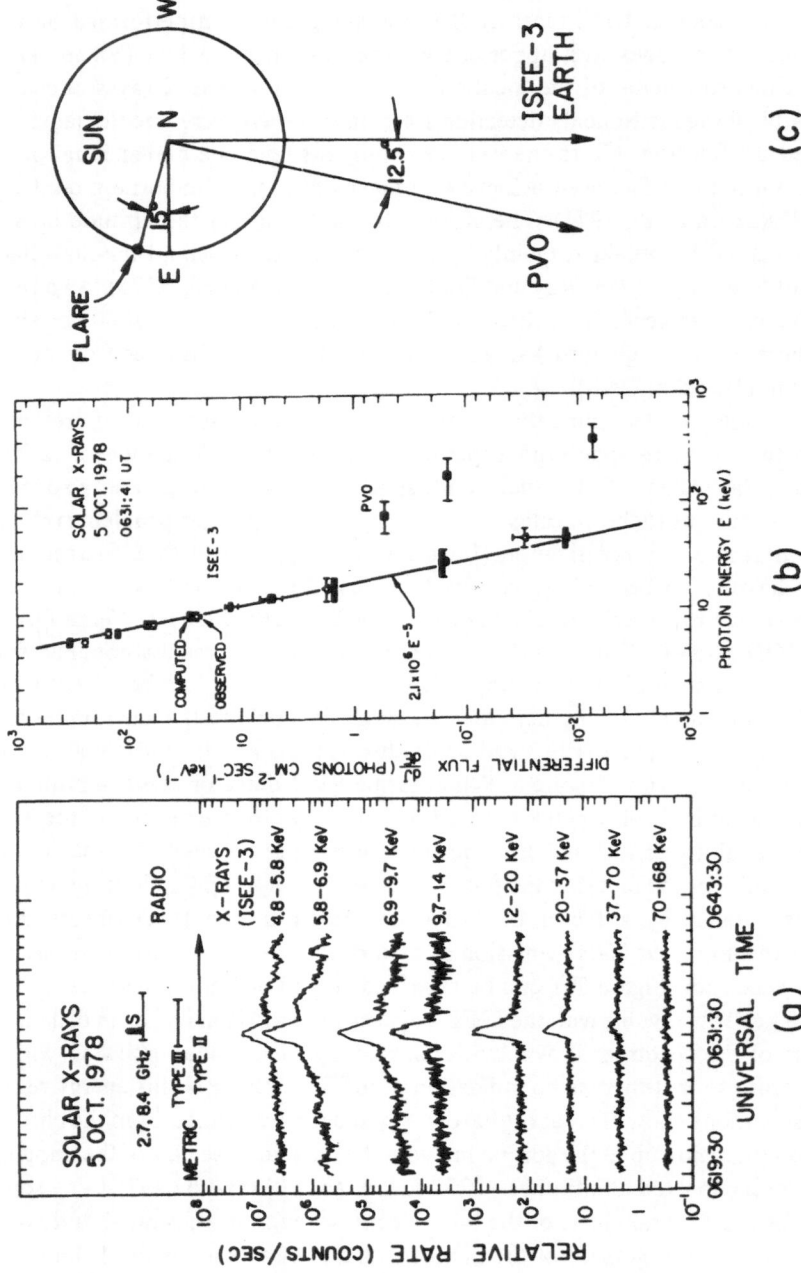

Fig. 5. Impulsive X-ray burst on 5 October, 1978 attributed to a solar flare located ≈ 15° behind the East limb of the Sun. (a) Intensity-time profiles at various energy channels of the ISEE-3 spectrometer. (b) X-ray spectra at the time of maximum. The X-ray flux observed by ISEE-3 is smaller by a factor ≈ 600 than that observed by PVO. (c) Estimated locations of the solar flare and the ISEE-3 and PVO spacecraft. (After Kane *et al.*, 1979).

In February 1980 the first spacecraft instrument able to image hard X-rays up to photon energies of 30 keV, the Hard X-Ray Imaging Spectrometer (HXIS), was launched aboard the Solar Maximum Mission (SMM) satellite (Van Beek *et al.*, 1980). The HXIS team represents a collaboration between the Astronomical Institute at Utrecht and the University of Birmingham. X-ray imaging is accomplished by a very complex multi-plate collimator containing 576 independent subcollimators. X-rays from

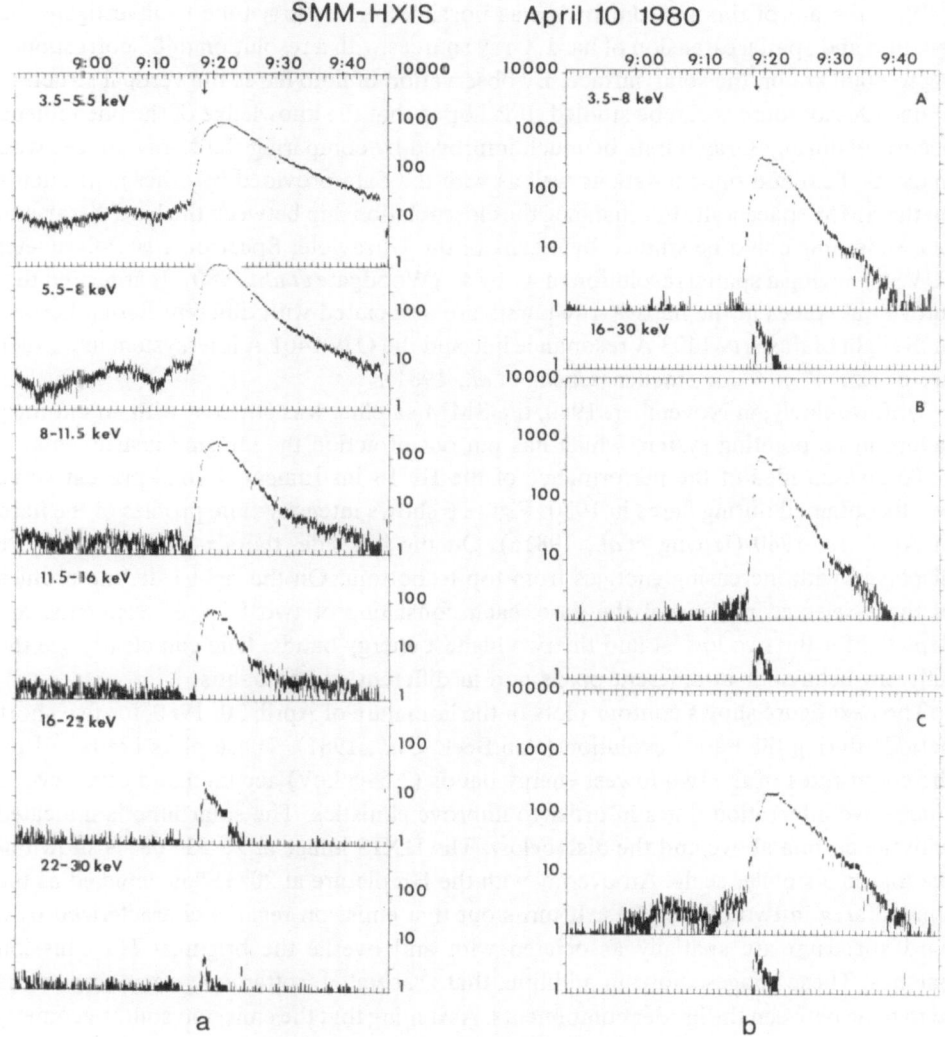

Fig. 6. Time evolution of the flare on 10 April, 1980 as observed by HXIS. (a) Intensity-time profiles in the six energy bands of HXIS, integrated over the whole flare area. (b) Intensity-time profiles in three different regions of the fine field of view, marked *A*, *B*, and *C*. Two 8″ × 8″ elements have been integrated in each case in the 3.5–8.0 keV (top) and 16.0–30.0 keV (bottom) energy ranges, respectively. The time of the hard X-ray peak is marked by an arrow. (After Hoyng *et al.*, 1981a.)

an individual point on the Sun are transmitted to one of an array of 900 mini-propor-
tional counters which serves as position sensitive detector system. The counter signals
are fed into six energy bands in the range 3.5 to 30 keV.

HXIS has two fields of view. The coarse field of view corresponds to $6'24''$ by $6'24''$
on the solar surface and contains 128 pixels with $32''$ spatial resolution. The fine field
of view covering $2'40''$ by $2'40''$ has 324 pixels with $8''$ resolution. The X-ray imaging
is continuous with a time resolution of 0.5 to 7 s depending on the mode of operation.

With the aid of this instrument it was possible for the first time to investigate the
location and spatial extension of hard X-ray sources with a resolution of $8''$ corresponding to 6000 km on the solar surface. By observation of limb flares the vertical structure
of hard X-ray sources can be studied. It is hoped that the knowledge of the phenomena
occurring during X-ray bursts be much improved by comparing the X-ray images with
magnetic field and optical data as well as with the data provided by other instruments
on the SMM spacecraft. For instance the close relationship between the hard X-ray and
UV emissions could be studied by means of the Ultraviolet Spectrometer Polarimeter
(UVSP) having a spatial resolution of $4''$ by $4''$ (Woodgate *et al.*, 1980). It appeared that
individual spikes in the hard X-ray bursts are associated with different flaring kernels
in the light of the Si IV 1403 Å resonance line and the O IV 1401 Å intersystem line which
are of size $4'' \times 4''$ or smaller (Cheng *et al.*, 1981).

Unfortunately, in November, 1980, the SMM satellite was stricken with an untimely
failure in its pointing system which has put out of action the imaging instruments.

To give an idea of the performance of the HXIS instrument, I shall present some
results obtained during flares in 1980. Figure 6 shows intensity-time profiles of the flare
of April, 10, 1980 (Hoyng *et al.*, 1981a). On the left side, the six energy bands are
displayed with increasing energies from top to bottom. On the right side, the profiles
of three selected regions of the flare, each consisting of two $8'' \times 8''$ elements, are
depicted for the two lowest and the two highest energy bands. One can clearly see the
different behaviour in different pixels and in different energy bands.

The next figure shows contour plots of the limb flare of April, 30, 1980, for five short
periods during the flare's evolution (Van Beek *et al.*, 1981). These plots are based on
the count rates of the two lowest energy bands (3.5–8 keV) accumulated over several
successive integration times in order to improve statistics. The solar limb is indicated
with the corona above and the disk below. The HXIS image at $20^{\rm h}31^{\rm m}08^{\rm s}$ is shown at
the top on a smaller scale. An overlay with the Hα picture at $20^{\rm h}31^{\rm m}$ is depicted as the
shaded area in two of the plots. It turns out that emission regions characterized by a
hard spectrum are spatially associated with and overlie the brightest Hα emission
patches. These images show, in addition, that associated, softer emission components
tend to lie between the harder components. Assuming that the emission source geometry
be that of a coronal loop, then a natural interpretation of these data is that the hard
X-rays are emitted at the loop footpoints, where the Hα emission is expected to take
place, whereas the softer X-rays are emitted at the top portion of the loop. It is
interesting to compare these findings with the results of recent interferometer observations of impulsive microwave bursts at high spatial resolution ($1''-3''$) (Marsh *et al.*,

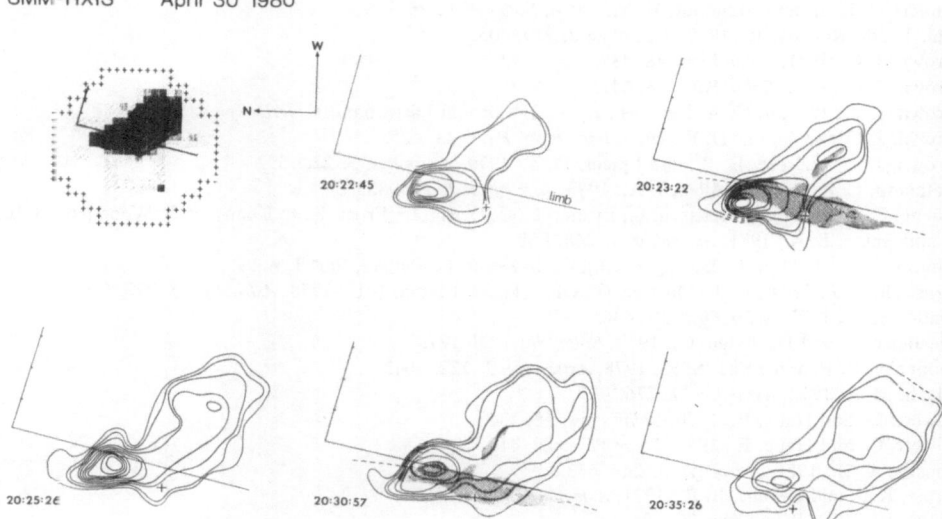

SMM-HXIS   April 30 1980

Fig. 7.   Contour plots of the limb flare of 30 April 1980 for five different times. An overlay with an Hα picture is shown in two of the images as shaded area. The fine field of view of HXIS is depicted at the top left, on a smaller scale, together with the X-ray image at $20^h31^m08^s$. The counting rate per $8'' \times 8''$ picture element is characterized by the degree of shading. The solar limb is indicated with the corona above and the disk below. The ordinate scale is in units of 10^4 km. (After Van Beek *et al.*, 1981.)

1980; Kundu *et al.*, 1981; Marsh *et al.*, 1981). The microwave sources are found to be also relatively compact ($\approx 10''$), they are, however, located *between* the Hα kernels, close to the magnetic neutral line, i.e., microwaves and hard X-rays were *not* produced by a common population of electrons, even though the energetic electrons may have a common acceleration site, namely the site of energy release.

A bright pointlike X-ray feature appearing at the beginning of the 30 April event may be interpreted to be the energy release site. Its position coincided to within $6''$, i.e., less than one picture element, with the position of the maximum intensity of radio emission at 10.6 GHz. Only 6 s later the X-ray tongue visible on the contour plots of Figure 7, had already a length of $\approx 30\,000$ km. The fact that the plasma did not diffuse any further shows that it was confined by the magnetic field of a pre-existing coronal structure.

These HXIS data illustrate the value of time-resolved imaging at higher photon energies. Although the analysis of the data is still in the preliminary stages, the HXIS observations combined with the measurements of other SMM instruments have shed new light on the problems associated with solar X-ray bursts. As a result the preponderance of evidence now points towards the non-thermal flare models (Hoyng *et al.*, 1981b). However, further investigation is required to arrive at a full understanding of the physical processes taking place during flares.

References

Anderson, K. A. and Mahoney, W. A.: 1974, *Solar Phys.* **35**, 419.

Bai, T. and Ramaty, R.: 1978, *Astrophys. J.* **219**, 705.

Brown, J. C.: 1971, *Solar Phys.* **18**, 489.

Brown, J. C.: 1972, *Solar Phys.* **26**, 441.

Brown, J. C.: 1974, in G. A. Newkirk, Jr. (ed.), 'Coronal Disturbances', *IAU Symp.* **57**, 395.

Brown, J. C. and Smith, D. F.: 1980, *Rep. Prog. Phys.* **43**, 125.

Brown, J. C., Melrose, D. B., and Spicer, D. S.: 1979, *Astrophys. J.* **228**, 592.

Catalano, C. P. and Van Allen, J. A.: 1973, *Astrophys. J.* **185**, 335.

Cheng, C.-C., Tandberg-Hanssen, E., Bruner, E. C., Orwig, L., Frost, K. J., Kenny, P. J., Woodgate, B. E., and Shine, R. A.: 1981, *Astrophys. J.* **248**, L39.

Chubb, T. A.: 1972, in C. De Jager (ed.), *Solar-Terrestrial Physics*, Part I, p. 99.

Crannell, C. J., Frost, K. J., Mätzler, C., Ohki, K., and Saba, J. L.: 1978, *Astrophys. J.* **223**, 620.

Datlowe, D.: 1971, *Solar Phys.* **17**, 436.

De Jager, C. and De Jonge, G.: 1978, *Solar Phys.* **58**, 127.

Donnelly, R. F. and Kane, S. R.: 1978, *Astrophys. J.* **222**, 1043.

Elcan, M. J.: 1978, *Astrophys. J.* **226**, L99.

Elwert, G. and Haug, E.: 1970, *Solar Phys.* **15**, 234.

Elwert, G. and Haug, E.: 1971, *Solar Phys.* **20**, 413.

Emslie, A. G.: 1981, *Astrophys. J.* **244**, 653.

Frost, K. J. and Dennis, B. R.: 1971, *Astrophys. J.* **165**, 655.

Haug, E.: 1972, *Solar Phys.* **25**, 425.

Haug, E.: 1975, *Solar Phys.* **45**, 453.

Hoyng, P., Machado, M. E., Duijveman, A., Boelee, A., De Jager, C., Fryer, R., Galama, M., Hoekstra, R., Imhof, J., Lafleur, H., Maseland, H. V. A. M., Mels, W. A., Schadee, A., Schrijver, J., Simnett, G. M., Švestka, Z., Van Beek, H. F., Van Tend, W., Van der Laan, J. J. M., Van Rens, P., Werkhoven, F., Willmore, A. P., Wilson, J. W. G., and Zandee, W.: 1981a, *Astrophys. J.* **244**, L153.

Hoyng, P., Duijveman, A., Machado, M. E., Rust, D. M., Švestka, Z., Boelee, A., De Jager, C., Frost, K. J., Lafleur, H., Simnett, G. M., Van Beek, H. F., and Woodgate, B. E.: 1981b, *Astrophys. J.* **246**, L155.

Hudson, H. S.: 1978, *Astrophys. J.* **224**, 235.

Kahler, S.: 1971a, *Astrophys. J.* **164**, 365.

Kahler, S.: 1971b, *Astrophys. J.* **168**, 319.

Kahler, S. W.: 1975, in S. R. Kane (ed.), 'Solar Gamma-, X-, and EUV Radiation', *IAU Symp.* **68**, 211.

Kane, S. R.: 1969, *Astrophys. J.* **157**, L139.

Kane, S. R.: 1974, in G. A. Newkirk, Jr. (ed.), 'Coronal Disturbances', *IAU Symp.* **57**, 105.

Kane, S. R.: 1981a, *Astrophys. Space Sci.* **75**, 163.

Kane, S. R.: 1981b, *Astrophys. J.* **247**, 1113.

Kane, S. R. and Anderson, K. A.: 1970, *Astrophys. J.* **162**, 1003.

Kane, S. R. and Donnelly, R. F.: 1971, *Astrophys. J.* **164**, 151.

Kane, S. R., Anderson, K. A., Evans, W. D., Klebesadel, R. W., and Laros, J.: 1979, *Astrophys. J.* **233**, L151.

Kosugi, T.: 1981, *Solar Phys.* **71**, 91.

Kundu, M. R.: 1961, *J. Geophys. Res.* **66**, 4308.

Kundu, M. R., Bobrowsky, M., and Velusamy, T.: 1981, *Astrophys. J.* **251**, 342.

Lin, R. P.: 1974, *Space Sci. Rev.* **16**, 189.

Lin, R. P. and Hudson, H. S.: 1971, *Solar Phys.* **17**, 412.

Marsh, K. A., Hurford, G. J., Zirin, H., and Hjellming, R. M.: 1980, *Astrophys. J.* **242**, 352.

Marsh, K. A., Hurford, G. J., Dulk, G. A., Dennis, B. R., Frost, K. J., and Orwig, L. E.: 1981, *Astrophys. J.* **251**, 797.

Mätzler, C., Bai, T., Crannell, C. J., and Frost, K. J.: 1978, *Astrophys. J.* **223**, 1058.

McKenzie, D. L.: 1975, *Solar Phys.* **40**, 183.

Novick, R.: 1975, *Space Sci. Rev.* **18**, 389.

Novick, R.: 1981, *Opt. Eng.* **20**, 31.

Peterson, L. E. and Winckler, J. R.: 1959, *J. Geophys. Res.* **64**, 697.

Roy, J. R. and Datlowe, D. W.: 1975, *Solar Phys.* **40**, 165.

Smith, D. F. and Auer, L. H.: 1980, *Astrophys. J.* **238**, 1126.

Smith, D. F. and Lilliequist, C. G.: 1979, *Astrophys. J.* **232**, 582.
Suri, A. N., Chupp, E. L., Forrest, D. J., and Reppin, C.: 1975, *Solar Phys.* **43**, 415.
Švestka, Z.: 1976, *Solar Flares*, D. Reidel Publ. Co., Dordrecht, Holland, Chapter VI.
Takakura, T., Ohki, K., Shibuya, N., Fujii, M., Matsuoka, M., Miyamoto, S., Nishimura, J., Oda, M., Ogawara, Y., and Ota, S.: 1971, *Solar Phys.* **16**, 454.
Tindo, I. P., Ivanov, V. D., Mandelstam, S. L., and Shurygin, A. I.: 1970, *Solar Phys.* **14**, 204.
Tindo, I. P., Ivanov, V. D., Mandelstam, S. L., and Shurygin, A. I.: 1972a, *Solar Phys.* **24**, 429.
Tindo, I. P., Ivanov, V. D., Valniček, B., and Livshits, M. A.: 1972b, *Solar Phys.* **27**, 426.
Tindo, I. P., Mandelstam, S. L., and Shurygin, A. I.: 1973, *Solar Phys.* **32**, 469.
Tindo, I. P., Shurygin, A. I., and Steffen, W.: 1976, *Solar Phys.* **46**, 219.
Vaiana, G. S., Krieger, A. S., and Timothy, A. F.: 1973, *Solar Phys.* **32**, 81.
Van Beek, H. F., De Feiter, L. D., and De Jager, C.: 1974, *Space Res.* **XIV**, 447.
Van Beek, H. F., Hoyng, P., Lafleur, B., and Simnett, G. M.: 1980, *Solar Phys.* **65**, 39.
Van Beek, H. F., De Jager, C., Fryer, R., Schadee, A., Švestka, Z., Boelee, A., Duijveman, A., Galama, M., Hoekstra, R., Hoyng, P., Imhof, J. P., Lafleur, H., Machado, M. E., Maseland, H. V. A. M., Mels, W. A., Schrijver, J., Simnett, G. M., Van der Laan, J. J. M., Van Rens, P., Van Tend, W., Werkhoven, F., Willmore, A. P., Wilson, J. W. G., and Zandee, W.: 1981, *Astrophys. J.* **244**, L157.
Woodgate, B. E., Tandberg-Hanssen, E. A., Bruner, E. C., Beckers, J. M., Brandt, J. C., Henze, W., Hyder, C. L., Kalet, M. W., Kenny, P. J., Knox, E. D., Michalitsianos, A. G., Rehse, R., Shine, R. A., and Tinsley, H. D.: 1980, *Solar Phys.* **65**, 73.

CORONAL INVESTIGATIONS WITH OCCULTED SPACECRAFT SIGNALS*

M. K. BIRD

Radioastronomisches Institut, Universität Bonn, Auf dem Hügel 71, 5300 Bonn, F.R.G.

Abstract. The radio telemetry links between Earth and a spacecraft near superior conjunction penetrate the corona at ranges well within the acceleration regime of the solar wind. Occultation experiments in the solar corona have been performed on many interplanetary missions beginning with the Mariner and Pioneer series and extending up to the more recent data on Helios, Viking, and Voyager. The changes in group and phase velocity of the radio signal are measured to determine the total electron content of the corona and its fluctuations. The broadening of the carrier signal may be used in combination with the electron content data to derive a solar wind velocity profile. The wave number spectrum of electron density fluctuations in the corona may be inferred from amplitude and phase scintillations of the received signal. Linearly polarized signals, which are rotated along the propagation path by the Faraday effect, can provide information on the coronal magnetic field and its variations.

1. Introduction

This paper concisely summarizes the results of those radio science experiments designed to investigate the structure and dynamics of the solar corona with spacecraft signals. Since the topic has many features common to other similar investigations, it is appropriate to delineate the scope of the present paper by briefly mentioning and including some references from those related fields of research that will not be treated in detail.

The motion of the Sun with respect to the celestial background provides an annual opportunity to observe solar occultations of natural sources located near the ecliptic plane. Perhaps the most famous of all solar occultation experiments is the empirical test of Einstein's predicted deflection of light rays from stars due to the gravitational field of the Sun (theoretical deviation at the limb: 1.75 arc sec). This experiment was performed at solar eclipses as early as 1919 (von Klüber, 1960). Optical observations of planetary occultations of stars have recently resulted in the discovery of the rings of Uranus (Elliot *et al.*, 1977) and a value for the minimum diameter of Charon (Walker, 1980). An extensive description of stellar occultation studies at visible wavelengths has been given by Elliot (1979).

The first solar occultation experiments at radio wavelengths were the source broadening measurements of Tau A (Crab Nebula) taken by Hewish (1958). These early efforts eventually evolved into a powerful radio scattering technique for coronal investigations (Hewish *et al.*, 1964), which is known generally as 'interplanetary scintillations' or IPS. The IPS observations of Dennison and Hewish (1967) were, in fact, the first evidence of a faster solar wind emanating from polar regions of the Sun than in the ecliptic. Reviews of IPS have been published by Hewish (1972), Jokipii (1973), Lotova (1975) and Coles (1978).

---

* Paper presented at the IX-th Lindau Workshop 'The Source Region of the Solar Wind'.

Space Science Reviews **33** (1982) 99–126. 0038–6308/82/0331–0099$04.20.

The Crab Nebula and its pulsar have been by far the most popular natural radio sources for coronal sounding, partially because of their fortuitous location close to the ecliptic plane, but also because of certain advantageous features of their radio emission. The shortest (perpendicular) distance from the Sun's center to the ray path from source to Earth, generally referred to as the 'solar offset', attains a minimum value of less than $5R_\odot$ on 15 June each year ($1R_\odot$ = solar radius $\simeq$ 16 arc min). The pulsed emission of the Crab Pulsar was monitored during the solar occultations from 1969–1973 at the Arecibo telescope to determine the total electron content of the corona from the arrival time delay of pulses at different frequencies (Counselman and Rankin, 1972, 1973; Weisberg et al., 1976). Furthermore, the high degree of linear polarization of the Crab Nebula has enabled measurements of coronal Faraday rotation (Sofue et al., 1976; Berlin et al., 1978). Measurements of coronal Faraday rotation have since been extended to other pulsars (Bird et al., 1980), which are the best suited natural sources for the experiment because of their high degree of linear polarization and their built-in calibration for solar noise during the emission pauses between pulses (Bird, 1981).

The first radio soundings of the solar corona with an artificial source were the ground-based experiments of Eshleman et al. (1960) and James (1966, 1970), which detected solar radar echoes at 26.5 and 38.2 MHz, respectively. More recent solar radar observations at a much higher frequency (2600 MHz) have been made by Benz and Fitze (1979). Their attempts to detect resonant echoes from plasma waves in the solar atmosphere have been thus far unsuccessful (Fitze and Benz, 1981).

With the advent of active reconnaissance of the solar system with spacecraft came the first opportunities to probe previously inaccessible regions of space not only with *in situ* and remote sensing instruments on board, but also with the radio telemetry links between the spacecraft and Earth. One of the more obvious regions available to radio sounding, the solar corona, is the subject of this review. It should also be mentioned, however, that many of the exciting discoveries of interplanetary exploration have resulted from radio investigations of planetary atmospheres, ionospheres and rings. The scope of the investigations performed by 'radio science teams' on recent NASA missions is exemplified by the experiment descriptions for Viking (Michael et al., 1972) and Voyager (Eshleman et al., 1977). Some selected examples of planetary encounters that have provided information from radio science occultations are: Mariner 10 at Mercury (Levy, 1977); Mariner 10 (Howard et al., 1974) and Pioneer Venus (Kliore et al., 1979) at Venus; Mariner 9 (Kliore et al., 1972) and Viking (Fjeldbo et al., 1977) at Mars; Pioneer 10 (Fjeldbo et al., 1975) and the Voyagers (Eshleman et al., 1979a, b) at Jupiter; Pioneer 11 (Kliore et al., 1980) and both Voyagers (Tyler et al., 1981a, 1982) at Saturn.

In the case of the solar corona, radio science investigations with spacecraft signals may not necessarily be aimed at determining the physical constitution of the propagation medium. Some examples of these, which will not be described in detail here, are the relativity experiments using time delay techniques as performed on Mariner 6 and Mariner 7 (Anderson et al., 1975) and Viking (Shapiro et al., 1977), or the search for gravitational deflection of linearly polarized radiation on Pioneer 6 (Harwit et al., 1974) and Helios (Dennison et al., 1978).

Having more precisely defined the boundaries of the material for this review with the above listing of topics that will *not* be addressed, it is time to turn to the business at hand, namely: coronal investigations with occulted spacecraft signals. The types of experiments are divided and discussed separately according to the measured quantity in the following section. Some examples of radio science observations of coronal dynamics are shown in Section 3. A few comments on possible coronal sounding experiments on future interplanetary missions will be given in the fourth section.

2. Coronal Sounding Experiments

The basic geometry for coronal investigations with occulted spacecraft signals is shown in Figure 1. The space probe is usually located at a point near superior conjunction so that the carrier signal (uplink and/or downlink) propagates through interplanetary space and the corona to and from ground stations on Earth. For small elongation angles ε, it is convenient to express R, the separation of the ray path from the Sun (solar offset), as

$$R = 1 \text{ AU} \sin \varepsilon \simeq 3.75\varepsilon \qquad (\text{in } R_\odot) \tag{1}$$

if ε is given in degrees.

The minimum solar offset at which useful radio science data can be taken is well below that at which telemetry reception becomes prohibitive, and is determined more or less by the signal-to-noise level (SNR). The SNR deteriorates rapidly at small R due to increased solar interference and carrier signal broadening, both of which are functions of carrier frequency. Much effort has been expended to study and model coronal effects for the purpose of improving spacecraft command and navigation capability at small solar offsets (Winn *et al.*, 1975; Rockwell, 1977; Taylor, 1977; Zygielbaum, 1977;

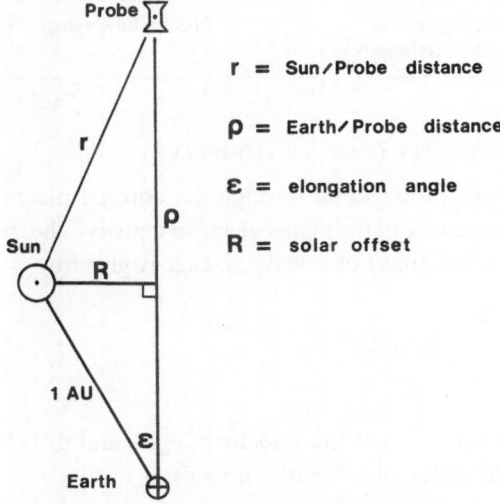

Probe

r = Sun∕Probe distance

ρ = Earth∕Probe distance

ε = elongation angle

R = solar offset

Sun

r

ρ

R

1 AU

ε

Earth

Fig. 1.   Basic geometrical configuration for coronal investigations with occulted spacecraft signals.

Layland, 1978). Radio science measurements have been conducted as close to the Sun as $R = 1.7R_{\odot}$ at S-band ($f \simeq 2.3$ GHz) on Helios 1 (Woo, 1978) and $R = 1.3R_{\odot}$ at X-band ($f \simeq 8.4$ GHz) on Viking (Tyler *et al.*, 1977).

The maximum solar offset at which coronal effects can still be detected is determined by the sensitivity of the system. Higher sensitivity is generally achieved at lower carrier frequencies and for dual-frequency systems as opposed to a single radio link. Definite coronal influence has been observed on Helios 2 (single S-band system) at solar offsets in excess of $20-30R_{\odot}$ (Edenhofer *et al.*, 1977). Most of the solar corona from the coronal base out to the Alfvén point, including the acceleration regime of the solar wind, is thereby accessible to sounding experiments with spacecraft radio signals.

Various types of radio science investigations in the solar corona are listed according to the measured signal parameter and the primary physical quantity inferred from the observations in Table I. Each will be discussed in turn in the following subsections.

TABLE I

Types of radio science investigations in the solar corona

	Signal parameter measured	Physical quantity inferred
1.	Group/phase velocity (ranging/Doppler)	Electron density N
2.	Faraday Rotation	Magnetic field B_L
3.	Spectral/angular broadening	Plasma flow velocity $V_{\perp}$
4.	Intensity/phase scintillations	Wave number spectrum of density fluctuations $P_N(\kappa)$
5.	Dual propagation path correlations	Plasma flow velocity $V_{\perp}$

2.1. GROUP/PHASE VELOCITY (RANGING/DOPPLER)

An electromagnetic wave propagating through the coronal plasma will travel slightly slower than in vacuum because of the higher electron density. The transit time from point '0' to point '*s*' (group delay time) of a wave packet is given by

$$t_g = \int_0^s \frac{\mathrm{d}y}{v_g}, \tag{2}$$

where $v_g = nc$ = group velocity, c is the velocity of light, and $\mathrm{d}y$ is the element of length along the ray path. The index of refraction n is given by

$$n = \sqrt{1 - (f_p/f)^2}, \tag{3}$$

where f_p is the plasma frequency (proportional to the square root of the electron density N). For wave frequencies well above the plasma frequency, i.e. $f^2 \gg f_p^2$, the index of refraction (3) is well approximated by

$$n = 1 - 40.3 \frac{N}{f^2} \tag{4}$$

with the electron density N given in m^{-3}. Therefore, (2) becomes

$$t_g = \frac{s}{c} + \frac{40.3}{cf^2} \int_0^s N \, dy. \tag{5}$$

The first term of (5) is the light time in free space and the second term is a small delay due to the presence of the plasma. Using 2 frequencies, one can elegantly eliminate the first term of (5) by measuring a differential group delay time Δt_g of the arrival of a ranging code (pulses, square wave, etc.) on the two carrier frequencies f_1 and f_2, that is

$$\Delta t_g = \frac{40.3}{c} (f_1^{-2} - f_2^{-2}) N_T \tag{6}$$

with $f_1 < f_2$. The quantity N_T is the physical parameter of interest, namely, the columnar electron content along the ray path as defined by

$$N_T = \int_0^s N \, dy. \tag{7}$$

The phase of an electromagnetic wave in a plasma increases faster than for propagation in a vacuum. Defining a phase delay time t_p by

$$t_p = \int_0^s \frac{dy}{v_p}, \tag{8}$$

where v_p is the phase velocity ($= c/n$) and the index of refraction n is again given by (4), one may proceed as before to obtain

$$t_p = \frac{s}{c} - \frac{40.3}{cf^2} \int_0^s N \, dy. \tag{9}$$

For a dual-frequency system one may therefore define a differential phase delay time given by

$$\Delta t_p = \frac{40.3}{c} (f_1^{-2} - f_2^{-2}) N_T \tag{10}$$

and the derived quantity is again the total electron content (7).

The differential phase delay time (10) is best expressed in terms of the number of cycles of phase shift at the lower frequency f_1. The measurement thus consists of a 'cycle count' m where

$$m = \frac{40.3 f_1}{c} (f_1^{-2} - f_2^{-2}) N_T. \tag{11}$$

In practice, the measured quantity m cannot be determined unambiguously because there is no possible way to distinguish how many integral cycles of phase difference have occurred along the ray path. The best one can do is to note the phase difference at the beginning of a tracking pass and then count cycles of phase difference to determine the change in the total electron content, but not its absolute value. Since the variation of m with time is a manifestation of a temporal change in the frequency difference, the phase measurements are often referred to simply as 'Doppler' data. The ambiguity associated with (11) can be resolved with a simultaneous ranging measurement (6).

The main advantage of the phase measurement (11) is its greater sensitivity to changes in the total electron content N_T. Whereas contemporary S/X-band systems are capable of determining group delay time differences (6) to an accuracy of ± 15 ns (Martin and Zygielbaum, 1977), phase jitter in standard phase-lock-loop receivers is typically of the order of $\pi/2$ radians (Winn *et al.*, 1975). This translates into the following typical accuracies for an S/X-band system:

$$\Delta N_T \text{(ranging)} \simeq 5 \times 10^{17} \text{ el m}^{-2} \tag{12}$$

$$\Delta N_T \text{(Doppler)} \simeq 5 \times 10^{15} \text{ el m}^{-2} \tag{13}$$

The phase measurements are thus about 100 times more accurate than ranging data on

TABLE II

Group/phase velocity measurements

Spacecraft	Frequency	Time interval	References
Pioneer 6		February–April 1966	Eshleman *et al.* (1966)
Pioneer 7	49.8 MHz	October 1966–January 1967	Koehler (1968)
Pioneer 8	and	August 1972	Croft (1973)
Pioneer 9	423 MHz	December 1965–June 1969	Landt (1974)
Mariner 5		February 1968–June 1971	Croft (1979)
Mariner 6	S-band	March–July 1970	Muhleman *et al.* (1977)
Mariner 7			
Helios 1	S-band	March–September 1975	Edenhofer *et al.* (1977)
Helios 2		March–June 1976	Esposito *et al.* (1980)
Viking	S/X-band	October–December 1976	Tyler *et al.* (1977)

the same radio system. To put these values in perspective, it could be recalled that a typical noontime ionosphere contains slightly more than 10^{17} el m^{-2}, so that an S/X-ranging experiment is virtually incapable of following ionospheric variations in N_T.

A summary of the ranging/Doppler measurements performed on various space probes is presented in Table II. The radio sounding experiment flown on the Pioneer 6–9 spacecraft is unique in that it is the only one with its own 'hardware' on board the spacecraft. The interplanetary plasma was probed only on the uplink, which was transmitted from a 300 kW command station with a 45 m parabolic antenna located at Stanford (Eshleman et al., 1966). The sounding frequencies were also considerably lower than the conventional carrier frequencies of their successors, thus providing an accuracy in N_T of $\pm 4 \times 10^{14}$ el m^{-2} (Koehler, 1968). As a result, the system was capable of mapping the electron content in interplanetary space, even at very large solar elongation angles. It was this feature of the low-frequency sounding system that prompted Croft (1978) to advocate such an experiment for the proposed Solar Probe Mission.

Were a map to be constructed of the total electron content between the Earth and various points in interplanetary space, it would probably resemble the plot in Figure 2. The x-axis in Figure 2 runs through the centers of the Earth and Sun, located at $x = 0$ and $x = 1$, respectively. The left side of the plot shows contours of constant N_T in the xy-plane (ecliptic), while the right side shows a cut through the solar pole in the xz-plane.

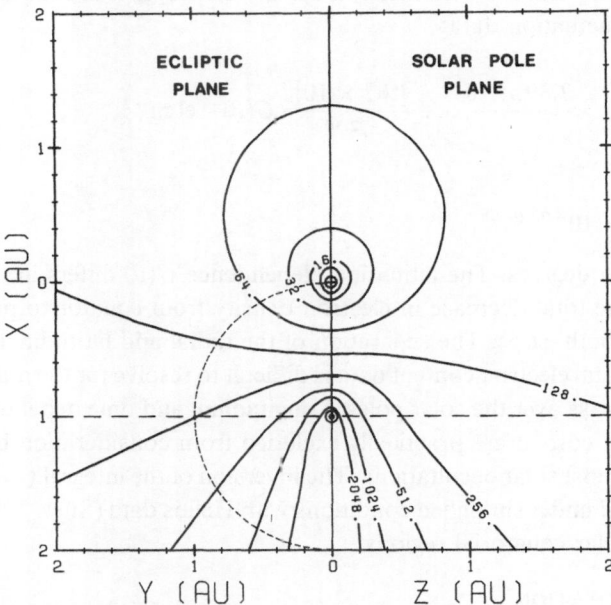

Fig. 2.   Contour map of columnar electron content N_T between Earth and points in the ecliptic plane (left half of diagram) and in the solar pole plane (right half). Earth is located at the center and the Sun is 1 AU below this. Lines of constant N_T are labeled in 'hexems', where 1 hexem = 10^{16} el m^{-2}.

The total electron content is given in 'hexems' (1 hexem $= 10^{16}$ el m^{-2}), and was derived from an assumed electron density in interplanetary space taken from a model from Tyler et al. (1977), i.e.

$$N(r, \theta) = \left[\frac{1.55 \times 10^{14}}{r^6} + \frac{3.44 \times 10^{11}}{r^2} \right] F(\theta) \quad \text{el m}^{-3} \tag{14}$$

with

$$F(\theta) = \left(\cos^2 \theta + \frac{1}{64} \sin^2 \theta \right)^{1/2}, \tag{15}$$

where θ is the heliographic latitude, assumed here to be identical with the ecliptic latitude, and r is given in solar radii. The density at the Earth's orbit from (14) is 7.4 el cm^{-3}.

The asymmetry of N_T from the ecliptic to the pole is readily apparent in Figure 2, particularly at intermediate values of N_T from 100–300 hexems. The electron content rises steeply near the Sun and appears to exceed 10^4 hexems at solar offsets inside of about $5R_\odot$ (Muhleman et al., 1977; Tyler et al., 1977; Esposito et al., 1980).

The exact radial dependence of N_T in the outer corona actually appears to be slightly closer to $r^{-1.3}$, indicating a falloff exponent of -2.3 for the electron density (Berman, 1979) in the case of spherical symmetry. Berman and Wackley (1977), for example, have derived the following model for the electron density distribution at solar minimum from Viking Doppler fluctuation data:

$$N(r, \theta) = \frac{2.39 \times 10^{14}}{r^6} + \frac{1.67 \times 10^{12}}{r^{2.30}} G(\theta) \quad \text{el m}^{-3} \tag{16}$$

with

$$G(\theta) = 10^{-0.90/90}, \tag{17}$$

where θ is given in degrees. The latitudinal dependence $G(\theta)$ differs in form from the model (15), but the total decrease in electron density from equator to pole is about a factor of eight in both cases. The separation of the radial and latitudinal effects in the interpretation of total electron content data is difficult to resolve for the many spacecraft occultations that pass over the solar poles. Longitudinal and time dependent variations of the background corona are practically excluded from consideration because of the short duration of most solar occultations. The inversion of the integral (7) has, however, been accomplished under simplified conditions with Helios data (Süss, 1980), which are confined to the solar equatorial regions.

2.2. FARADAY ROTATION

If the probe signal is linearly polarized, a measurement of the incident polarization angle on the ground will determine the Faraday rotation of the signal along its ray path. This information may then be used to infer the strength and polarity of the magnetic field in the propagation medium. A good approximation for the Faraday rotation Ω (Stelzried, 1970) is given by

$$\Omega = \frac{K}{f^2} \int_0^s N \mathbf{B} \cdot \mathbf{dy} \quad \text{radians} \tag{18}$$

where

$$K = 2.36 \times 10^4 \text{ (SI units)}.$$

The angle Ω is related to the locally determined polarization angle q of the incident radiation as shown in Figure 3. The ray path is viewed end-on in Figure 3 and is assumed to be coincident with the intersection of the ecliptic and the plane of the horizon. The initial electric vector $\mathbf{E}_0$ is oriented perpendicular to the ecliptic plane. If the magnetic field is directed toward the observer as indicated, the Faraday rotation Ω occurs in the mathematically positive direction. The angle q is measured with respect to the antenna structure and is usually referred to local zenith. The only missing information is the angle p between the horizon and the ecliptic, which is a well defined function of the given observation coordinates such as right ascension and declination of the spacecraft, latitude and longitude of the ground station and Universal Time (Stelzried et al., 1972). An isolated measurement of q (and therefore Ω) is possible only in the interval from 0 to 180 deg. The multicycle ambiguity can be resolved by continually tracking the polarization angle for the duration of the solar occultation.

The mechanical measurement of the polarization angle q with S-band automatic tracking polarimeters is described in Ohlson et al. (1974). Typical accuracies at large solar offset are $\Delta\Omega \simeq \pm 1°$. The terrestrial ionosphere contributes a small positive

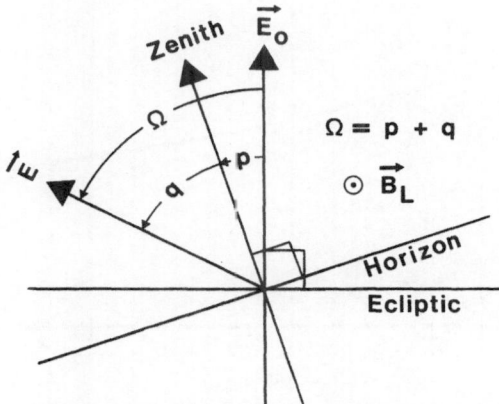

Fig. 3.   View along line-of-sight to spacecraft showing definition of angles used for measurement of the signal Faraday rotation Ω. The mean magnetic field along the ray path is directed toward the observer in this schematic example.

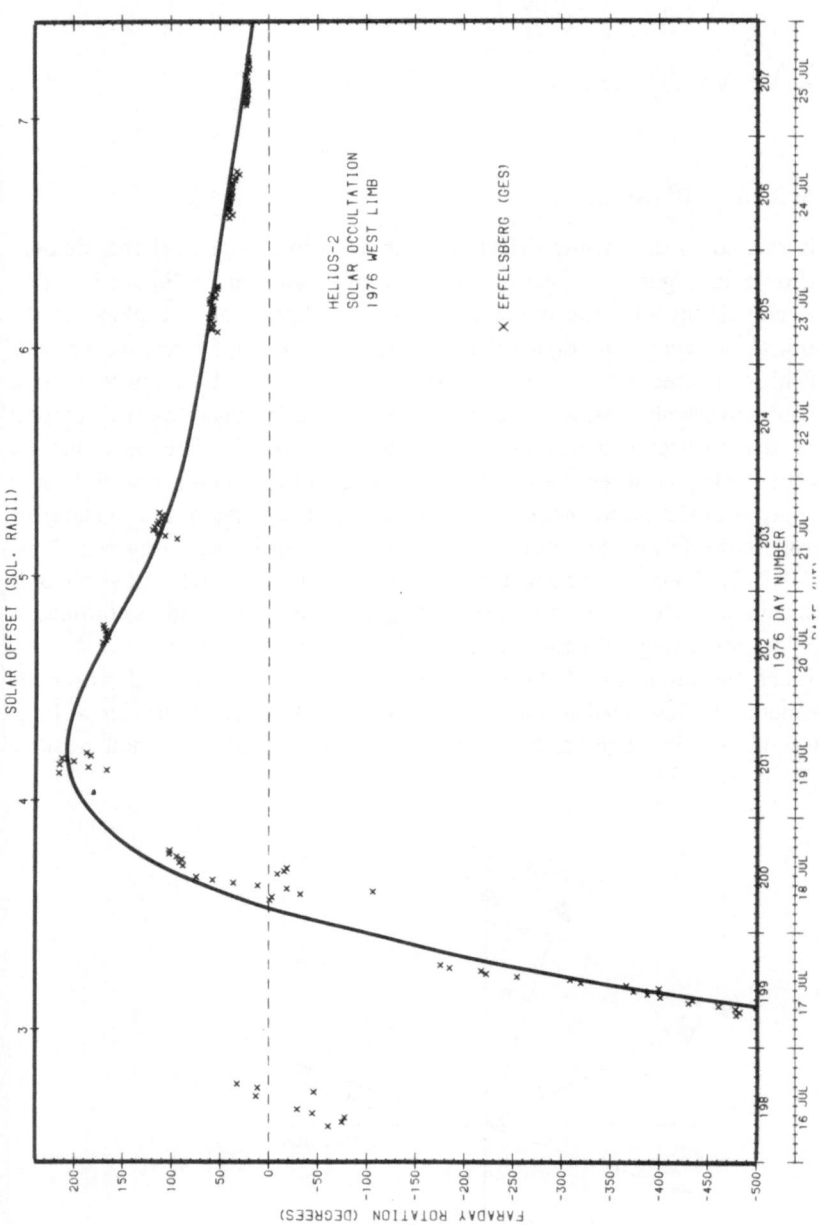

Fig. 4. Faraday rotation of Helios 2 signal during west limb occultation in July 1976. Only the Effelsberg station was available for Helios coverage due to the landing of Viking on Mars, which occupied the NASA Deep Space Network. The data on 16 July (DOY 198) are ambiguous by ± nπ, where n is an integer. The solid curve is the simulated Faraday rotation profile calculated from the coronal model of Figure 5.

TABLE III

Faraday rotation measurements

Spacecraft	Frequency	Time interval	References
Pioneer 6	S-band	November–December 1968	Stelzried et al. (1970)
Pioneer 9	S-band	December 1970–January 1971	Cannon (1976)
Helios 1	S-band	April–September 1975	Volland et al. (1977)
Helios 2		May–June 1976	Bird et al. (1978)

Faraday rotation at northern tracking stations of about 2–5 deg. Coronal Faraday rotation begins to exceed that of the ionosphere at about $10R_\odot$ and has been measured as close to the Sun as $2.0R_\odot$ with Helios 2 (Bird et al., 1978). The coronal Faraday rotation experiments performed with linearly polarized spacecraft signals are shown in Table III.

Measurements of coronal Faraday rotation taken at the 100 m radio telescope in Effelsberg during the solar occultation of Helios 2 in July 1976 are displayed in Figure 4. The mean coronal magnetic field on the west limb is seen to switch from predominantly negative (away from observer) to predominantly positive (toward observer) on 18 July (day of year = 200) when the solar offset was $3.5R_\odot$. The measurements on 16 July (DOY 198) are ambiguous by an unknown number of half-rotations ($\pm n\pi$, with n an integer) due to the lack of continuous tracking. The data from the remaining passes, however, are correctly positioned to a high degree of confidence, because the radial gradient of the coronal Faraday rotation was smaller.

The solid curve of Figure 4 is the theoretical Faraday rotation obtained from the coronal model of Figure 5, which was derived from the Faraday rotation data in the manner described by Volland et al. (1977). Figure 5 shows the corona as viewed from the north ecliptic pole at time $t_0 = 200.0 = 18.0$ July 1976. The direction to Earth is downwards. The coronal magnetic field is assumed to be radial and the product $N \cdot B_r$ (electron density times magnetic field) is assumed to decrease with distance from the Sun as $r^{-(\gamma + 1)}$. The longitudinal structure of the corona is modelled by 5 sectors with different values of $N \cdot B_r$. No models of the corona with less than 5 different sectors could be found to adequately simulate the Faraday rotation data and at the same time be consistent with observations of the mean solar magnetic field as registered at Stanford (Scherrer et al., 1977) and published in Solar-Geophysical Data. The Helios occultations of 1975 even required 6 different sectors to meet these conditions. The polarity and strength of $N \cdot B_r$ at the solar surface (in $10^8 \ T \ m^{-3}$) is indicated in each sector of the model on Figure 5. A large segment of the corona is marked 'uncertain', since these longitudes were not probed by the signal. Around the perimeter of the diagram are the daily measurements of the polarity of the mean solar field mentioned above, with which the model sectors were required to be compatible. With the exception of five days of very weak mean solar magnetic field, the Sun was, for all practical purposes, a positive magnetic monopole during most of July 1976.

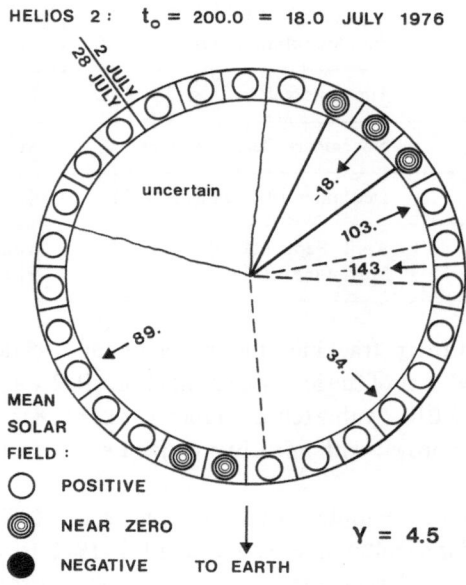

Fig. 5.   Model of the corona derived from the Faraday rotation data of Figure 4. The diagram shows a view of the corona from the north ecliptic pole at time $t_0 = 200.0$ (Earth downwards). The radial decrease of the product $N \cdot B_r$ (electron density × radial magnetic field) is a power law with exponent $\gamma + 1 = 5.5$. The longitudinal structure is modeled by 5 sectors of different $N \cdot B_r$ with widths, polarities and strengths (in $10^8 \, T \, m^{-3}$) as indicated. Six days of helio-longitude are marked 'uncertain', because they were not adequately probed by the Helios signal on the west limb. The polarities of the model sectors were required to agree in the large with the polarity of the mean solar magnetic field as given around the perimeter of the diagram.

The contributions from electron density and magnetic field in the integrand of (18) cannot be separated without supplementary information such as group/phase delay measurements or white light coronagraph observations to determine the total electron content alone. Using a combination of Faraday rotation and electron content measurements, Hollweg *et al.* (1982) were able to show that fluctuations of the Faraday rotation data are primarily due to oscillations of the coronal magnetic field and not the electron density, thus providing evidence for coronal Alfvén waves.

2.3. Spectral/angular broadening

The scattering of radio waves from inhomogenieties in the refractive index of the coronal plasma causes the signals to broaden in frequency and apparent angular width. Since the fluctuations of the refractive index (4) are just proportional to the fluctuations in electron concentration, the extent of these broadening effects is controlled by the strength and spectral characteristics of the coronal 'turbulence'. Comprehensive discussions of the theory of angular and spectral broadening can be found in Hollweg (1970) and Woo *et al.* (1977). The reported observations of spectral and/or angular broadening of spacecraft signals are listed in Table IV.

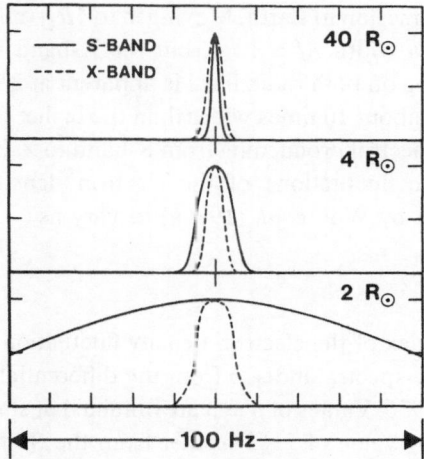

Fig. 6. Spectral broadening of the unmodulated carrier signals at S-band (2.3 GHz) and X-band (8.4 GHz) for various ray path offsets from the Sun.

The effect of spectral broadening is schematically displayed in Figure 6, which shows spectrograms of the carrier signals at various solar offsets for two conventional transmission frequencies at S-band (2.3 GHz) and X-band (8.4 GHz). The spectrograms are constructed by a Fourier transform technique first implemented by Goldstein (1969) for the solar occultation of Pioneer 6. The frequency resolution across the bandwidth of 100 Hz is of the order of 0.2 Hz. At large solar offsets (upper panel) the two radio links are essentially unaffected by the corona and the spectral width is governed by the stability of the spacecraft oscillator. The intrinsic width of the X-band carrier is wider than at S-band since it is derived from the same oscillator and frequency multiplied by

TABLE IV

Spectral/angular broadening measurements

Spacecraft	Frequency	Time interval	References
Pioneer 6	S-band	November–December 1968	Goldstein (1969) Woo *et al.* (1976a)
Mariner 10	S/X-band	May–June 1974	Goldstein and Stelzried (1975)
Helios 1 Helios 2	S-band	April–September 1975 May–September 1976	Woo (1977) Woo (1978)
Venera 10	928 MHz	April–June 1976	Efimov *et al.* (1977) Razmanov *et al.* (1979)
Voyager 2	S-band	August 1979	Bradford and Routledge (1980)

a factor 11/3 before transmission to Earth. Moving in to $4R_\odot$ (middle panel), the S-band signal has broadened to a width $\Delta f \simeq 8$ Hz, but the X-band is still very close to its intrinsic width. Broadening on both radio links is apparent at $2R_\odot$ (lower panel), where the S-band profile is now about 10 times wider than the higher frequency carrier signal.

The dispersion of the spectral broadening from S-band to X-band is dependent upon the spectral index of the fluctuations of the electron density in the corona. The bandwidth Δf was shown by Woo et al. (1976a) to vary as

$$\Delta f \sim f^{-2/(p-2)}, \tag{19}$$

where p is the spectral index of the electron density fluctuation spectrum. Woo (1981) has derived values for the spectral index p from the differential broadening of the two Viking carrier signals in 1976. Values of $p \simeq 3$ are obtained at small solar offset and tend to rise to the Kolmogorov value of 11/3 further from the Sun ($r > 20R_\odot$).

As pointed out by Woo et al. (1977), simultaneous measurements of spectral and angular broadening of a monochromatic signal can be employed for a determination of the plasma flow speed perpendicular to the line-of-sight $v_\perp$, because spectral broadening is directly proportional to $v_\perp$ and angular broadening is independent of this quantity. If $\Delta\theta$ is the measured angular broadening of the signal, than Woo (1978) has shown that

$$v_\perp = \text{const} \times \frac{1}{f}\frac{\Delta f}{\Delta\theta}L, \tag{20}$$

where

$$L = \frac{\rho}{\rho - \cos\varepsilon} \tag{21}$$

and

ρ = distance from Earth to spacecraft in AU (Figure 1);

ε = solar elongation angle;

f = signal frequency.

The constant in (20) depends on such things as direction of the coronal magnetic field, ellipticity of the angular broadening pattern (axial ratio) etc., but it is independent of the strength or spectral index p of the coronal turbulence. Woo (1978) made the first calculation of $v_\perp$ using (20) to derive a value of 24 km s^{-1} at a solar offset of only $1.7R_\odot$. This projected velocity is compatible with those obtained by James (1970) with radar echoes and by Ekers and Little (1971) with dual-path scintillation time delay measurements.

Typical radial variations of spectral and angular broadening are shown in Figure 7. Most of the data for Figure 7 was taken at S-band. The remaining data have been adjusted with a scaling factor $\sim f^{-2}$ for both diagrams, corresponding to $p \simeq 3$. The vertical dashed line at $R = 1.36R_\odot$ represents the theoretical occulting disk of the Sun at S-band (Bracewell et al., 1969). The rather copious spectral broadening data cluster

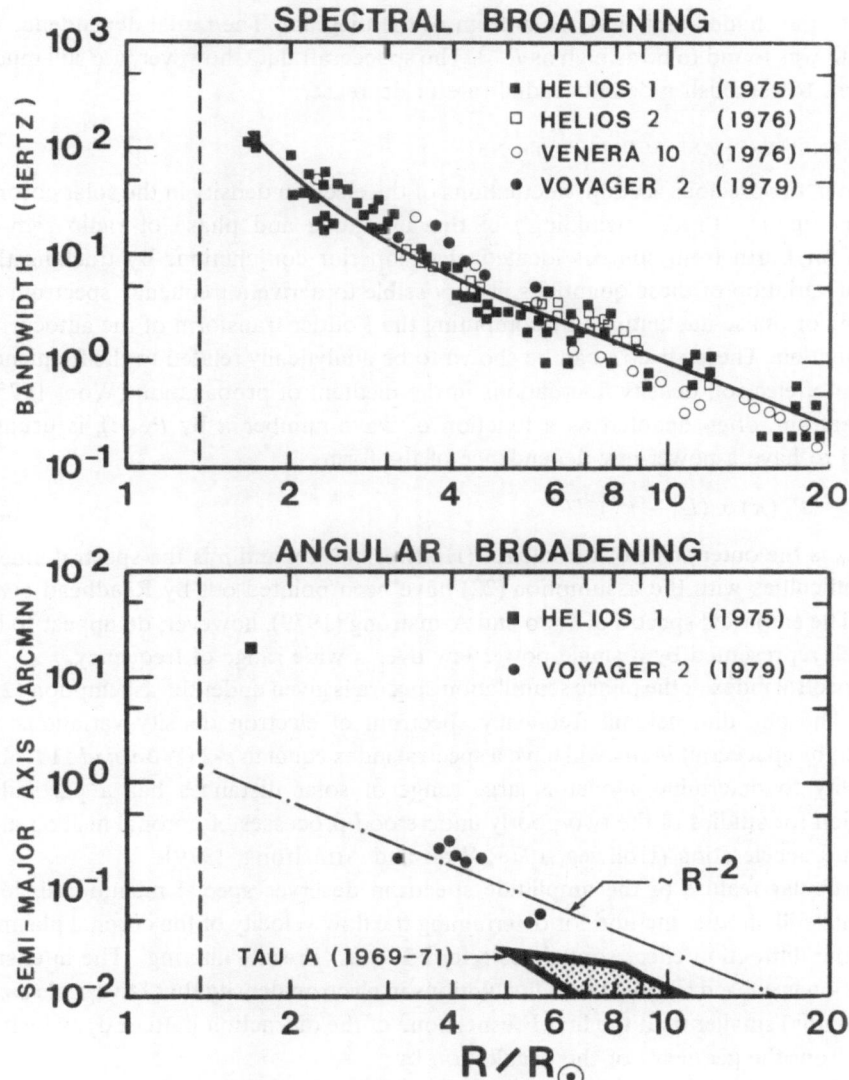

Fig. 7. Measurements of the radial dependence of spectral broadening (upper panel) and angular broadening (lower panel).

about an empirical model of Rockwell (1977) denoted by the solid curve. In contrast, there are very few angular broadening data from spacecraft (lower diagram). For comparison with measurements on natural sources at 'infinity', the observed angular broadening of the Helios and Voyager signals has been multiplied by the factor (21) to convert these to the plane wave values. The Helios 1 measurement at $1.7R_{\odot}$ does not appear to lie on the extrapolated R^{-2} line fitted to the Voyager data. The 80 MHz observations of Tau A (Blesing and Dennison, 1972) in the years 1969–71 translate at

S-band to the shaded region at the bottom of the diagram. The radial dependence of these data was found to be as high as R^{-3}. The spacecraft data, however, are still much too sparse to establish a definite radial rate of decrease.

2.4. INTENSITY/PHASE SCINTILLATIONS

As noted in the precious section, fluctuations of the electron density in the solar corona cause scintillation ('radio twinkling') of the amplitude and phase of radio signals received on Earth from sources located near superior conjunction. By tracking the temporal variation of these quantities, it is possible to derive a frequency spectrum of amplitude or phase fluctuations by computing the Fourier transform of the autocorrelation function. These spectra can be shown to be analytically related to the frequency spectrum of electron density fluctuations in the medium of propagation (Woo, 1975). This spectrum, often denoted as a function of wave number κ by $P_N(\kappa)$, is usually assumed to have a power-law dependence of the form

$$P_N(\kappa) \propto (L_0^2 + \kappa^2)^{-p/2}, \tag{22}$$

where L_0 is the outer scale of turbulence (Hollweg, 1970), and p is the spectral index. Some difficulties with the assumption (22) have been pointed out by Readhead et al. (1978). The empirical spectra of Woo and Armstrong (1979), however, do appear to be fairly well represented by a single power law over a wide range of frequency.

The spectral index of the phase scintillation spectra is given under the assumption (22) by p-1. The one dimensional frequency spectrum of electron density variations as measured by spacecraft in situ will have a spectral index equal to p-2 (Woo et al., 1976b). The ability to determine p over a large range of solar distances has a particular application for studies of the two poorly understood processes of coronal heating and solar wind acceleration (Hollweg, 1978; Woo and Armstrong, 1979).

A particular feature of the amplitude spectrum deserves special mention, since it represents still another method for determining the flow velocity of the coronal plasma. This is the diffraction effect sometimes referred to as 'Fresnel filtering'. The intensity scintillation is caused by only those fluctuations in electron density that have scale sizes (wavelengths) smaller than the first Fresnel zone of the diffraction pattern d_F, which is defined from the geometry of the occultation by

$$d_F = 773 \, (\lambda_0/L)^{1/2} \, \text{km} \tag{23}$$

with L given by (21) and λ_0, the wavelength of the radio signal, given in meters. The contributions to the amplitude power spectrum thus increase with increasing scale size until d_F, at which point the spectrum will form a 'knee' and remain flat for still lower frequencies. The break frequency f_b will occur when irregularities of the scale size (23) are transported past the Fresnel zone at the solar wind speed, i.e.

$$f_b = v_\perp/d_F. \tag{24}$$

Typical break frequencies are 0.1 to 1.0 Hz at S-band for the anticipated solar wind velocities between 30–300 km s^{-1} in the solar corona. Since phase scintillations are sensitive to coronal irregularities of any scale size, the phase spectra do not contain the characteristic knee at f_b (Woo, 1975).

Some amplitude and phase spectra from Tyler *et al.* (1981b) are reproduced here in Figure 8. These examples were taken at S-band from the Viking spacecraft on 29 December, 1976 at a solar offset of $36.4R_\odot$. The break frequency in the amplitude spectrum converts to a projected velocity of 229 km s^{-1}. The slope of the spectrum for $f > f_b$ is -1.63, yielding a spectral index p of 2.63. The second flat section of the amplitude spectrum for $f > 5$ Hz is noise associated with the radio system and is not of coronal origin. Phase spectra covering a much larger frequency range have been published by Woo and Armstrong (1979).

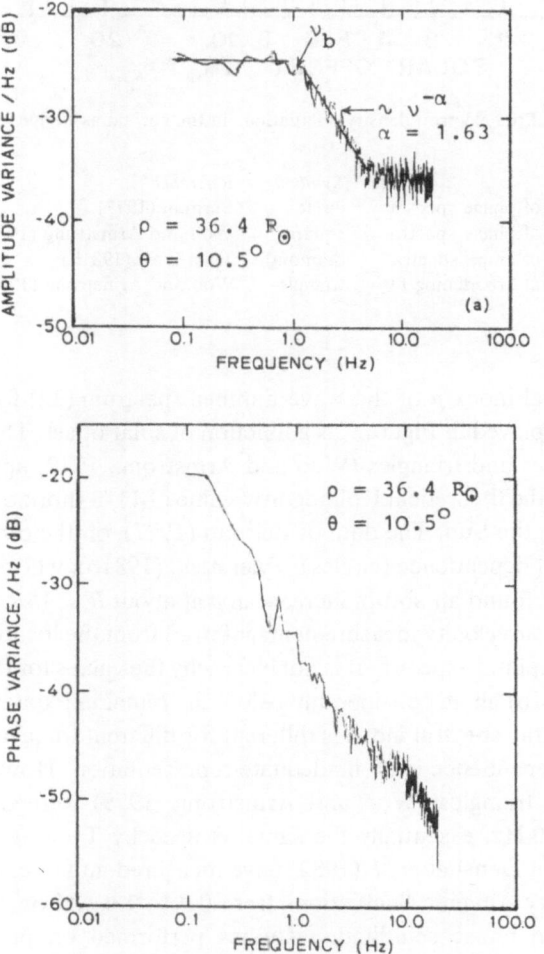

Fig. 8. Examples of frequency spectra of amplitude scintillations (upper plot) and phase scintillations (lower plot) as taken from Tyler *et al.* (1981b). The amplitude plot has a knee at the frequency ν_b, which is proportional to the projected flow velocity of the coronal plasma fluctuations.

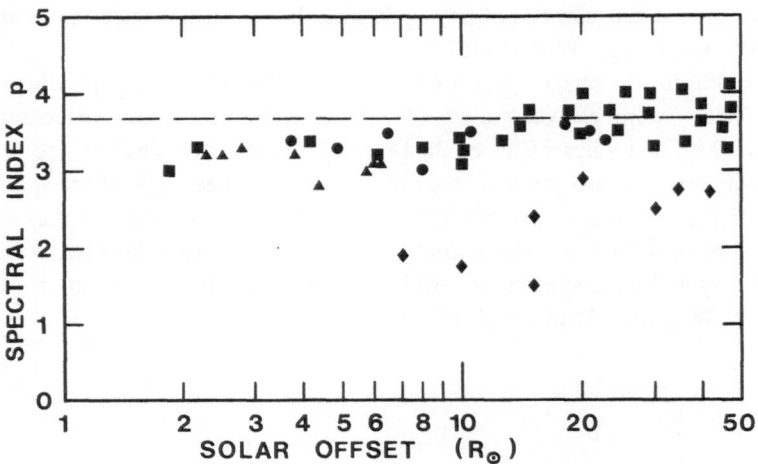

Fig. 9.   Spectral index p of the electron density fluctuations in the corona as derived from various data
types:

Type	Symbol	Reference
1. slope of phase spectra	circle	Berman (1977)
2. slope of phase spectra	square	Woo and Armstrong (1979)
3. slope of ampl. spectra	diamond	Tyler *et al.* (1981b)
4. spectral broadening fit	triangle	Woo and Armstrong (1979)

The derived spectral index p of the wave number spectrum (22) for some selected
Viking data sets is displayed in Figure 9 as a function of solar offset. The measurements
denoted by the squares and triangles (Woo and Armstrong, 1979) appear to increase
with R and approach the theoretical Kolmogorov value of 11/3 (horizontal dashed line)
at large distance from the Sun. The data of Berman (1977), on the other hand, do not
exhibit a strong radial dependence (circles). Tyler *et al.* (1981b), whose data points are
marked by diamonds, found an abrupt increase in p at about $R = 15R_\odot$, that coincided
with an increase in their velocity measurements inferred from the locations of the break
frequency f_b in the amplitude spectra. It is not clear why the values for the spectral index
from Tyler *et al.* (1981b) all lie considerably below the remaining data. One possibility
might have been that the spectral index is different for different frequency domains, i.e.
a single power-law dependence is an inadequate representation. However, the data in
Figure 9 denoted by triangles (Woo and Armstrong, 1979) were derived from the
frequency range 1–10 Hz, essentially the same as used by Tyler *et al.* (1981b). It is
interesting to note that Denskat *et al.* (1982) have measured an increase in the spectral
index of interplanetary Alfvénic fluctuations from 0.3 to 1.0 AU on Helios.

The amplitude and phase scintillation studies performed on past interplanetary
missions are listed in Table V. This and all the previous tables of radio science
investigations in the corona are quite possibly incomplete, but are felt to represent a
broad-based survey of the literature.

TABLE V

Intensity/phase scintillation measurements

Spacecraft	Frequency	Time interval	References
Mariner 6	S-band	May–June 1970	
Mariner 7		May–June 1970	Callahan (1975)
Mariner 9		August–October 1972	
Mariner 10	S/X-band	April–July 1974	Woo et al. (1976b)
Mars 2	1000 MHz	June–September 1974	Yakovlev et al. (1977)
Pioneer 10	S-band	March–June 1975	Berman and
Pioneer 11			Rockwell (1975)
Helios 1	S-band	July–October 1975	Berman (1976)
Venera 10	928 MHz	April–June 1976	Kolosov et al. (1978)
Viking	S/X-band	August 1976–February 1977	Callahan (1978)
		April 1976–March 1978	Woo and Armstrong (1979)
		October 1976–January 1977	Tyler et al. (1981b)
Voyager 1	S-band	June–August 1978	Berman and Conteas (1978)
		July–September 1979	Berman and Conteas (1979)

2.5. DUAL PROPAGATION PATH CORRELATIONS

In addition to the two schemes already discussed for determining the plasma flow velocity in the corona (i.e. spectral/angular broadening and the location of the break frequency of amplitude scintillation spectra), there is a more obvious method available whenever two radio links are simultaneously probing the corona. The measurement in its simplest form is merely to record the elapsed time of a distinguishing wave form in the time profile of some signal parameter in its transit from one ray path to the next. Instead of actually timing individual pulses, the time lag is usually determined by computing the cross-correlation function of the two time series. More than two receiving stations can improve the accuracy of the procedure and help eliminate projection effects. Solar wind velocity estimates using IPS are based on this concept (Coles and Kaufman, 1978).

A schematic representation of possible dual propagation path experiments is shown in Figure 10. The geometrical configurations from top to bottom are:

(a) One spacecraft P_1 and two (or more) ground stations R_1 and R_2: The receivers could be widely separated (intercontinental baseline), for which the projected separation of the ray paths in the corona is of the order of an Earth radius. With coronal velocities of over 100 km s^{-1}, this results in probable correlation time lags of a minute or even less. Tyler et al. (1981b) were unable to detect significant cross-correlation from Viking data, which they attribute to non-radial alignment of the solar offsets of the two ray paths

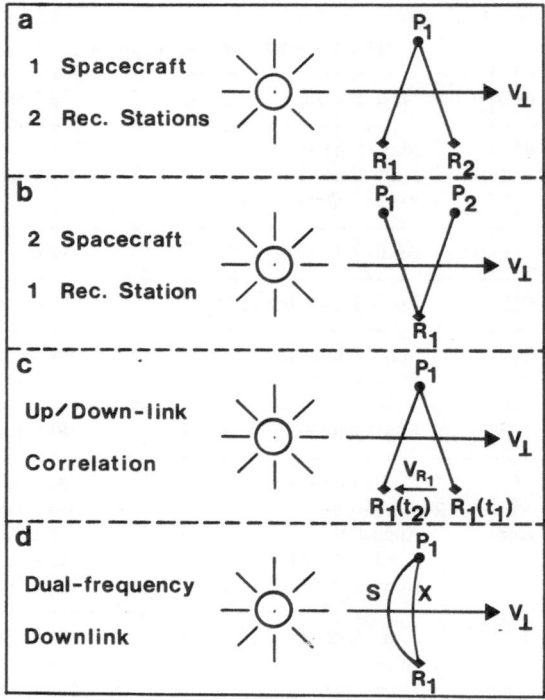

Fig. 10.   Various geometrical configurations for a determination of coronal velocities from a dual propagation path correlation analysis.

as well as their large separation. It is also possible to use two or more receivers at essentially one longitude (e.g. one DSN site), for which the correlation time lags are considerably shorter (fractions of a second). Woo and Armstrong (1980) have used this technique to determine solar wind speeds from S-band intensity scintillations using two DSN stations at the Goldstone site. Their data clearly indicate a solar wind acceleration over the range from $10-30R_\odot$. A subsequent study by Armstrong and Woo (1981) revealed that the fractional random velocity $\Delta v/v$ tended to increase dramatically at solar offsets $R < 20R_\odot$. This corresponds to the solar range where the coronal electron density spectrum is evolving, as evidenced by the change in spectral index (Figure 9).

   (b) Two spacecraft P_1, P_2 and one ground station R_1: This configuration could be extended, of course, to more than one ground station if the sources could not be simultaneously tracked at one ground station. Solar occultations of spacecraft are rare enough that the chance of catching two of them in superior conjunction at the same time would at first seem to be miniscule. On the other hand, spacecraft are very often launched in pairs and also tend to accumulate in the same general region of the solar system close to their primary objects of investigation. This trend resulted in the 'spacecraft parade' of August-September 1979, in which both Voyagers, Pioneer 11 and Pioneer Venus all underwent superior conjunction within a short time interval. Radial line-ups of the ray paths occurred between Pioneer Venus, the only spacecraft moving from west-to-east, and the other three probes. The original dual-spacecraft version of the International Solar Polar Mission (ISPM) would have featured a continuous two spacecraft solar occultation that could have been tracked at one DSN station, since both probe signals would have been located well within the antenna beam. Since the typical ray path separations here are of the order of solar radii rather than Earth radii, the possibility of detecting correlation is much more remote than for (a).

(c) Correlation between up- and downlink at one ground station R_1: The relative motion of the Earth with respect to the Sun/spacecraft system results in a separation d of the uplink ray path from the downlink ray path given approximately by

$$d = v_p t_c / L,\tag{25}$$

where L ($L > 1$) is given by (21), v_p is the apparent velocity of the spacecraft seen from Earth with respect to the Sun, and t_c is the round-trip light time to the spacecraft, i.e.

$$t_c = 2\rho/c.\tag{26}$$

The separation d for the Helios spacecraft at recent occultations, for which $L \simeq 4$ and $v_p \simeq 5R_\odot/\text{day}$, is of the order of an Earth radius. For more distant spacecraft the separation gets progressively greater. The solar wind velocity is determined from the time lag τ_p of the peak in the autocorrelation function from the relation

$$v_\perp = \frac{d}{\left| t_c/L - \tau_p \right|}.\tag{27}$$

Some preliminary results using this technique have been presented by Lüneburg and Esposito (1979).

(d) Dual-Frequency downlink, one spacecraft P_1 and one ground station R_1: The ray paths of radio signals are deflected away from the Sun by an amount inversely proportional to the square of the frequency. If the corona is probed with a dual-frequency downlink, this dispersive refraction produces a small separation between the ray paths at the solar proximate distance that could be exploited for a velocity determination via a cross-correlation analysis. For an S/X-band combination, the separation increases from about 10 km at $8R_\odot$ to over 10^3 km inside $3R_\odot$ (Tyler et al., 1977).

Since only the very few results cited above are available as examples of dual propagation path correlations, a table of references for this subsection was deemed unnecessary.

3. Coronal Dynamics

Radio science experiments with signals from spacecraft in solar occultation can provide valuable supplementary observations of abrupt temporal variations such as the coronal mass ejection event. Even prior to the first optical observations of coronal transients from the orbiting coronagraphs on OSO-7 (Howard et al., 1975) and Skylab (MacQueen et al., 1974), solar associated transient effects were recorded in the Faraday rotation (Levy *et al.*, 1969) and spectral broadening (Goldstein, 1969) of the Pioneer 6 carrier signal. The August 1972 flare events produced massive travelling interplanetary waves that were detected by the radio sounding experiment on Pioneer 9 (Croft, 1973). Additional Faraday rotation events were registered on Helios by Bird *et al.* (1977) at solar minimum.

An example of the types of data that can be analyzed to infer the physical characteristics of a dynamic coronal event is shown in Figure 11 (as taken from Woo and Armstrong, 1981). The ray path from the Voyager 1 spacecraft was probing the east solar limb on 18 August 1979 at about $13R_\odot$ from the Sun. Starting at 15 : 01 UT, the bandwidths of both carrier signals at S- and X-band broadened dramatically, the strength of the amplitude scintillation increased, and the difference between the carrier phases at S- and X-band began to grow at an abnormally fast rate.

Using two independent methods to determine the velocity of the leading edge of the disturbance from the spectral broadening and amplitude scintillation data, Woo and Armstrong (1981) compute a plasma flow speed of over 2500 km s^{-1}. Considering the disturbance to be an expanding spherical shell, the differential phase data was used to derive a profile of the electron density. The initial increase in columnar electron content N_T was about twice the typical ambient value at $13R_\odot$, but the increase in N from pre-event values to the peak of the disturbence was a factor of 3.7, consistent with

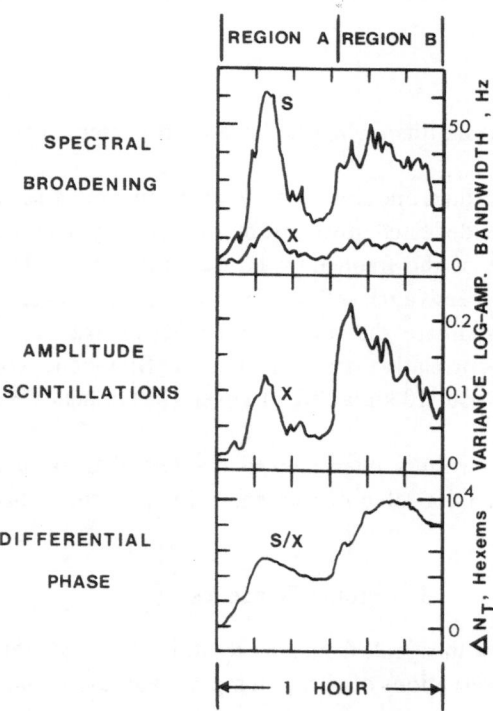

Fig. 11.   Measurements of spectral broadening at S-band and X-band, variance of the log-amplitude at X-band, and S/X-band differential phase using the dual-frequency radio system of Voyager 1 during the passage of an interplanetary shock wave on 18 August 1979 (after Woo and Armstrong, 1981).

the interpretation that this was a strong shock. Under this condition, the inferred shock speed was found to be over 3500 km s^{-1}.

Two regions are clearly distinguished in the data of Figure 11. The region A displays a high velocity from the spectral broadening and a relatively high electron density from the differential phase data, probably characteristic for a post-shock plasma. The region B, however, is much more turbulent (very strong amplitude scintillations relative to region A) and is much slower (reduced signal broadening), corresponding perhaps to the 'driver gas' of the shock wave.

4. Future Investigations

Very few spacecraft are presently available for coronal sounding experiments and the situation is likely to deteriorate further in the next few years. Only Helios 1 is still routinely tracked at its annual superior conjunctions, the eighth one having occurred in December 1981 during its 14th perihelion passage.

The mission plans of the two primary interplanetary ventures of this decade, ISPM and Galileo, are not especially well-suited for extensive radio science experiments in the solar corona. The ISPM spacecraft, for example, will pass behind the Sun only once during its flight to Jupiter before being injected into its final trajectory inclined almost perpendicular to the ecliptic. The situation with Galileo is somewhat better for two reasons: (a) the S-band probe signal is linearly polarized (Faraday rotation) and stronger than that on ISPM, and (b) the new DELTA-VEGA trajectory, which circles the Sun one extra time before returning to Earth for a gravitational assist, provides a total of three solar occultations enroute to Jupiter. Both missions will feature the dual-frequency capability.

The Solar Probe Mission (renamed 'Starprobe' recently to avoid confusion with the Solar Polar Mission), which proposes to send a spacecraft to within $4R_\odot$ of the solar surface, would provide some interesting new possibilities for coronal sounding from a position in the region of investigation. There would also be opportunities for radio science prior to perihelion (Croft, 1978; Coles et al., 1978). Unfortunately, Starprobe has been relegated along with many other space projects to an uncertain status because of the presently unfavorable fiscal climate.

These difficulties notwithstanding, Porsche et al. (1980) have proposed an interplanetary mission designed to provide continuous sounding of the solar corona. This project, dubbed COSIMI (Coronal Sounding Interplanetary Mission) by the proposers, would represent the ultimate in coronal investigations with occulted probe signals. The spacecraft would be inserted via a transfer orbit of about 2.5 y duration into the 'anti-Earth' position at perpetual superior conjunction. The many possible orbits available upon reaching superior conjunction all have pros and cons, but an overriding prerequisite would be the desire to keep the probe signal monitored as continuously as possible.

Examples of orbits of interplanetary spacecraft for radio sounding in the corona are given in Figure 12. The left-hand diagrams show the orbit in the ecliptic plane in a fixed

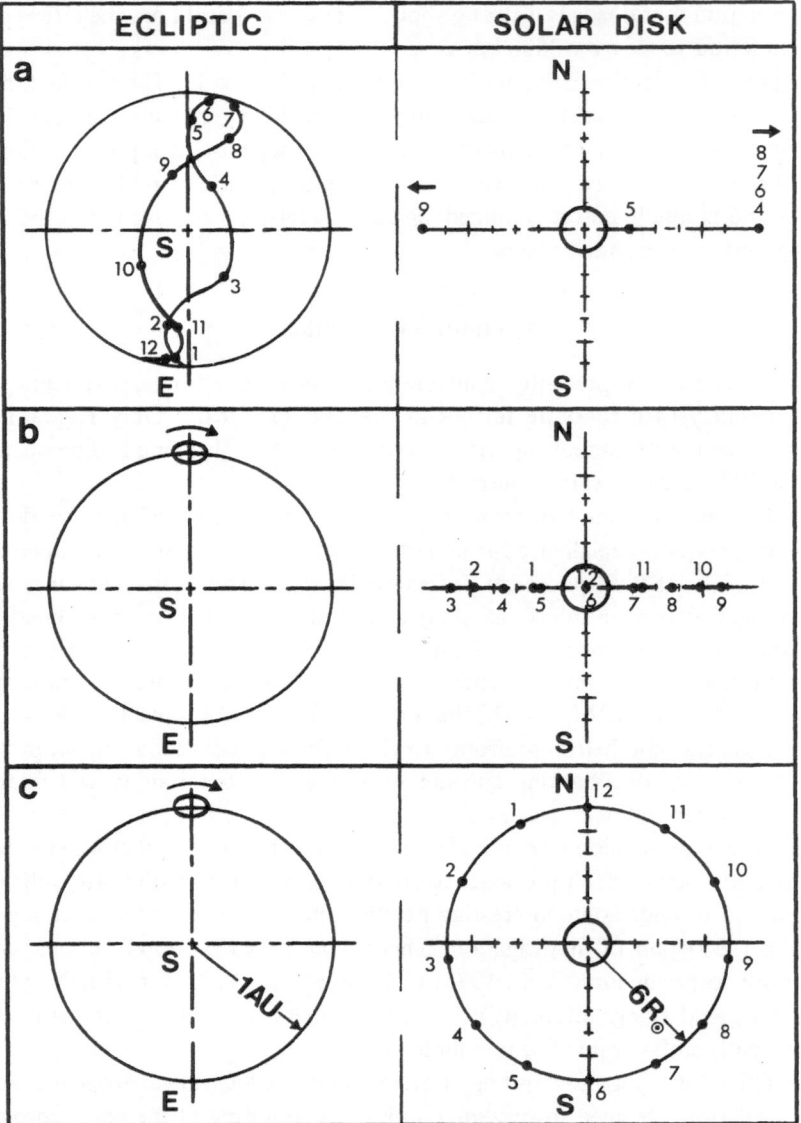

Fig. 12.   Various orbital configurations for a coronal sounding mission. The plots show a view of the ecliptic plane in a fixed Sun-Earth reference system (left), and a view of the solar disk with the north ecliptic pole directed upwards (right). See text for description of the orbits and the orbital elements.

Sun-Earth reference system. The solar disk views are drawn on the right. The upper panel (a) is the first year trajectory of the Helios 1 spacecraft (inclination $i \simeq 0°$, eccentricity $e \simeq 0.52$) that demonstrates its curious 'Figure 8' form in the rotating coordinate system. Only a small fraction of the orbital period was spent in solar occultation. The orbit in the middle panel (b) is also confined to the ecliptic ($i = 0°$), but has a small eccentricity ($e \simeq 0.03$) so that the probe signal moves back and forth along a radial from east limb to west limb. The spacecraft is too close to the Sun for coronal sounding for about three months of its one year period. If a small inclination ($i \simeq 3.2°$) is also added to the trajectory, the virtual orbit from Earth will look like the lower panel (c) with the ray path probing all position angles in the corona at a constant solar offset over the course of one year.

Stationing an interplanetary spacecraft like COSIMI at superior conjunction would be of benefit not only for coronal sounding, but also for solar backside observations. The evolution of active regions could be followed without interruption by combining Earth-based measurements with those from solar disk monitors on the COSIMI spacecraft. It would also be much easier to separate spatial and temporal variations with continuous coronal sounding over several solar rotations. *In situ* measurements from a spacecraft like Starprobe, which streaks through an entire solar hemisphere in less than one day, can only render a 'snapshot' of the coronal structure along the trajectory, saying very little about dynamic variations.

5. Summary

Coronal sounding investigations with occulted spacecraft signals have been shown to be a powerful tool for determining the structure of the important acceleration region of the solar wind, which is presently and will probably remain for some time unexplored by *in situ* spacecraft. The electron density N, the longitudinal component of the mean coronal magnetic field B_L, the coronal plasma flow velocity perpendicular to the line-of-sight $v_\perp$, and the wave number spectrum of electron density fluctuations $P_N(\kappa)$ are examples of physical quantities derived from radiometric data from spacecraft located near superior conjunction.

Measurements of electron content from ranging and Doppler data yield a radial decrease in electron density according to $r^{-2.3}$ (Muhleman *et al.*, 1977; Edenhofer *et al.*, 1977; Berman, 1979; Esposito *et al.*, 1980). Since the product $N \cdot B_r$ obeys a power law $\sim r^{-5.5}$ for $r \simeq 2{-}10R_\odot$ (Volland *et al.*, 1977), it is inferred that the magnetic field of the corona in this range decreases with a power closer to 3 (dipole field) and not the Parker spiral field exponent of 2. The radial velocity of the solar wind can be determined using many various radio techniques that clearly indicate the acceleration in the corona, even if the speeds are somewhat lower than expected for expansion from coronal holes (Woo, 1978; Woo and Armstrong, 1980; Tyler *et al.*, 1981b). The spectral index of electron density fluctuations tends to increase with distance from the Sun, indicating a relative loss in power at the higher frequencies (Woo and Armstrong, 1979; Tyler *et al.*, 1981b). The flatter spectra at small solar distances are most likely associated with an energy deposition that contributes to the acceleration of the solar wind.

Radio sounding techniques have been used on occasion to derive the time profiles of important physical parameters during abrupt coronal disturbences (Woo and Armstrong, 1981), thus offering a valuable contribution to the studies of travelling interplanetary phenomena.

Future investigations on the interplanetary missions ISPM and Galileo will provide some opportunities for coronal sounding with spacecraft signals. However, only a mission with continuous tracking of a stationary source at superior conjunction (Porsche *et al.*, 1980) would be capable of separating the various temporal and spatial variations of the solar corona.

References

Anderson, J. D., Esposito, P. B., Martin, W., Thornton, C. L., and Muhleman, D. O.: 1975, *Astrophys. J.* **200**, 221.

Armstrong, J. W. and Woo, R.: 1981, *Astron. Astrophys.* **103**, 415.

Benz, A. O. and Fitze, H. R.: 1979, *Astron. Astrophys.* **76**, 354.

Berlin, A. B., Korol'kov, D. V., Pariiskii, Y. N., Soboleva, N. S., and Timofeeva, G. M.: 1978, *Sov. Astron. Letters* **4**, 102.

Berman, A. L.: 1976, *DSN-PR* **42–32**, Jet Propulsion Laboratory, Pasadena, p. 262.

Berman, A. L.: 1977, *DSN-PR* **42–41**, Jet Propulsion Laboratory, Pasadena, p. 135.

Berman, A. L.: 1979, *DSN-PR* **42–50**, Jet Propulsion Laboratory, Pasadena, p. 124.

Berman, A. L. and Conteas, A. D.: 1978, *DSN-PR* **42–48**, Jet Propulsion Laboratory, Pasadena, p. 59.

Berman, A. L. and Conteas, A. D.: 1979, *DSN-PR* **42–54**, Jet Propulsion Laboratory, Pasadena, p. 71.

Berman, A. L. and Rockwell, S. T.: 1975, *DSN-PR* **42–30**, Jet Propulsion Laboratory, Pasadena, p. 231.

Berman, A. L. and Wackley, J. A.: 1977, *DSN-PR* **42–38**, Jet Propulsion Laboratory, Pasadena, p. 152.

Bird, M. K.: 1981, in H. Rosenbauer (ed.), *Solar Wind Four*, MPAE–W–100–81–31, p. 78.

Bird, M. K., Volland, H., Stelzried, C. T., Levy, G. S., and Seidel, B. L.: 1977, in M. A. Shea, D. F. Smart, and S. T. Wu (eds.), *Contributed Papers for the Study of Traveling Interplanetary Phenomena/1977*, D. Reidel Publ. Co., Dordrecht, Holland, p. 63.

Bird, M. K., Volland, H., Hirth, W., and Fürst, E.: 1978, *Space Res.* **18**, 377.

Bird, M. K., Schrüfer, E., Volland, H., and Sieber, W.: 1980, *Nature* **283**, 459.

Blesing, R. G. and Dennison, P. A.: 1972, *Proc. Astron. Soc. Australia* **2**, 84.

Bracewell, R. N., Eshleman, V. R., and Hollweg, J. V.: 1969, *Astrophys. J.* **155**, 367.

Bradford, H. M. and Routledge, D.: 1979, *Monthly Notices Roy. Astron. Soc.* **190**, 73P.

Callahan, P. S.: 1975, *Astrophys. J.* **199**, 227.

Callahan, P. S.: 1978, *DSN-PR* **42–44**, Jet Propulsion Laboratory, Pasadena, p. 75.

Cannon, A. R.: 1976, Ph. D. Thesis, U. California, Berkeley.

Coles, W. A.: 1978, *Space Sci. Rev.* **21**, 411.

Coles, W. A. and Kaufman, J. J.: 1978, *Radio Sci.* **13**, 591.

Coles, W. A., Rickett, B. J., and Scott, S. L.: 1978, in M. Neugebauer and R. W. Davies (eds.), *A Close-up of the Sun*, JPL Publ. 78–70, Jet Propulsion Laboratory, Pasadena, p. 388.

Counselman, C. C. and Rankin, J. M.: 1972, *Astrophys. J.* **175**, 843.

Counselman, C. C. and Rankin, J. M.: 1973, *Astrophys. J.* **185**, 357.

Croft, T. A.: 1973, *J. Geophys. Res.* **78**, 3159.

Croft, T. A.: 1978, in M. Neugebauer and R. W. Davies (eds.), *A Close-up of the Sun*, JPL Publ. 78–70, Jet Propulsion Laboratory, Pasadena, p. 397.

Croft, T. A.: 1979, *J. Geophys. Res.* **84**, 439.

Dennison, P. A. and Hewish, A.: 1967, *Nature* **213**, 343.

Dennison, B., Melnick, G., Harwit, M., Sato, T., Stelzried, C. T., and Jauncey, D.: 1978, *Nature* **273**, 33.

Denskat, K. U. and Neubauer, F. M.: 1982, *J. Geophys. Res.* **87**, 2215.

Edenhofer, P., Esposito, P. B., Hansen, R. T., Hansen, S. F., Lüneburg, E., Martin, W. L., and Zygielbaum, A. I.: 1977, *J. Geophys.* **42**, 673.

Efimov, A. I., Yakovlev, O. I., Razmanov, V. M., and Shtrykov, V. K.: 1977, *Sov. Astron. Letters* **3**, 172.

Ekers, R. D. and Little, L. T.: 1971, *Astron. Astrophys.* **10**, 310.

Elliot, J. L.: 1979, *Ann. Rev. Astron. Astrophys.* **17**, 445.

Elliot, J. L., Dunham, E., and Mink, D.: 1977, *Nature* **267**, 328.

Eshlemar, V. R., Barthle, R. C., and Gallagher, P. B.: 1960, *Science* **131**, 329.

Eshlemar, V. R., Garriott, O. K., Leadabrand, R. L., Peterson, A. M. *et al.*: 1966, *J. Geophys. Res.* **71**, 3325.

Eshlemar, V. R., Tyler, G. L., Anderson, J. D., Fjeldbo, G., Levy, G. S., Wood, G. E., and Croft, T. A.: 1977, *Space Sci. Rev.* **21**, 207.

Eshlemar, V. R., Tyler, G. L., Wood, G. E., Lindal, G. F., Anderson, J. D., Levy, G. S., and Croft, T. A.: 1979a, *Science* **204**, 976.

Eshlemar, V. R., Tyler, G. L., Wood, G. E., Lindal, G. F., Anderson, J. D., Levy, G. S., and Croft, T. A.: 1979b, *Science* **206**, 959.

Esposito, P. B., Edenhofer, P., and Lüneburg, E.: 1980, *J. Geophys. Res.* **85**, 3414.

Fitze, H. R. and Benz, A. O.: 1981, *Astrophys. J.* **250**, 782.

Fjeldbo, G., Kliore, A., Seidel, B., Sweetnam, D., and Cain, D.: 1975, *Astron. Astrophys.* **39**, 91.

Fjeldbo, G., Sweetnam, D., Brenkle, J., Christensen, E., Farless, D., Mehta, J., Seidel, B., Michael, Jr., W., Wallio, A., and Grossi, M.: 1977, *J. Geophys. Res.* **82**, 4317.

Goldstein, R. M.: 1969, *Science* **166**, 598.

Goldstein, R. M. and Stelzried, C. T.: 1975, *DSN-PR* **42–27**, Jet Propulsion Laboratory, Pasadena, p. 81.

Harwit, M., Lovelace, R. V. E., Dennison, B., Jauncey, D. L., and Broaderick, J.: 1974, *Nature* **249**, 230.

Hewish, A.: 1958, *Monthly Notices Roy. Astron. Soc.* **118**, 534.

Hewish, A.: 1972, in C. P. Sonett, P. J. Coleman, Jr., and J. M. Wilcox (eds.), *Solar Wind*, NASA SP-308, U.S. Gov. Print. Office, Washington, p. 477.

Hewish, A., Scott, P. F., and Wills, D.: 1964, *Nature* **203**, 1214.

Hollweg, J. V.: 1970, *J. Geophys. Res.* **75**, 3715.

Hollweg, J. V.: 1978, *Rev. Geophys. Space Phys.* **16**, 689.

Hollweg, J. V., Bird, M. K., Volland, H., Edenhofer, P., Stelzried, C. T., and Seidel, B. L.: 1982, *J. Geophys. Res.* **87**, 1.

Howard, H. T., Tyler, G. L., Fjeldbo, G., Kliore, A., Levy, G. S., Brunn, D. L., Dickinson, R., Edelson, R. E., Martin, W. L., Postal, R. B., Seidel, B. L., Sesplaukis, T. T., Shirley, D. L., Stelzried, C. T., Sweetnam, D. N., Zygielbaum, A. I., Esposito, P. B., Anderson, J. D., Shapiro, I. I., and Reasonberg, R. D.: 1974, *Science* **183**, 1297.

Howard, R. A., Koomen, M. J., Michels, D. J., Tousey, R., Detweiler, C. R., Roberts, D. E., Seal, R. T., Whitney, J. D., Hansen, R. T., Garcia, C. J., and Yasukawa, E.: 1975, Report UAG-48, World Data Center A, NOAA, Boulder.

James, J. C.: 1966, *Astrophys. J.* **146**, 356.

James, J. C.: 1970, *Solar Phys.* **12**, 143.

Jokipii, J. R.: 1973, *Ann. Rev. Astron. Astrophys.* **11**, 1.

Kliore, A. J., Cain, D. L., Fjeldbo, G., Seidel, B. L., Sykes, M. J., and Rasool, S. I.: 1972, *Icarus* **17**, 484.

Kliore, A. J., Woo, R., Armstrong, J. W., Patel, I. R., and Croft, T. A.: 1979, *Science* **203**, 765.

Kliore, A. J., Patel, I. R., Lindal, G. F., Sweetnam, D. N., Hotz, H. B., Waite, Jr., J. H., and McDonough, T. R.: 1980, *J. Geophys. Res.* **85**, 5857.

Klüber, H. von: 1960, in A. Beer (ed.), *Vistas in Astronomy*, Permagon Press, London, p. 47.

Koehler, R. L.: 1968, *J. Geophys. Res.* **73**, 4883.

Kolosov, M. A., Yakovlev, O. I., Rogal'skii, V. I., Efimov, A. I., Razmanov, V. M., and Shtrykov, V. K.: 1978, *Sov. Phys. Dokl.* **23**, 440.

Landt, J. A.: 1974, *J. Geophys. Res.* **79**, 2761.

Layland, J. W.: 1978, *DSN-PR* **42–44**, Jet Propulsion Laboratory, Pasadena, p. 54.

Levy, G. S.: 1977, *Mariner Venus Mercury 1973 S/X-band Experiment*, JPL Publ. 77–17, Jet Propulsion Laboratory, Pasadena.

Levy, G. S., Sato, T., Seidel, B. L., Stelzried, C. T., Ohlson, J. E., and Rusch, W. V. T.: 1969, *Science* **166**, 596.

Lotova, N. A.: 1975, *Sov. Phys. Usp.* **18**, 292.

Lüneburg, E. and Esposito, P. B.: 1979, paper presented at Nat. Radio Sci. Meet. URSI, 18–21 June, Seattle.

MacQueen, R. M., Eddy, J. A., Gosling, J. T., Hildner, E., Munro, R. H., Newkirk, Jr., G. A., Poland, A. I., and Ross, C. L.: 1974, *Astrophys. J.* **187**, L85.

Martin, W. L. and Zygielbaum, A. I.: 1977, *Mu-II Ranging, TM* **33–768**, Jet Propulsion Laboratory, Pasadena.

Michael, Jr., W. H., Cain, D. L., Fjeldbo, G., Levy, G. S., Davies, J. G., Grossi, M. D., Shapiro, I. I., and Tyler, G. L.: 1972, *Icarus* **16**, 57.

Muhleman, D. O., Esposito, P. B., and Anderson, J. D.: 1977, *Astrophys. J.* **211**, 943.

Ohlson, J. E., Levy, G. S., and Stelzried, C. T.: 1974, *Trans, IEEE* **IM–23**, 167.

Porsche, H., Volland, H., Bird, M. K., and Edenhofer, P.: 1980, in M. Dryer and E. Tandberg-Hanssen (eds.), *Solar and Interplanetary Dynamics*, D. Reidel Publ. Co., Dordrecht, Holland, p. 541.

Razmanov, V. M., Efimov, A. I., and Yakovlev, O. I.: 1979, *Radiophys. Quant. Electronics* **22**(9), 728.

Readhead, A. C. S., Kemp, M. C., and Hewish, A.: 1978, *Monthly Notices Roy. Astron. Soc.* **185**, 207.

Rockwell, S. T.:: 1977, *DSN-PR* **42–43**, Jet Propulsion Laboratory, Pasadena, p. 216.

Scherrer, P. H., Wilcox, J. M., Svalgaard, L., Duvall, Jr., T. L., Dittmer, P. H., and Gustofson, E. K.: 1977, *Solar Phys.* **54**, 353.

Shapiro, I. I., Reasonberg, R. D., MacNeil, P. E., Goldstein, R. B., Brenkle, J. P., Cain, D. C., Komarek, T., Zygielbaum, A. I., Cuddihy, W. F., and Michael, Jr., W. H.: 1977, *J. Geophys. Res.* **82**, 4329.

Sofue, Y., Kawabata, K., Takahashi, F., and Nawajiri, N.: 1976, *Solar Phys.* **50**, 465.

Stelzried, C. T.: 1970, *A Faraday Rotation Measurement of a 13-cm Signal in the Solar Corona*, TR 32-1401, Jet Propulsion Laboratory, Pasadena.

Stelzried, C. T., Levy, G. S., Sato, T., Rusch, W. V. T., Ohlson, J. E., Schatten, K. H., and Wilcox, J. M.: 1970, *Solar Phys.* **14**, 440.

Stelzried, C. T., Sato, T., and Abreu, A.: 1972, *J. Spacecraft* **9**, 69.

Süss, H.: 1980, *Kleinheubacher Ber.* **23**, 275 (in German).

Taylor, F. H. J.: 1977, *DSN-PR* **42–40**, Jet Propulsion Laboratory, Pasadena, p. 57.

Tyler, G. L., Brenkle, J. P., Komarek, T. A., and Zygielbaum, A. I.: 1977, *J. Geophys. Res.* **82**, 4335.

Tyler, G. L., Eshleman, V. R., Anderson, J. D., Levy, G. S., Lindal, G. F., Wood, G. E., and Croft, T. A.: 1981a, *Science* **212**, 201.

Tyler, G. L., Vesecky, J. F., Plume, M. A., Howard, H. T., and Barnes, A.: 1981b, *Astrophys. J.* **249**, 318.

Tyler, G. L., Eshleman, V. R., Anderson, J. D., Levy, G. S., Lindal, G. F., Wood, G. E., and Croft, T. A.: 1982, *Science* **215**, 553.

Volland, H., Bird, M. K., Levy, G. S., Stelzried, C. T., and Seidel, B. L.: 1977, *J. Geophys.* **42**, 659.

Walker, A. R.: 1980, *Monthly Notices Roy. Astron. Soc.* **192**, 47P.

Weisberg, J. M., Rankin, J. M., Payne, R. R., and Counselman, C. C.: 1976, *Astrophys. J.* **209**, 252.

Winn, F. B., Reinbold, S. R., Yip, K. W., Koch, R. E., and Lubeley, A.: 1975, *DSN-PR* **42–30**, Jet Propulsion Laboratory, Pasadena, p. 88.

Woo, R.: 1975, *Astrophys. J.* **201**, 238.

Woo, R.: 1977, in M. A. Shea, D. F. Smart, and S. T. Wu (eds.), *Study of Travelling Interplanetary Phenomena/1977*, D. Reidel Publ. Co., Dordrecht, Holland, p. 81.

Woo, R.: 1978, *Astrophys. J.* **219**, 727.

Woo, R.: 1981, in H. Rosenbauer (ed.), *Solar Wind Four*, MPAE–W–100–81–31, p. 66.

Woo, R. and Armstrong, J. W.: 1979, *J. Geophys. Res.* **84**, 7288.

Woo, R. and Armstrong, J. W.: 1980, paper presented at STIP Workshop, 10-15 June, Smolenice, Czechoslovakia.

Woo, R. and Armstrong, J. W.: 1981, *Nature* **292**, 608.

Woo, R., Yang, F. C., and Ishimaru, A.: 1976a, *Astrophys. J.* **210**, 593.

Woo, R., Yang, F. C., Yip, K. W., and Kendall, W. B.: 1976b, *Astrophys. J.* **210**, 568.

Woo, R., Yang, F. C., and Ishimaru, A.: 1977, *Astrophys. J.* **218**, 557.

Yakovlev, O. I., Molotov, Y. P., Kruglov, Y. M., Yefimov, A. I., Razmanov, V. M., Timofeyeva, T. S., and Shtrykov, V. F.: 1977, *Radio Engg. Elec. Phys.* **22**(2), 29.

Zygielbaum, A. I.: 1977, *DSN-PR* **42–41**, Jet Propulsion Laboratory, Pasadena, p. 43.

MEASUREMENTS OF THE PROPERTIES OF SOLAR WIND PLASMA RELEVANT TO STUDIES OF ITS CORONAL SOURCES*

MARCIA NEUGEBAUER

Jet Propulsion Laboratory, California Institute of Technology, Pasadena, CA 91109, U.S.A.

Abstract. Interplanetary measurements of the speeds, densities, abundances, and charge states of solar wind ions are diagnostic of conditions in the source region of the solar wind. The absolute values of the mass, momentum, and energy fluxes in the solar wind are not known to an accuracy of 20%. The principal limitations on the absolute accuracies of observations of solar wind protons and alpha particles arise from uncertain instrument calibrations, from the methods used to reduce the data, and from sampling biases. Sampling biases are very important in studies of alpha particles. Instrumental resolution and measurement ambiguities are additional major problems for the observation of ions heavier than helium. Progress in overcoming some of these measurement inadequacies is reviewed.

1. Introduction

Direct measurements of the solar wind plasma have, so far, been limited to the region beyond 0.3 AU from the Sun where the flow is already supersonic. By this distance, much information about coronal processes has already been lost or masked by interplanetary effects. This review focusses on the measurements of those plasma properties which are most diagnostic of coronal conditions and least affected by inter-planetary propagation. The review considers the accuracy with which these properties can be measured and the development of instrumentation to decrease the observational uncertainties.

The plasma parameters considered are:

(1) *Bulk flow speed.* One of the first discovered and most striking features of the solar wind is its organization into high- and low-speed streams. High-speed streams have been associated with flow from coronal holes and solar flare sites. Some of the low-speed wind originates in coronal streamers. One aim of present solar wind research is to identify other solar wind sources, especially of the low-speed wind. The speed observed at 1 AU has been used to map the plasma streamlines back to the Sun (Snyder and Neugebauer, 1966; Nolte and Roelof, 1973) in an attempt to identify the solar longitude of their sources.

(2) *Mass, momentum, and energy fluxes.* The mass, momentum, and energy fluxes observed at 1 AU serve as a boundary condition for theories or models of the acceleration of the solar wind from various solar sources. A successful model must be able to explain these fluxes as well as the solar wind speed.

(3) *Elemental abundances.* The abundance of helium relative to hydrogen in the solar wind is highly variable (see Neugebauer (1981) for a recent review). Studies of variations of helium as well as other ion abundances may help elucidate the relative importance

---

* Paper presented at the IX-th Lindau Workshop 'The Source Region of the Solar Wind'.

Space Science Reviews 33 (1982) 127–150. 0038–6308/82/0332–0127$03.45.

of various processes such as thermal diffusion, turbulent mixing, Coulomb drag, and wave acceleration in the regions of origin of the solar wind.

(4) *Charge states*. The distribution of charge states of heavy ions (e.g., the relative amounts of O^{5+}, O^{6+}, O^{7+}, ions) carries information on the energy spectrum of electrons at the maximum altitude in the corona at which ion-electron collisions occur (Hundhausen *et al.*, 1968; Fenimore, 1980). Because of their different ionization potentials, different ion species (iron and oxygen, for example) yield information about different heights in the corona.

(5) *Magnetic field*. Comparison of the direction of the interplanetary magnetic field and photospheric magnetic fields is another useful tool for identifying possible source regions of the solar wind (e.g., Burlaga *et al.*, 1978; Levine, 1978). Furthermore, the field strength at 1 AU is another boundary condition which must be met by solar wind models (e.g., Hundhausen, 1977).

With the exception of the magnetic field, all these parameters relevant to studies of the solar wind soure region involve the measurement of positive ions. It is the purpose of this paper to comment on the accuracy with which it has been possible to measure these particular properties of solar wind ions, to point out some of the experimental difficulties, and to discuss ways of avoiding some of them in the future. Not all properties of solar wind ions are included in the discussion. Specifically, the ion and electron bulk flow directions, temperatures, and anisotropies at 1 AU are largely controlled by interplanetary rather than coronal processes. In this review, they are therefore considered only as they affect the accuracy of the measurements of the parameters of principal interest to studies of solar wind sources.

2. Solar Wind Bulk Flow Properties

King (1977) has made a thorough comparison of hourly averages of near-Earth solar wind properties measured by a number of different instruments on a number of spacecraft. For each pair of instruments, he performed a regression analysis of selected solar wind properties for those hours during which both members of the pair observed the solar wind. His results for bulk speed observations are shown in Figure 1. Each line in this Figure represents the fit to the speeds observed by a single pair, while the error bars on each line show the standard deviation about the regression line. Random errors between pairs of hourly averages of solar wind speed are, on the average, about 12 km s^{-1}, while systematic differences can be as large as 60 km s^{-1}.

The first three columns of Table I show how these velocity uncertainties translate into uncertainties of mapping streamlines back to the Sun. To understand this table, consider its first row. Suppose the speed of a particular feature observed near 1 AU by Explorers 34 and 33 was roughly 300 km s^{-1}. Using Explorer-34 data, one would calculate that this particular bit of plasma came from a solar longitude of $\phi_{34} = \phi_E + \Omega r/v_{34}$, where ϕ_E is the solar longitude of the Earth, Ω is the solar rotation rate, $r = 1$ AU, and v_{34} is the solar wind speed observed by Explorer 34. Similarly, for Explorer 33, $\phi_{33} = \phi_E + \Omega r/v_{33}$. The difference in solar longitude listed in Table I is

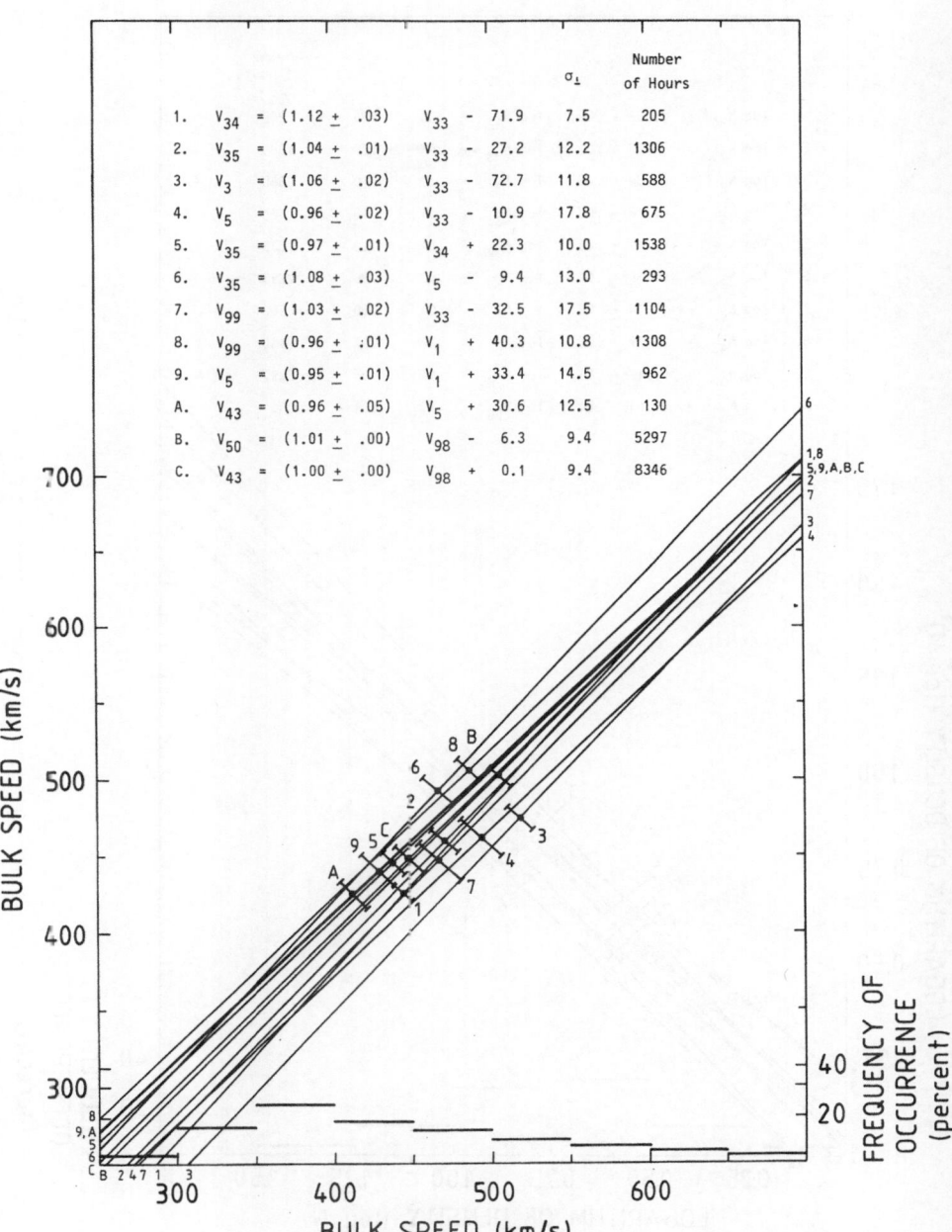

Fig. 1.   Regression lines for hourly averages of the solar wind bulk speed for several pairs of spacecraft which made simultaneous observations of the solar wind. After King (1977).

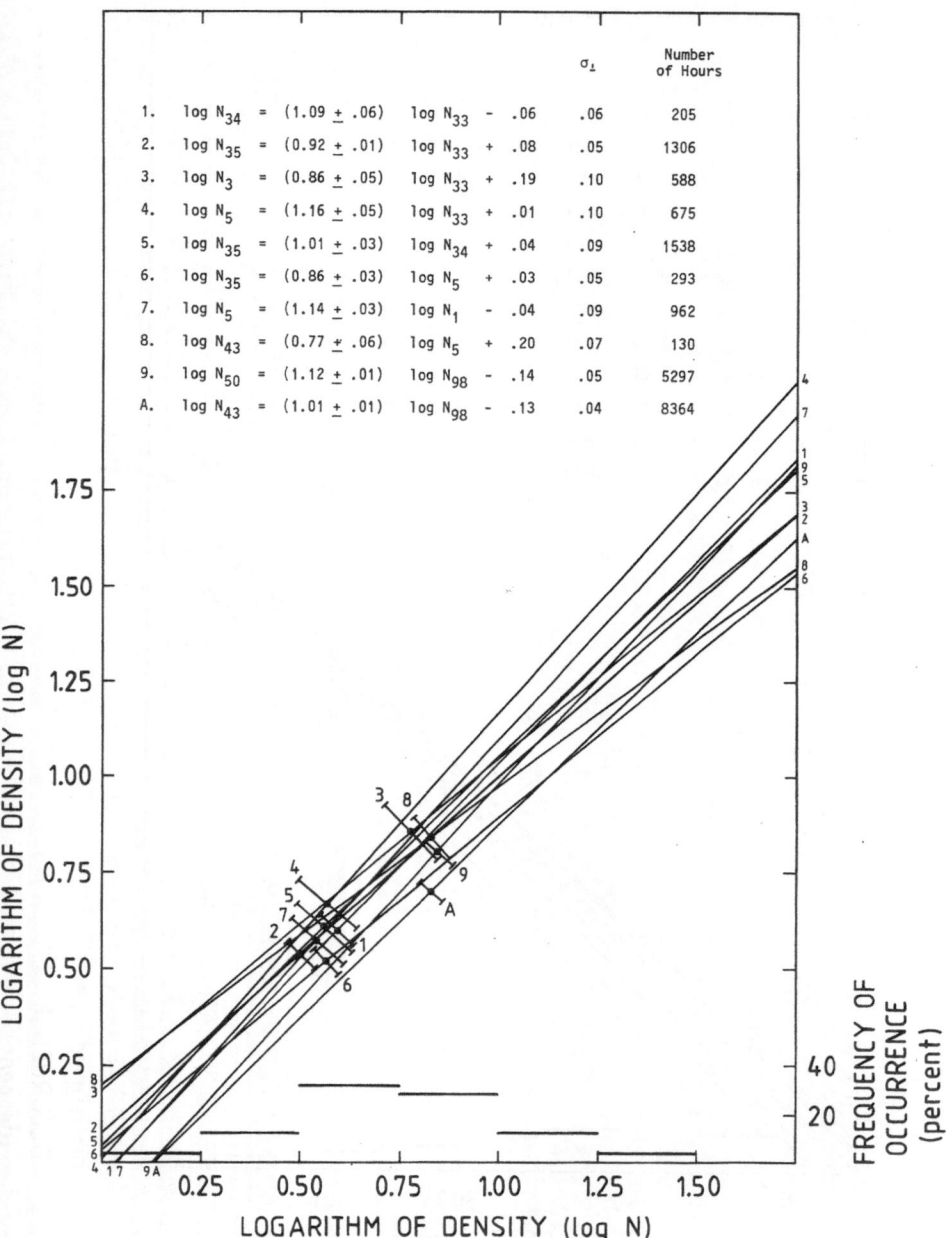

Fig. 2.   Regression lines for the logarithm of hourly averages of the number density of the solar wind for several pairs of spacecraft which made simultaneous observations of the solar wind. After King (1977).

TABLE I

Illustration of uncertainties arising in determination of the longitude of a solar source region and of the number flux and energy flux of the solar wind using sets of simultaneous data obtained by different spacecraft

Comparison	Difference in solar longitude		% difference in flux (nv)		% difference in energy flux (nv^3)	
	@300 km s⁻¹	@750 km s⁻¹	@300 km s⁻¹ and 10 cm⁻³	@750 km s⁻¹ and 6 cm⁻³	@300 km s⁻¹ and 10 cm⁻³	@750 km s⁻¹ and 6 cm⁻³
E34 vs E33	10.4ª	0.8ᶜ	6	5	31	9
E35 vs E33	4.2	0.2	51	25	61	5
V3 vs E33	17.2	1.1	9	15	48	7
O5 vs E33	6.4	1.7	31	25	15	14
E35 vs E34	3.2	0.0	16	11	24	11
E35 vs O5	3.6	2.1	21	12	11	2
O5 vs HEOS	4.5	0.1	31	17	43	16
E43 vs O5	4.5	0.1	1	5	11	5
I8 vs I6–8	0.8	0.1	6	11	8	10
E43 vs I6–8	0.0	0.0	27	28	27	28
Average			20	15	23	11

$2|\phi_{34} - \phi_{33}|/(\phi_{34} + \phi_{33}) = 10.4°$. That is, the calculated solar origin of a particular low-speed feature, such as a sector boundary, would differ by 10.4°, depending on whether Explorer-34 or Explorer-33 data were used. A few of the longitude differences in the table, especially at low speeds, are greater than the sizes of or spacings between possibly relevant solar features. The conclusion is that, in addition to the instrinsic uncertainly in the mapping method because of its underlying assumptions that the flow is constant and radial, significant further uncertainty can arise from the inaccuracy of the velocity measurements.

King's results for proton density are shown in Figure 2. Note that the scales in this Figure are logarithmic. For densities in the range $3-6$ cm^{-3}, systematic differences between instruments are as large as 44%. We note in passing the most notoriously divergent solar wind density measurements are not included in King's analysis.

The combined effects of systematic differences in velocity and density are given in the center and right sections of Table I. Suppose, for example, one wished to compute the energy flux in a certain low-speed structure observed by both OGO 5 and HEOS. The answers would probably differ by about 43%. There is no way to know which result, if either, is correct.

Hundhausen (1977) has noted that the proton flux in the high-speed streams from three large coronal holes were only $15-50\%$ above average values observed earlier in the solar cycle. The discrepancies in flux measurements shown in Table I are large enough to raise some doubt about the reality of any enhanced proton flux in these streams. Hundhausen's calculated factor of two increase in energy flux is probably beyond question, however.

There are many possible causes for the observed discrepancies between data sets.

(1) *Solar wind variability.* First, there is the possibility that some of the random differences are real. Spatial gradients do exist in the solar wind, but they are usually small over distances comparable to the separation of the spacecraft in King's sample. When dealing with hourly averages, differences can also arise from the propagation time of a solar wind disturbance from one spacecraft to the other.

(2) *Terrestrial and lunar effects.* In a time varying solar wind, differences in hourly averages can also be caused by one spacecraft spending a greater or lesser fraction of the hour beyond the bow shock in a time varying solar wind. Such effects should contribute only to the random errors and not to the systematic discrepancies, however. These errors could be avoided if hours containing bow shock crossings were excluded from comparative studies such as King's.

Discrepancies can also be caused by undetected passages into the magnetosheath. There is certainly some magnetosheath data masquerading as solar wind data in King's compilation and in others. How much is not known. It can be minimized by extremely careful and laborious examination of simultaneous plasma and field data at high time resolution. Systematic differences could arise between data sets if one group of experimenters was consistently more careful than another about identifying magneto-sheath data. There is no evidence that this is the case (perhaps because no one has looked for such evidence).

(3) *Instrument calibrations and stability.* Solar wind ion spectrometers are usually calibrated preflight by measurements of their responses to an ion beam as functions of ion energy and angle of incidence. It is difficult to obtain ion beams which are uniform and parallel over an area as large as the detector aperture. Quoted measurement accuracies range from 1 to 10% for speed and from 8 to 30% for density. The fraction of these uncertainties attributable to the calibration process is usually not stated.

Temporal changes in instrument calibration or sensitivity may also be fairly common. Some such changes are indicated by changes in the regression lines between instruments from one year to the next. Some experiments carried two sensors, such as a channel electron multiplier and an electrometer, which could be compared and intercalibrated during the flight. On Pioneers 10 and 11, the sensitivity of the channel multipliers decreased relative to the electrometers, by a large factor (D. Intriligator, Personal Communication). On Helios 1 and 2, however, the intercalibrations of the same two types of sensors have remained constant, within 1%, over a period of almost seven years, to date (H. Rosenbauer, Personal Communication).

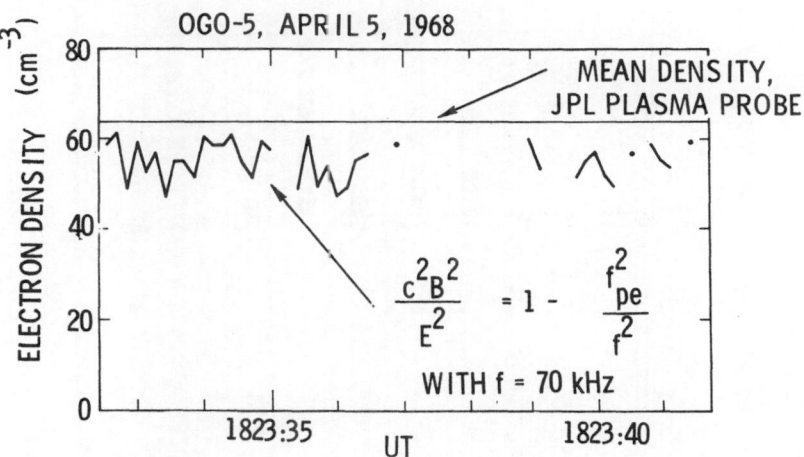

Fig. 3. Electron density as measured by the plasma spectrometer and by an analysis of plasma waves observed by OGO 5. From Scarf *et al.* (1970).

What is needed is a method of in-flight calibration, especially of the absolute value of plasma density. It is possible to obtain occasional in-flight calibration through the observation of plasma waves. Figure 3 shows data from one period when this could be done with the instrumentation on OGO 5. The plasma wave instruments on this spacecraft could observe both electric and magnetic oscillations at 70 kHz. Several bursts were observed when the electron density was near 63 cm^{-3}, which corresponds to a local plasma frequency of 70 kHz. In this case it was possible to identify the presence of a single wave mode (an electromagnetic wave with frequency slightly greater than the local plasma frequency) and to calculate the corresponding local electron density (Scarf *et al.*, 1970). The results are plotted in Figure 3; the wave observations indicated a plasma density about 10% below the density calculated from the plasma probe data with much poorer time resolution. More of this type of cross-correlation between instruments would be possible in the future if sufficiently capable wave instrumentation is included in the payload.

(4) *Data reduction methods.* Besides possible miscalibrations, perhaps the greatest source of discrepancy arises during the data reduction process.

First, there is the problem of what to do about unmeasured quantities. The problem is illustrated, in an exaggerated way, by the spectrum in the top of Figure 4. This is a slightly better than typical spectrum obtained by the curved-plate analyzer on Mariner 2 (Neugebauer and Snyder, 1966). Because there was only a single detector, which always

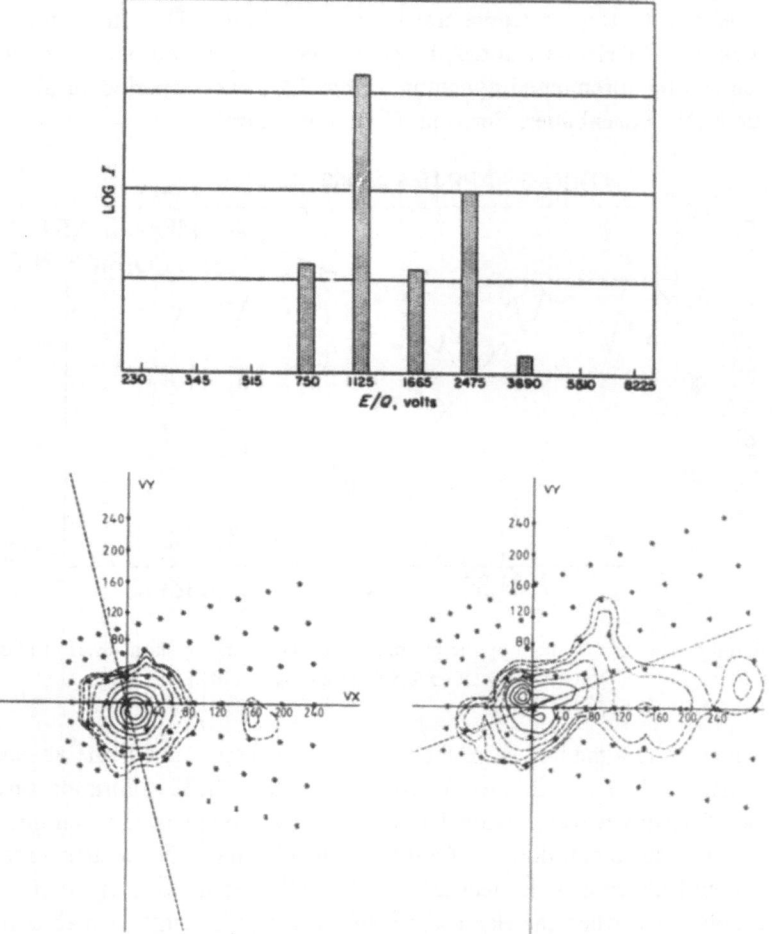

Fig. 4.   (Top) An energy/charge spectrum obtained with the electrostatic analyzer on Mariner 2. From Neugebauer and Snyder (1966). (Bottom) Two-dimensional cuts through three-dimensional distribution functions measured by Helios 2. The contours have been calculated from data obtained at the values marked by dots, as well as similar data in other planes. From Marsch *et al.* (1982).

faced the Sun, the measurement was one dimensional. Even in the best spectra, only five data points defined two spectral peaks. The data were reduced by assuming (1) only protons and alphas contributed to the spectra, (2) the protons and alphas had the same vector velocities, (3) the direction of the flow was determined by radial flow of the plasma from the Sun and the known tangential speed of the spacecraft, (4) both the proton and alpha distribution functions were convected isotropic Maxwellians, and (5) the alpha temperature was four times the proton temperature (the data were clearly inconsistent with equal temperatures). Four data points were used to calculate the four parameters v, T_p, n_p, and n_α on the basis of these assumptions, and when a fifth point was available, it was used to determine the goodness of fit. If any of the assumption changed, so did the answers. If any of the assumptions were wrong, so were the answers. More recent, more detailed measurements have shown that, in fact, none of the assumptions were probably consistently correct.

As stated above, the Mariner-2 data represent an extreme case of unmeasured parameters. But thermal anisotropies could not be detected with many of the early instruments. Users of solar wind data also tend to forget that there is a whole generation of two-dimensional plasma data obtained with hemispherical electrostatic analysers which measured only the projection of the distribution function onto the spacecraft equator.

The obvious cure for this problem is to make better measurements, and this is now possible to do. The bottom of Figure 4 shows, for example, two two-dimensional cuts through three-dimensional distribution functions obtained on Helios 2. Very little is left to the imagination or the realm of assumption.

Even with fairly complete measurements, however, many data reduction problems remain. For studies of solar wind acceleration and dynamics, we are interested in knowing the ion density n, where

$$n = \int f(v) \, \mathrm{d}^3 v$$

the flux $n\mathbf{v}$

$$n\mathbf{v}_0 = \int \mathbf{v} \, f(v) \, \mathrm{d}^3 v$$

the pressure tensor $\mathbf{P}$

$$\mathbf{P} = \int (\mathbf{v} - \mathbf{v}_0)(\mathbf{v} - \mathbf{v}_0) \, f(v) \, \mathrm{d}^3 v$$

and, in some cases, the heat flux vector $\mathbf{q}$

$$\mathbf{q} = (m/2) \int (\mathbf{v} - \mathbf{v}_0) \, |\mathbf{v} - \mathbf{v}_0|^2 \, f(v) \, \mathrm{d}^3 v \,,$$

where $f(v)$ is the three-dimensional distribution function of the ions. The problem is that the measured quantities are series of counting rates (or currents) obtained over a range of instrument look directions and a series of voltage settings of an electrostatic analyzer. Each voltage setting corresponds to a limited range or window of energy per unit charge. In general, the counting rate C_{ijk} is given by

$$C_{ijk} = A \int \mathbf{v} \cdot \hat{n}\, G(\theta, \phi, v/v_i)\, f(v)\, \mathrm{d}^3 v \,,$$

where A and $\hat{n}$ are the area of and unit normal vector to the detector aperture and G is the instrument transmission or resolution function. Contours of equal values of G are shown in Figure 5 for each of four different types of electrostatic analyzers. In each diagram in Figure 5, the vertical and horizontal axes are proportional to the velocity components normal to and in the plane of the instrument aperture, respectively. The four types shown are (a) cylindrical curved plate analyzers as flown on Mariner 2 (Neugebauer and Snyder, 1966) and OGO 5 (Neugebauer, 1970), (b) hemispherical curved plate analyzers as flown on the Vela satellites and on the Los Alamos instruments in IMPs 6 and 7 (Hundhausen *et al.*, 1967) (c) quadrispherical curved plate analyzers as flown on the Pioneer 6–11 (Smith and Day, 1971; McKibben *et al.*, 1977) and Helios (Rosenbauer *et al.*, 1977) spacecraft, and (d) modulated Faraday cup analyzers as flown on IMP 1 (Bridge *et al.*, 1965), Mariners 4 and 5 (Lazarus *et al.*, 1967), Explorers 33

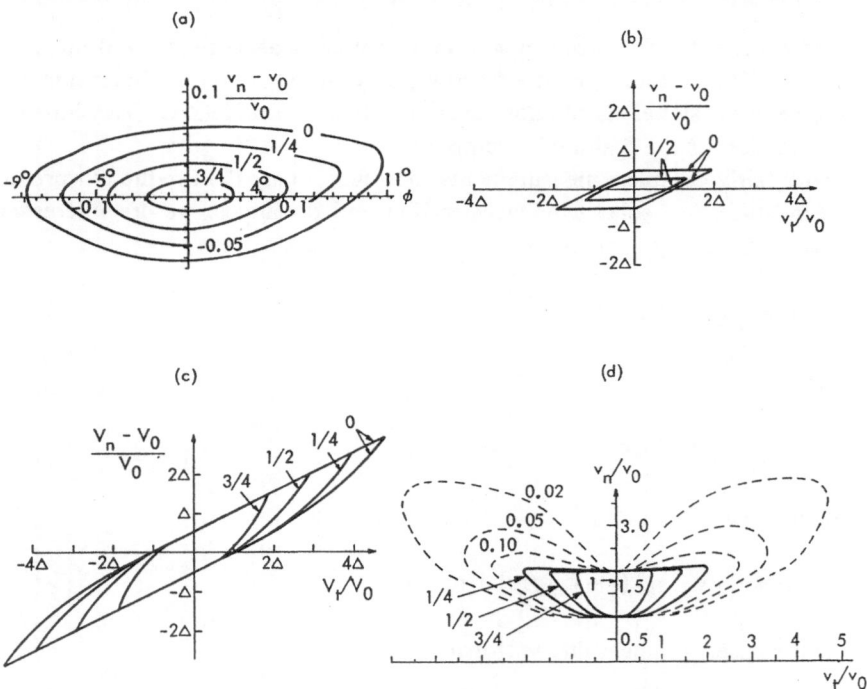

Fig. 5. Representative transmission functions for curved-plate electrostatic analyzers with (a) cylindrical, (b) hemispherical, and (c) quadrispherical geometry and (d) for a modulated Faraday cup detector. The notation used is: v_0 = ion velocity; v_n, v_t = components of v_0 normal and tangential to the detector aperture; $\phi = \sin^{-1}(v_t/v_0)$; and $\Delta = 2(r_2 - r_1)/r_0$, where r_2 and r_1 are the radii of curvature of the outer and inner electrodes, respectively, and $r_0 = (r_2 + r_1)/2$.

and 35 (Lyon *et al.*, 1967) and, in a modified form, on Voyager (Bridge *et al.*, 1977). The basic properties of these different types of electrostatic analyzers have been reviewed by Vasyliunas (1971).

The problem is how best to deconvolve the data to obtain n, $\mathbf{v}_0$, $\mathbf{P}$ and $\mathbf{q}$ from a set of C_{ijk}. Three classes of methods are commonly used: these are the methods of moments, fits to model distributions, and interpolation (Vasyliunas, 1971).

The method of moments is often based on the assumption that the distribution function f is much wider than the transmission function G in all dimensions. Then

$$C_i = f(v_i) \left[A \int \mathbf{v} \cdot \hat{n} \, G \, \mathrm{d}^3 v \approx Q v_i^4 \right],$$

where Q is a constant, and

$$n = \sum C_i / Q v_i^4$$
$$n\mathbf{v}_0 = \sum C_i \mathbf{v}_i / Q v_i^4$$

etc., for the higher moments.

The difficulty with this method of calculation is that, for solar wind ions, the two functions f and G are often approximately equally broad, with neither function remaining nearly constant while the other varies rapidly. The use of this method is therefore largely restricted to the analysis of low Mach number plasmas such as solar wind electrons and magnetospheric ions.

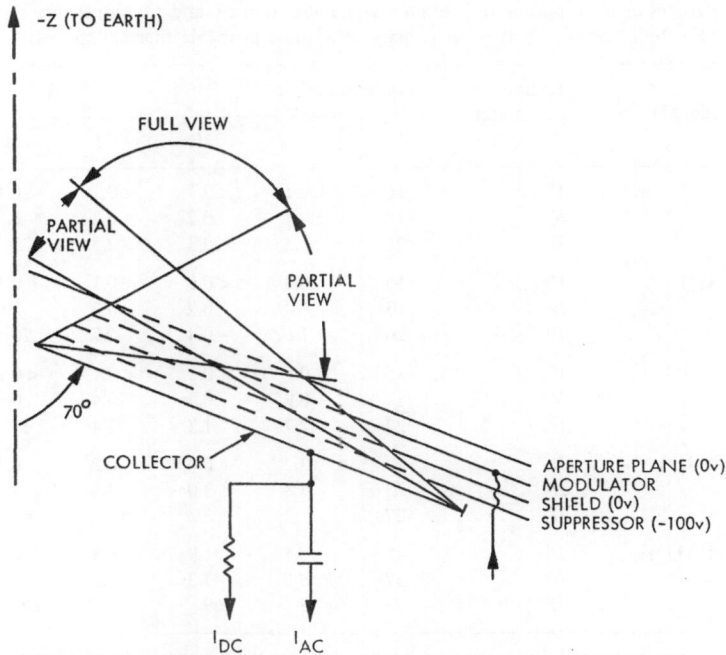

Fig. 6.   Faraday cup geometry used on the Voyager mission. From Bridge *et al.* (1977).

The solar wind experiment on the Voyager spacecraft had an unusual transmission function which allowed the successful application of the method of moments to solar wind ions (Bridge et al., 1977). Figure 6 shows the field of view of the instrument. Over an angle of about 70°, G is nearly independent of angle of incidence. For flows totally within this field of view, the current to the collector can be estimated by

$$I_i = eA\Delta u_i \int \mathbf{v} \cdot \hat{n} \, \delta(u_i - \mathbf{v}_0 \cdot \hat{n}) \, f(v) \, \mathrm{d}^3 v \, ,$$

where $u_i = \mathbf{v} \cdot \hat{n}$ at the center of the ith window, Δu_i is the width of this window, and $\delta(u_i - \mathbf{v}_0 \cdot \hat{n})$ is the Dirac delta function. Then,

$$n = \sum (I_i/(eAu_i))$$

$$n\mathbf{v}_0 \cdot \hat{n} = \sum I_i/(eA)$$

etc., for the higher moments. Since Voyager had three separate detectors oriented in three different directions, the vector velocity $\mathbf{v}_0$ could be calculated together with three independent estimates of the density n. Table II shows the approximate accuracy of the plasma parameters estimated in this way for convected isotropic Maxwellian distributions with different densities and Mach numbers. This computer simulation included a model of the noise in the instrument's electrometer. The percent accuracy achieved

TABLE II

Percent error in parameter estimate vs number density and Mach number
(V = bulk speed, N = number density, W = most probable thermal speed)

Number density (cm^{-3})	Estimated parameter	Mach number				
		1	3	10	30	100
10	V	16	< 0.1	< 0.1	< 0.1	< 0.1
	N	13	< 0.1	0.2	0.2	0.5
	W	21	0.2	0.3	2.5	27
1 (2.4 AU[a])	V	15	0.3	< 0.1	< 0.1	< 0.1
	N	13	0.4	0.2	0.2	0.6
	W	20	1.6	0.4	2.7	26
0.1 (7.7 AU[a])	V	15	0.5	< 0.1	< 0.1	< 0.1
	N	14	0.9	0.4	0.4	0.4
	W	22	2.7	1.2	3.4	31
0.01 (24 AU[a])	V	20	3.2	0.4	0.2	0.1
	N	23	6.8	3.0	2.4	2.8
	W	27	1.2	3.2	18	41
0.003 (44 AU[a])	V	48	6.5	1.9	0.3	0.3
	N	32	8.0	7.3	6.3	7.7
	W	35	13	69	26	131

[a] Note: Distance at which the number density represents the average solar wind conditions assuming a value of 6 cm^{-3} at 1 AU, and an inverse r^2 dependence on distance.

with this method is very good at Mach numbers corresponding to average solar wind conditions; it is poorer at low Mach numbers for which the beam is not entirely within the 70° field of view and at very high Mach numbers for which the delta function approximation for the energy dependence of G is not very accurate.

The second commonly used method of data reduction is to assume that the plasma distribution function has a certain form, such as a convected isotropic Maxwellian, a convected bi-Maxwellian, or a distribution function with a high energy tail such as the κ distribution (Vasyliunas, 1968). The free parameters such as n, v_0, and T are selected to give the best fit (often in the least squares sense) to the data. This is often done by comparing the data to precomputed tables of integrals of G convolved with the assumed distribution function. The results achieved with this method can be erroneous if the actual distribution function is very different from the one assumed. For example, if a single convected Maxwellian function is used to fit a double stream, for which two convected Maxwellians would be more appropriate, the computed temperature would be anomolously high, the computed speed would be intermediate to the two true speeds, and the computed density could also be in error. This situation must be fairly common in several solar wind data sets.

The third method of data reduction is useful for modern data with many three-dimensional data points C_{ijk} to work with. The goal is to remove the instrumental effects while preserving as much detail about the distribution function as possible. This is accomplished (e.g., Hundhausen et al., 1967; Feldman et al., 1973) by fitting the observations at a small number of contiguous points (usually three) in velocity space to a function such as a cubic or a Gaussian, similarly approximating the local value of

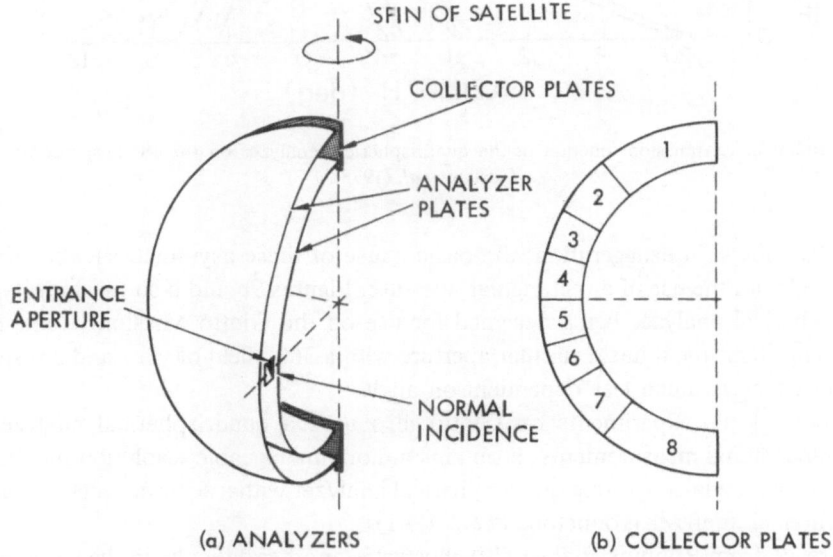

Fig. 7.   Conceptual diagram of a quadrispherical curved-plate electrostatic analyzer.

G, integrating the resulting equation, and then solving for the distribution function at the central point. The data in the lower part of Figure 4 were fit in this way. If desired, the quantities n, v_0, **P**, and **q** can be found by numerical integration of the resulting corrected distribution functions.

Some resolution functions are more troublesome than others, no matter what method of data reduction is used. Perhaps the most infamous is that of the quadrispherical analyzer, one example of which is shown in Figure 7. This type of analyzer has the advantage that the elevation-angle distribution can be measured while azimuth is mapped by the spacecraft spin, thus allowing three-dimensional measurements. But, as shown in the bottom left diagram of Figure 5, the resolution function is long and narrow, with a correlation between energy and angle. Figure 8 (Gosling *et al.*, 1978) shows how the response can shift and broaden at large angles of incidence. At large elevation angles, the transmission function is much broader than the angular spread of the solar wind.

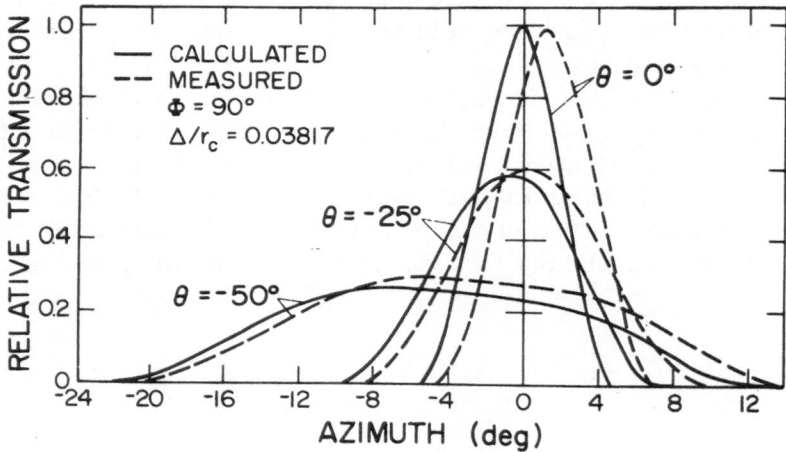

Fig. 8.   Relative transmission function of the quadrispherical analyzer on the ISEE spacecraft. From Gosling *et al.* (1978).

Figure 9a shows, in exaggeration, the basic cause of these asymmetries; the effect is associated with the use of a rectangular aperture. Figures 9b and c show a new type of quadrispherical analyzer being designed for use on the Giotto Mission (Rème *et al.*, 1981). This instrument has a circular aperture with a 360° field of view and a response function which is much less dependent on angle.

Another of the experiments on Giotto also uses a quadrispherical analyzer for three-dimensional measurements; it obtains a more manageable resolution function by preceding the angle-separating quadrispherical analyzer with a separate energy-defining hemispherical analyzer (Johnstone *et al.*, 1981).

Figure 10, from Robbins *et al.* (1970), illustrates another difficulty in the reduction of data obtained with electrostatic analyzers. One must devise a method of determining

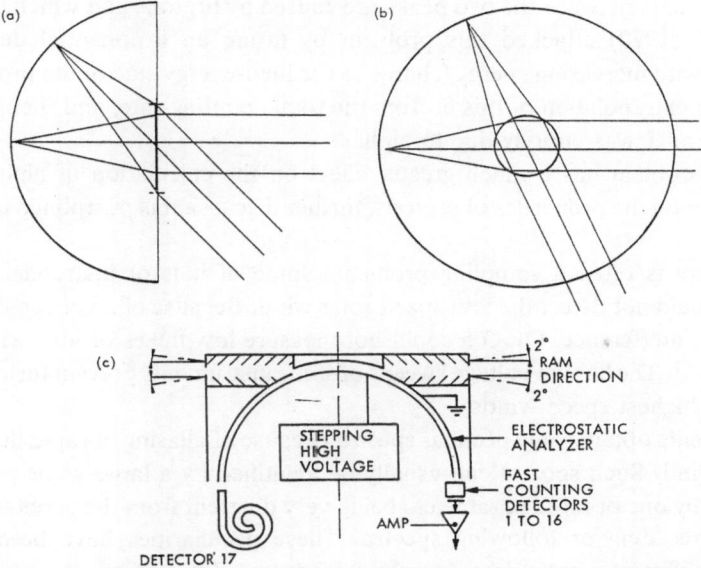

Fig. 9. Exaggerated projected ion ray paths through (a) a conventional quadrispherical analyzer and (b) the 360° analyzer to be flown on the Giotto mission. This new detector concept is shown in cross section in panel (c). After Rème *et al.* (1981).

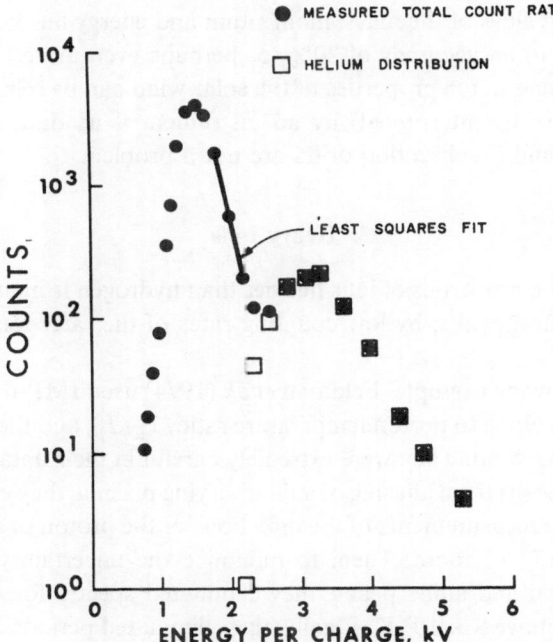

Fig. 10. A typical solar wind ion spectrum obtained by Vela 3. The method of separating proton from alpha particle counts is illustrated. From Robbins *et al.* (1970).

which of the counts between the two peaks are caused by protons and which by alphas. Robbins *et al.* (1970) attacked this problem by fitting an exponential decrease of counting rate with increasing energy/charge to the high-energy side of the proton peak, subtracting an extrapolation of this fit from the total counting rate, and then assuming that the remainder was entirely due to alphas.

Since this problem has a much greater effect on the calculation of alpha-particle properties than on the properties of protons, further discussion is postponed to the next section.

Finally, there is often a sampling problem. Some of it is of instrumental origin. Explorer 35 could not detect the low-speed solar wind. Because of a voltage-dependent photoelectron interference, OGO 5 could not measure low fluxes of ions with speeds near 350 km s^{-1}. The limited voltage sweeps of some instruments prevent their detecting the lowest or highest speed winds.

All instruments obtain some peculiar spectra because of aliasing of rapid fluctuations in the solar wind. Such spectra can usually be identified by a large χ^2 or variance to a model fit or by one or more parameters being very different from those obtained from analysis of preceding or following spectra. These peculiarities have been handled differently in different sets of data, thereby introducing different biases.

The following inferences can be drawn about the reliability of measurements of the bulk flow properties of the solar wind:

(1) Later, more modern measurements are, on the whole, more thorough and reliable than the early measurements.

(2) The absolute values of the mass, momentum and energy fluxes in the solar wind are still not known to an accuracy of 20%, or perhaps even more.

(3) Relative changes in the properties of the solar wind can be reliably detected (but not necessarily correctly interpreted) by an instrument if its data are reduced in a consistent manner and if calibration drifts are not a problem.

3. Heavy Ions

The calculation of the properties of ions heavier than hydrogen is hampered by overlap of neighboring spectral peaks, by low counting rates of the rarer ions, and by severe sampling biases.

Consider the following example. Feldman *et al.* (1974) used IMP-6 data to study the relation between the alpha to proton temperature ratio, T_α/T_p, and the rate of Coulomb collisions in the solar wind. They were extremely careful in their data reduction. First, to eliminate faulty results from aliasing of a time varying plasma, they eliminated spectra for which successive measurements of the directions of the proton or electron heat flux vectors varied by 15° or more. Then, to minimize the uncertainty associated with overlap of the proton and alpha peaks, they eliminated spectra for which the proton temperatures were above 3×10^5 K. Finally, they eliminated periods of double stream-ing or of unusual shaped spectra by requiring that the χ^2 value associated with the fit of the proton spectra to the model be $\leqslant 15$; the average value of χ^2 for the entire data set was 27. The result: 'most of the IMP 6 ion data were eliminated from this study'. Their results are summarized by the line labelled 'I6' in Figure 11. This study demonstrated the operation of a physical process, but with a set of data that was strongly biased in favor of low temperature, low speed, low wave activity conditions.

The data labelled 'O5' in Figure 11 were obtained by OGO 5. These data were treated quite differently. The energy resolution of this instrument usually precluded the identification of double streams. A process similar to that illustrated in Figure 10 was used to separate the two peaks; spectra with $T_p > 3 \times 10^5$ K were included if the ratio n_α/n_p was high enough to yield a well defined peak. As stated above, the OGO data also had an undersampling of plasma conditions near 350 km s^{-1}.

Both data sets showed an inverse relation between T_α/T_p and the ratio of expansion time scale τ_e to the Coulomb collision energy equipartition time τ_c, with approximately the same slope. But the different sampling biases of the two data sets have produced a sizeable offset between the two. Probably neither of the results is correct for the 'average' solar wind. Although alpha-particle temperatures have been discussed in this example, the same problems arise in the calculation of alpha particle densities and abundances relative to protons.

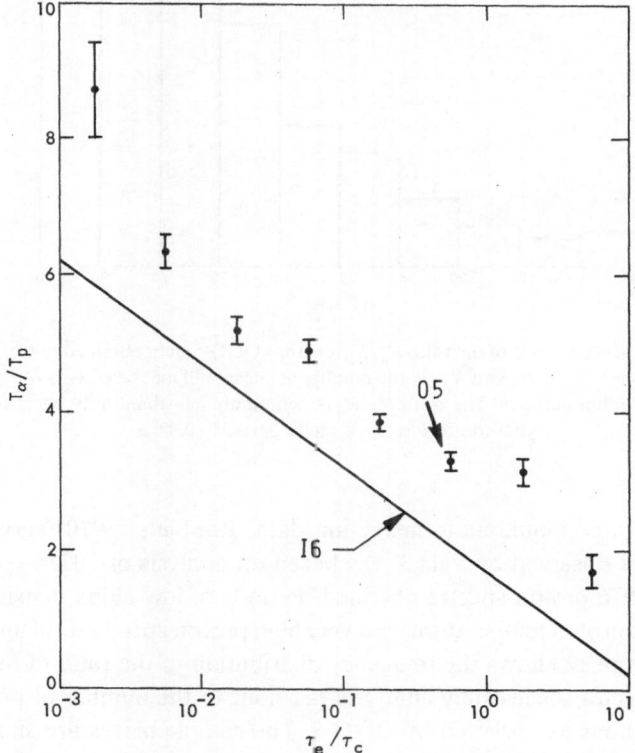

Fig. 11.   Alpha to proton temperature ratio versus the ratio of the solar wind expansion time to the Coulomb collision energy partition time. The line labelled I6 summarizes IMP-6 data while the data points are from OGO 5. From Neugebauer (1976).

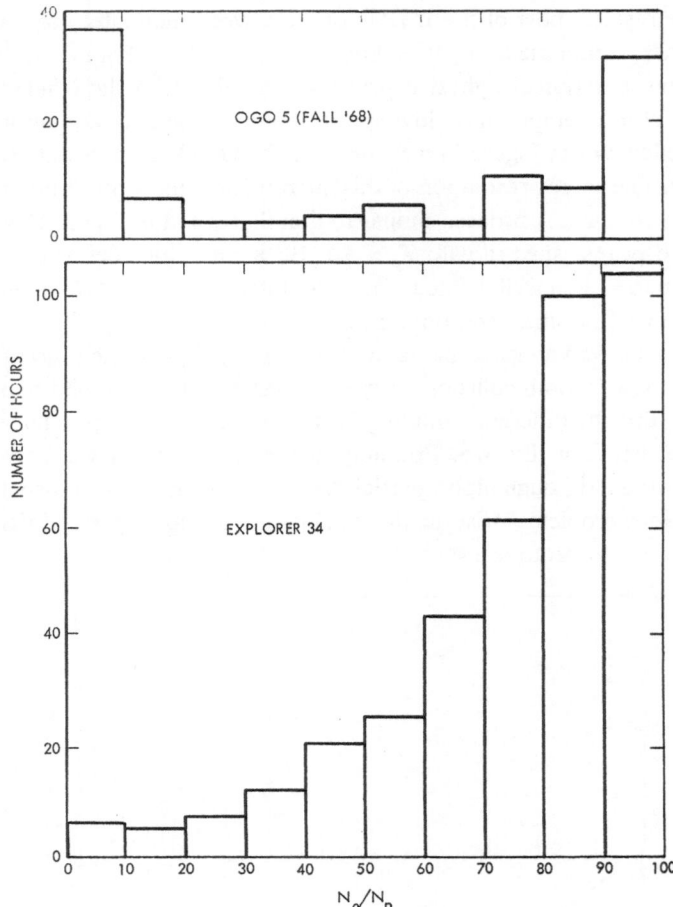

Fig. 12.   Frequency of occurrence of the ratio N_α/N_p for (top) OGO-5 data obtained during the fall of 1968 and (bottom) Explorer-34 data. N_α and N_p are the number of successful measurements of alpha and proton properties obtained within an hour. The ratio N_α/N_p is commonly less than unity because of the poorer signal-to-noise ratio of alpha particle spectra.

Sample biases are ubiquitous in heavy ion data. Robbin's (1970) survey of alpha-particle properties observed by Vela 3 was based on analysis of 10314 spectra, which is 74% of the 13976 proton spectra obtained. Periods of low alpha density, low alpha abundances, high proton temperature, and very high proton speed are all undersampled.

   The top of Figure 12 shows the frequency distribution of the ratio of the number of alpha particle spectra successfully analyzed per hour to the number of proton spectra during the same hour as observed by OGO 5. The sample biases are similar to those encountered by Robbins *et al.* (1970) except that the OGO instrument could detect alphas at higher speeds than could Vela 3, but speeds of 300–400 km s^{-1} were under-represented.

The bottom of Figure 12 shows a similar frequency distribution for the GSFC experiment on Explorer 34 (K. Ogilvie, personal communication). (The bottom of the figure does not include hours for which no alpha spectra were successfully analyzed while the top of the figure does.) The principal bias of the Explorer-34 experiment is the undersampling of periods of low helium flux.

Electrostatic analyzer observations of ions heavier than helium are limited to periods of low ion thermal speeds (Bame *et al.*, 1970). In 21 months of Vela data, Fenimore (1980) found only 156 usable spectra, each of which had been accumulated over approximately one hour.

The study of heavy ions is also sometimes limited to low or moderate speed winds by the maximum deflection voltage of the analyzer. For example, the Vela 5–6 analyzers could not detect ions with energy/charge greater than 8 kV; this would limit detection of Fe^{9+}, for example, to speeds below 500 km s^{-1}.

In recent years, quite a lot of effort has gone into the development of compound analyzers which add a new dimension to the measurement of solar wind plasma in order to identify the ions contributing to overlapping spectral peaks. These compound analyzers combine electrostatic analysis (which measures energy/charge) with either magnetic analysis (for momentum/charge) or a time of flight measurement of speed. The details of the implementation vary greatly. The magnetic analysis can consist of variable deflection in a purely magnetic field of ions of differing momenta, as on Prognoz 1 (Bosqued *et al.*, 1977) and as planned for Galileo, of a balance between electric and magnetic forces as in the Wien filters on Explorer 34 and ISEE-3 (Coplan *et al.*, 1978), or of a net curvature in superimposed electric and magnetic fields as on GEOS (Balsiger *et al.*, 1976) and ISEE (Shelley *et al.*, 1978). Two different configurations for time-of-flight measurements are planned to be flown on ISPM (Gloeckler and Hsieh, 1979) and Giotto (Johnstone *et al.*, 1981).

Figure 13 shows how useful such compound analyzers can sometimes be. The curves marked 'ESA' are energy/charge spectra of the solar wind obtained with an electrostatic

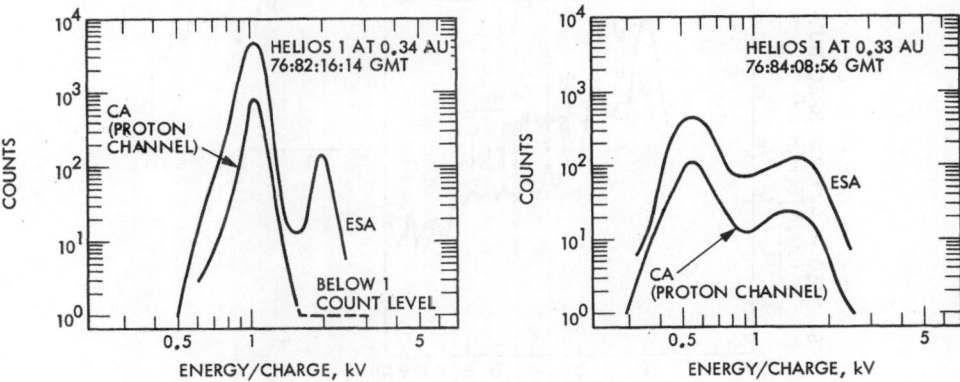

Fig. 13.   Energy/charge spectra obtained by the Helios-1 spacecraft. The curves labeled ESA were measured with an electrostatic analyzer while those labeled CA were measured with a compound analyzer which responded only to protons. After Rosenbauer *et al.* (1977).

analyzer on Helios 1. They look like typical, or nearly typical, solar wind proton and alpha peaks. Helios also carried a compound electrostatic and time-of-flight analyzer, which gave the curves labeled CA in Figure 13. Only protons can be detected in this channel. On the left, as expected, only the lower peak, the proton peak, is present. During the period on the right, however, both peaks were proton peaks! The second peak would have been erroneously handled as alpha particle data by any experimenter working with electrostatic analyzer data alone. Fortunately, occasions when a second proton stream masquerades as a stream of helium occur only rarely (H. Rosenbauer, personal communication).

The most important use of the new generation of compound analyzers will probably be associated with the measurement of ions heavier than helium. Figure 14 shows three heavy ion spectra, obtained by summing Vela electrostatic-analyzer data over periods of approximately one hour. Feldman *et al.* (1981) have estimated the relative abundances of O^{7+} and O^{6+} ions in the solar wind from the ratio of the heights of the labeled

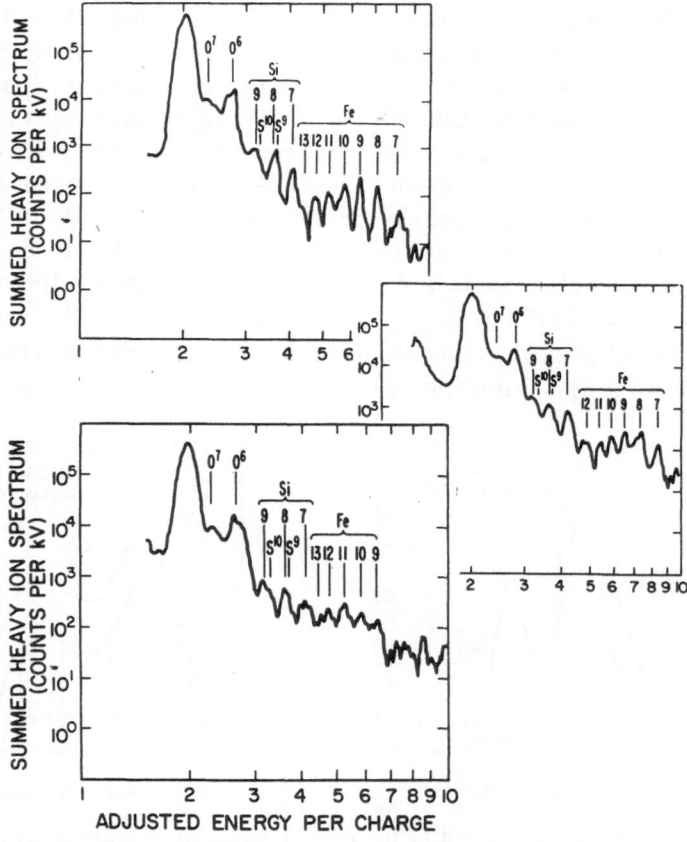

Fig. 14.   Three summed heavy ion spectra obtained with the electrostatic analyzers on the Vela spacecraft. After Feldman *et al.* (1981).

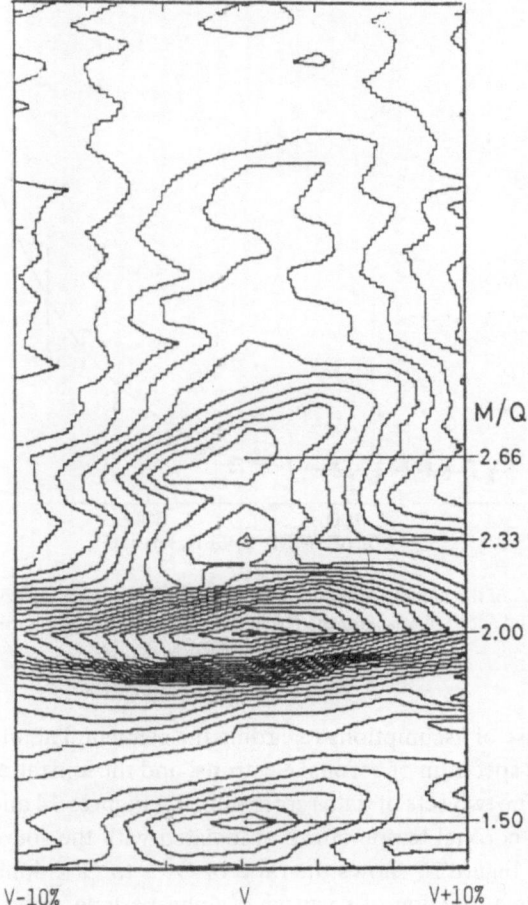

Fig. 15.   Mass per unit charge versus velocity distributions obtained with the heavy-ion analyzer on ISEE 3.
After Ogilvie *et al.* (1980).

peaks. The peaks are often not well defined and separated from neighboring spectral peaks.

Figure 15 shows a two dimensional distribution obtained with a compound electrostatic-magnetic analyzer on ISEE-3 (Coplan *et al.*, 1978). In this figure, the ordinate is mass/charge while the abscissa is the speed of solar wind alpha particles, which presumably is the speed of the heavier ions as well. Even though Figure 15 clearly contains more information than Figure 14, which can be used to constrain the interpretation of the data, some ambiguities about the relative contributions of different ion species remain. Ogilvie and Vogt (1980) have used the ISEE-3 data illustrated by Figure 15, and Feldman *et al.* (1981) have used Vela data illustrated by Figure 14, to calculate relative abundances of O^{6+} ($m/q = 2.67$) and O^{7+} ($m/q = 2.29$) in the solar wind. Calculations of the coronal temperatures at which the oxygen charge states were

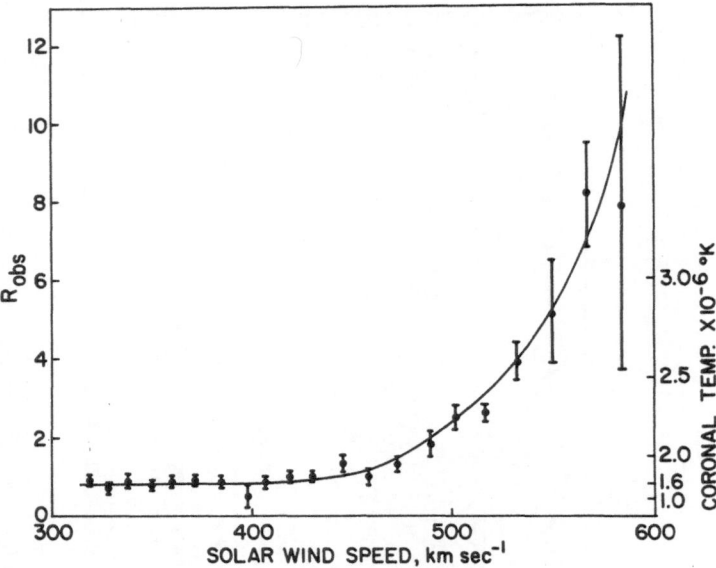

Fig. 16.   The ratio R_{obs} of the flux observed at $M/Q = 2.29$ (O^{7+}) to that observed at $M/Q = 2.67$ (O^{6+}) plotted versus solar wind speed. A scale for the coronal temperature is given on the right. From Ogilvie and Vogt (1980).

frozen in require use of assumptions regarding the elemental abundances present in the corona, the energy spectrum of coronal electrons, and the ionization and recombination rates of the ions. The two sets of data represented by Figures 14 and 15 give significantly different values of coronal temperatures associated with the source regions of the low speed solar wind. Figure 16 shows the ratio of O^{7+} to O^{6+} ions reported by Ogilvie and Vogt (1980) as a function of solar wind alpha-particle speed. Their conversion of abundances to coronal temperatures is shown by the scale on the right of Figure 16. At low speeds, they find the average oxygen freezing-in temperature to be near 1.6×10^6 K. From their electrostatic analyzer data shown by the examples in Figure 14, however, Feldman *et al.* (1981) find an average coronal temperature for the low-speed wind of $(2.1 \pm 0.2) \times 10^6$ K. The disagreement actually exceeds the difference between 1.6 and 2.1 million degrees, because Feldman et al. have removed from their sample all examples of unusually high-ionization-state, low-speed wind as sometimes found, for example, following interplanetary shocks. The source of this disagreement could be either the measurement of the ratios of O^{7+} to O^{6+} or their interpretation in terms of coronal temperature. Future experiments, such as that to be flown on ISPM (Gloeckler and Hsieh, 1979) may help resolve some of the ambiguities because measurement of ionic charge as well as energy and mass can be used to remove the contributions of ions such as He^{2+}, C^{5+}, Ne^{8+}, and Mg^{10+} so that the O^{7+} and O^{6+} ions can be more clearly determined. Continued progress in coronagraph and EUV observations will probably lead to better models of the elemental abundances and of the distribution of electron density and energy distributions with solar distance so that ion charge-state distributions at 1 AU can be less ambiguously interpreted and better related to coronal phenomena.

Acknowledgements

This research was conducted at the Jet Propulsion Laboratory of the California Institute of Technology, Pasadena, California, and was supported by the Space Plasma Physics Program of NASA under contract NAS 7–100.

References

Balsiger, H., Eberhardt, P., Geiss, J., Ghielmetti, A., Walker, H. P., Young, D. T., Loidl, H., and Rosenbauer, H.: 1976, *Space Sci. Instr.* **2**, 499.
Bame, S. J., Asbridge, J. R., Hundhausen, A. J., and Montgomery, M. D.: 1970, *J. Geophys. Res.* **75**, 6360.
Bosqued, J. M., D'Uston, C., Zertzalov, A. A., and Vaisberg, O. L.: 1977, *Solar Phys.* **51**, 231.
Bridge, H. S., Egidi, A., Lazarus, A., Lyon, E., and Jacobson, L.: 1965, *Space Res.* **5**, 969.
Bridge, H. S., Belcher, J. W., Butler, R. J., Lazarus, A. J., Mavretic, A. M., Sullivan, J. D., Siscoe, G. L., and Vasyliunas, V. M.: 1977, *Space Sci. Rev.* **21**, 259.
Burlaga, L. F., Behannnon, K. W., Hansen, S. F., Pneuman, G. W., and Feldman, W. C.: 1978, *J. Geophys. Res.* **83**, 4177.
Coplan, M. A., Ogilvie, K. W., Bochsler, P. A., and Geiss, J.: 1978, *IEEE Trans. Geosci. Electronics* **GE–16**, 185.
Feldman, W. C., Asbridge, J. R., Bame, S. J., and Montgomery, M. D.: 1973, *J. Geophys. Res.* **78**, 2017.
Feldman, W. C., Asbridge, J. R., and Bame, S. J.:1974, *J. Geophys. Res.* **79**, 2319.
Feldman, W. C., Asbridge, J. R., Bame, S. J., Fenimore, E. E., and Gosling, J. T.: 1981, *J. Geophys. Res.* **86**, 5408.
Fenimore, E. E.: 1980, *Astrophys. J.* **235**, 245.
Gloeckler, G., and Hsieh, K. C.: 1979, *Nuclear Instrum. and Methods* **165**, 537.
Gosling, J. T., Asbridge, J. R., Bame, S. J., and Feldman, W. C.: 1978, *Rev. Sci. Instrum.* **49**, 1260.
Hundhausen, A. J.: 1977, in J. B. Zirker (ed.), *Coronal Holes and High Speed Wind Streams*, Colorado Assoc. Univ. Press, p. 225.
Hundhausen, A. J., Asbridge, J. R., Bame, S. J., Gilbert, H. E., and Strong, I. B.: 1967, *J. Geophys. Res.* **72**, 87.
Hundhausen, A. J., Gilbert, H. E., and Bame, S. J.: 1968, *J. Geophys. Res.* **73**, 5485.
Johnstone, A., Bryant, D., Edwards, T., Hultquist, B., Formisano, V., Biermann, L., Luest, R., Schmidt, H., Feldman, W., Cerulli-Irelli, P., Dobrowclny, M., Egidi, A., Terenzi, R., Jockers, K., Rosenbauer, H., Studemann, W., Wilken, B., Wallis, M., Haerendel, G., Paschmann, G., Winningham, J. D., and Reme, H.: 1981, *Scientific and Experimental Aspects of the Giotto Mission*, European Space Agency, SP-169, p. 17.
King, J. H.: 1977, *Interplanetary Medium Data Book*, NSSDC/WDS-A-R 77-04, NASA, Greenbelt, MD.
Lazarus, A. J., Bridge, H. S., Davis, J. M., and Snyder, C. W.: 1967, *Space Res.* **7**, 1296.
Levine, R. H.: 1978, *J. Geophys. Res.* **83**, 4193.
Lyon, E. F., Bridge, H. S., and Binsack, J. H.: 1967, *J. Geophys. Res.* **72**, 6113.
Marsch, E., Muhlhauser, K.-H., Schwenn, R., Rosenbauer, H., Pilipp, W., and Neubauer, F. M.: 1982, *J. Geophys. Res.* **87**, 52.
McKibben, D. D., Wolfe, J. H., Collard, H. R., Savage, H. F., and Molari, R.: 1977, *Space Sci. Instr.* **3**, 219.
Neugebauer, M.: 1970, *J. Geophys. Res.* **75**, 717.
Neugebauer, M.: 1976, *J. Geophys. Res.* **81**, 78.
Neugebauer, M.: 1981, *Fund. Cosmic Phys.* **7**, 131.
Neugebauer, M. and Snyder, C. W.: 1966, *J. Geophys. Res.* **71**, 4469.
Nolte, J. T. and Roelof, E. C.: 1973, *Solar Phys.* **33**, 241.
Ogilvie, K. W. and Vogt, C.: 1980, *Geophys. Res. Letters* **8**, 577.
Ogilvie, K. W., Bochsler, P., Coplan, M. A., and Geiss, J.: 1980, *J. Geophys. Res.* **85**, 6069.

Rème, H., Cotin, F., d'Uston, C., Sauvaud, J. A., Korth, A., Richter, A. K., Anderson, K. A., Carlson, C. W., Lin, R. P., Wekhof, A., Mendis, D. A., and Johnstone, A.: 1981, *Scientific and Experimental Aspects of the Giotto Mission*, European Space Agency, SP-169, p. 29.
Robbins, D. F., Hundhausen, A. J., and Bame, S. J.: 1970, *J. Geophys. Res.* **75**, 1178.
Rosenbauer, H., Schwenn, R., Marsch, E., Meyer, B., Miggenrieder, H., Montgomery, M. D., Mühlhäuser, K. H., Pilipp, W., Voges, W., and Zink, S. M.: 1977, *J. Geophys.* **42**, 561.
Scarf, F. L., Fredricks, R. W., Green, I. M., and Neugebauer, M.: 1970, *J. Geophys. Res.* **75**, 3735.
Shelley, E. G., Sharp, R. D., Johnson, R. G., Geiss, J., Eberhardt, P., Balsiger, H., Haerendel, G., and Rosenbauer, H.: 1978, *IEEE Trans. Geosci. Electronics* **GE–16**, 266.
Smith, Z. K. and Day, J. R.: 1971, *Rev. Sci. Instrum.* **42**, 968.
Snyder, C. W. and Neugebauer, M.: 1966, in R. J. Mackin and M. Neugebauer (eds.), *The Solar Wind*, Pergamon Press, p. 25.
Vasyliunas, V. M.: 1968, *J. Geophys. Res.* **73**, 2839.
Vasyliunas, V. M.: 1971, in R. H. Lovbergs (ed.), *Methods of Experimental Physics*, Vol. 9B of *Plasma Physics*, Academic Press, p. 49.

FLOW OF MATERIAL AT THE CHROMOSPHERE-CORONA
TRANSITION ZONE OF THE SUN*

W. M. GLENCROSS

Dept of Physics and Astronomy, University College London, Gower Street, London WC1E 6BT

Abstract. Early models of the Chromosphere-Corona transition zone of the Sun considered it to be a static plane-parallel region. From these it became clear that the layer was extremely thin and had an important role in the conduction of energy from the Corona. More recent observations show mass motions of order 10 km s^{-1}, which means that heating and cooling of the moving plasma has an important affect on the energy balance, while transient effects producing far higher velocities are also common. Studies of plasma motion through the zone are clearly relevant to the initial heating of material which enters the solar wind.

1. Introduction

Since it is an extremely thin layer sandwiched between the chromosphere and corona, a comprehensive discussion of the characteristics of the transition zone can only be attempted in conjunction with a description of the rest of the atmosphere. The aim of this present review is more limited however: rather than up-date all aspects of the problem discussed in earlier reviews (for example, Frisch, 1972; Whithbroe and Noyes, 1977), it is planned to concentrate mainly on the characteristics arising from any divergence from hydrostatic equilibrium at the base of the solar atmosphere. This clearly has considerable relevance to the formation of the solar wind, which is the topic of the accompanying papers.

Even with such a restricted aim, it is impossible to avoid some discussion of the general problem of heating of the solar atmosphere, since the approach to be taken depends critically on where the deposition of energy to heat the corona occurs. The early view was that mechanical waves propagating upwards through the chromosphere were dissipated particularly rapidly at the steep density gradient of the transition layer (see Kuperus, 1969). The characteristics of the layer in such a case would clearly be dominated by the development of steep pressure gradients required to ensure momentum balance (Delache, 1969). We shall see that considerable mechanical activity is indeed observed on various short timescales, but the concensus of opinion at present appears to be that heating of the corona is distributed rather more evenly throughout that region. The arguments for such a view are put forward elsewhere (Kuperus *et al.*, 1981).

Thus, we accept for this discussion that the dissipation of mechanical waves, which heats the chromosphere, is essentially complete below the base of the corona which is heated by some other process. It might be argued that the transition region becomes of far less interest if it is not the main seat of coronal heating: the fact is that its main fascination lies in its being so thin, which is due to its more passive role associated with transfer of heat between the corona and chromosphere. The problems which then arise

---

* Paper presented at the IX-th Lindau Workshop 'The Source Region of the Solar Wind'.

Space Science Reviews **33** (1982) 151–160. 0038–6308/82/0332–0151$01.50.

concern the geometrical configuration of the region, its energy balance, and the way in which it is affected by the flow of plasma in either direction between the chromosphere and corona. Of particular interest is the limitation placed on the speed of outflowing atoms, which subsequently contribute to the solar wind, if these are to be heated to coronal temperatures during the brief passage through the thin zone.

2. Geometrical Configuration

We are concerned with a layer of solar atmosphere within which the temperature rises by an order of magnitude from roughly 5×10^4 K. Since the whole change occurs within a region which is no thicker than the spatial resolution element of a telescope, the detailed structure cannot be ascertained directly. The state of ionization of suitable chemical elements varies significantly throughout this large temperature range however, so the relative intensities of spectral lines produced at different stages contribute some information on the geometry. Since early observations in the extreme ultraviolet region of the spectrum provided relatively poor spectral and spatial resolution, modelling based on these data was restricted to considering a series of plane-parallel layers of varying temperature. Details of the comparison between measured and predicted (for which

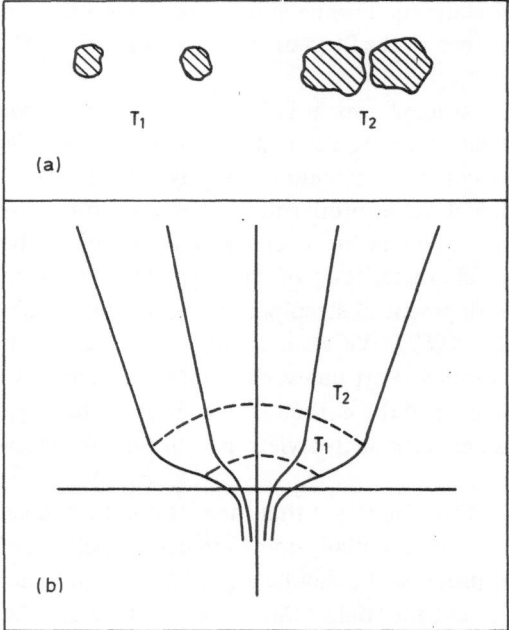

Fig. 1.   Diagram (a) provides a schematic impression of two emitting features on the solar disk viewed at different wavelengths. The size of each element is shown to increase with the temperature of the emitting region ($T_1 < T_2$). It is generally assumed that divergence of the magnetic field associated with each feature accounts for these changes: the relevant isothermal contours are seen in diagram (b). An alternative explanation is illustrated in Figure 4.

atomic data are required) spectral-line intensities are discussed in a number of articles, such as Pottasch(1963) and Frisch (1972). In fact, the intensity of a spectral line depends on both the volume V of the emitting region and the density of material (characterised by the electron number density N_e), so this modelling procedure only provides information on the variation of emission measure $\int_V N_e^2 \, dV$ with temperature. Other measurements (see Frisch, 1972) are needed to determine the density and thickness of each temperature layer, and thus provide the temperature gradient. Obviously such modelling requires care to ensure that the necessary physical restrictions are observed: in particular, the pressure variation must comply with assumptions about gravitational and velocity fields, and the temperature gradient at each level must be consistent with energy-balance considerations involving conduction of heat between layers.

Whereas modelling tends to be simplified by considering plane parallel layers, the assumption is not in keeping with observation that the horizontal area of any discrete solar feature varies with height in the transition zone. Thus, Figure 1 provides a schematic impression in which structures are smaller when the 'cool' component is viewed, but the size increases with temperature to the extent that details are 'washed out' at coronal temperatures. A possible explanation (considered by Gabriel, 1976 and others) is that the magnetic field enclosing the feature diverges into the corona, as shown in the diagram. Even this assumption, which still maintains approximately horizontal layering, must be a poor representation of the situation in some parts of the solar atmosphere however: spicules are narrow columns of material at chromospheric temperature (typically 10 000 K) projecting to an altitude of order 10 000 km into the corona, so the transition to coronal temperatures probably occurs throughout a series of concentric vertical cylinders.

3. Energy Balance

When the analysis just referred to is carried out, it becomes clear that the temperature change of a factor ten occurs along a path length which varies with location on the Sun, but generally lies in the range 10–1000 km. The greatest temperature gradients are found above plage, while thicker layers are found above sunspots (regions of particularly intense magnetic fields) and at 'quiet' regions having low average magnetic fields. Even in the less extreme cases, the temperature gradient of order 1 K m^{-1} leads to considerable heatflow because the thermal conductivity of fully ionized hydrogen is of order $10^{-6} \times T^{5/2}$ erg cm^{-1} s^{-1} deg^{-1} (Spitzer, 1962). This highlights one of the important characteristics of the transition zone – that it provides a conduction path for considerable heat loss from the base of the corona to the chromosphere. Numerous studies of the energy balance of the zone have been carried out in the hope that any limits placed on some of the relevant parameters might shed light on the processes involved – this would be particularly valuable if the need for local heating of the plasma to explain energy balance could be demonstrated. Such studies are discussed by Jordan (1977): although they help explain general trends in the observations, it has not proved possible to reach precise detailed conclusions. This probably reflects the fact that conditions vary so rapidly with position on the Sun that no set of observations can provide a true representtation of conditions.

Even the information which is obtained most directly, the variation of emission measure with temperature, is only available with sufficient accuracy to provide some general trends. Thus, following a suggestion by Athay (1966), there has been widespread acceptance that the emission measure varies with temperature in proportion to $T^{3/2}$ over much of the Sun. Such a trend is significant since it corresponds to a constant thermal flux (that is, $T^{5/2} \, dT/dh = $ constant, where the thermal conductivity is proportional to $T^{5/2}$) through a layer where the pressure remains essentially constant. Jordan (1980) demonstrates clearly how conclusions can be drawn from any measured dependence of emission measure on temperature by considering models involving the variation with height of heating, heat-loss, and conductive flow. It seems unlikely that the radiation process considered is the dominant heat-loss mechanism in the transition region however: the reason for this doubt is that considerable movement of transition-region plama can now be detected from the Doppler shifting of spectral lines. The heating or cooling of this material as it moves in the appropriate direction along the temperature gradient can become an important term in the energy-balance equation. This is demonstrated in Figure 2 which applies to a specific case where a steady flux of material is being evaporated from the chromosphere and rises into the corona. It is seem that

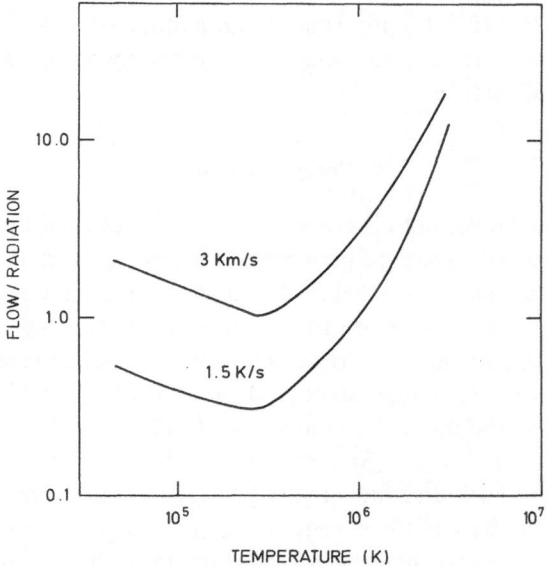

Fig. 2.   This diagram, taken from Glencross (1981), deals with evaporation of material from the chromo-sphere. It is assumed that the thermal energy deposited at each level of the transition zone from the conducted flow goes partly towards heating the rising plasma (flow term) while the rest is radiated away (radiation term). The ratio of the two quantities is shown here as a function of temperature for two flow rates. Since a constant flux of material through the region is assumed in each case, these velocities apply to the 10^6 K level, and became correspondingly small where the temperature is lower.

the energy required to heat the expanding gas as it passes through a narrow horizontal layer exceeds the radiation loss from the layer even when the upward flow velocity is small. The effect of such motions on the energy balance of the transition zone and corona has been included in studies of outflow of plasma in the solar wind (Elzner and Elwert, 1980), while the reverse process – the role of spicule material falling back onto the chromosphere after heating in the corona – has also been examined (Pneuman and Kopp, 1977, 1978; Athay, 1981). Following earlier work by Meyer and Schmidt (1968), there has been interest recently in the motion of coronal material along magnetic arches due to pressure differences between the feet (Glencross, 1980, 1981; Cargill and Priest, 1980; Noci, 1981): such flow will tend to be directed towards the region of more intense magnetic field. It has not proved possible to observe the complete flow pattern, since this involves studies of the Doppler shift of coronal X-ray emission lines in addition to those from the transition zone, though similar motions have been seen along lower arches containing cooler medium (Ellison, 1944). Nevertheless, the inclusion of steady flow in the models provides explanations for a number of observed characteristics of the transition zone. For example, the motion of material along arches connecting sunspots to the surrounding plage in active regions can account for the considerable difference in temperature gradient of the association transition layers. Since pressure gradients between the feet of these arches tend to favour upflow at plage, the transition layer has to be relatively thin there to accommodate the thermal conduction from the corona necessary to heat the rising plasma. Glencross (1981) has examined the energy balance of the region and shown how the thickness of the zone has to adjust to maintain specific rates of plasma flow. Even so, such studies can only demonstrate likely trends since there are, once again more free parameters in the calculations than can be determined from current observations. As Figure 3 shows, the density of material and its velocity have to be determined, as well as the fraction of the energy conducted from the corona which reaches the chromosphere rather than contributing to heating the rising plasma. What is clear however, is that conditions associated with the downflow towards sunspots are not so critically dependent upon the development of a steep transition zone. The energy balance associated with downflow arising in various ways has been discussed by Pneuman and Kopp (1977, 1978) and Athay (1981). On the observational side, it has been known for some time (Lites et al., 1976) that red-shifts of transition region lines are dominant: the explanation usually given is that spicules account for the upflow (involving some unknown mechanism) of material, which is too cool to produce emission characteristic of the transition zone, so that the red shift arises as the medium falls back after heating in the corona (Pneuman and Kopp, 1977, 1978). An attractive alternative explanation is provided if flow of material along magnetic arches is important however. As just explained, the transition zone tends to be particularly thin where upflow takes place, so the appropriate emission measure is low. Observations will then be biased towards the thicker regions where downflow dominates. This is in agreement with another conclusion reached by Lites et al. (1976) who detected weak blue shifted lines (upflow) above plage, and strong red-shifted lines from more-intense magnetic fields. In fact, the considerable vertical extension of material at

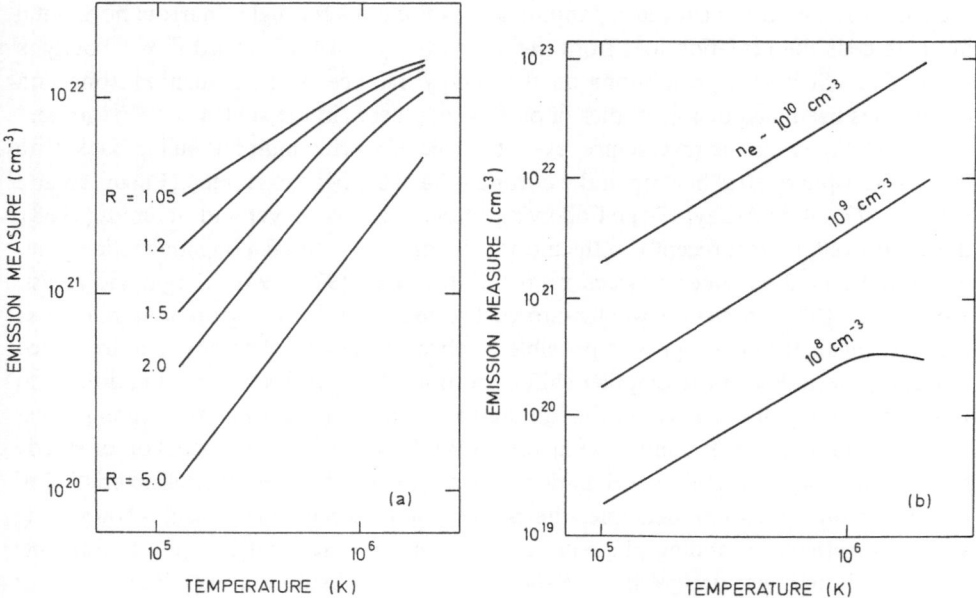

Fig. 3. These two graphs illustrate the difficulty in reaching unambiguous conclusions by modelling the transition zone to explain observations of the variation of emission measure with temperature. Evaporation of material is assumed such that an upflow velocity of 10 km s^{-1} is reached where the temperature is 10^6 K, and the emission measure is calculated for a column of 1 cm^2 cross-section through the zone. Diagram (a) applies to the case where the electron number density is 2×10^9 cm^{-3} where $T = 10^6$ K. It is seen that the parameter R = (conducted energy reaching chromosphere)/(conducted energy heating rising plasma) plays an important role. Similarly, diagram (b) shows the result of altering the electron number-density when $R = 3$.

transition-region temperatures above sunspots is clearly recognised in pictures present-ed by Foukal (1975, 1976). Though motion of material cannot be observed directly, it is clear that there must be downflow because each column is too cool to be in static equilibrium.

A comment is also necessary on the height of the transition 'layer' at this stage. It is recognised from the great variety of chromospheric structure observed on the limb that the vertical position of the transition zone must vary considerably, though the mean is taken to be around 2000 km above the photosphere. Simple modelling (Glencross, 1980) shows how the level has to fall in order to accommodate any increase of density within the overlying corona, whereas the steeper temperature gradients tend to be lower in the atmosphere if upflow of material is to be sustained by thermal conduction from the corona.

It should also be noted that this variability in thickness of the transition layer can account for the changing cross-section of features on the disk (see Figure 1) without involving large divergence of the magnetic field: this is important because coronal material generally appears to be embedded within reasonably collimated flux tubes. Furthermore, the fairly uniform section of spicules also points to that material being

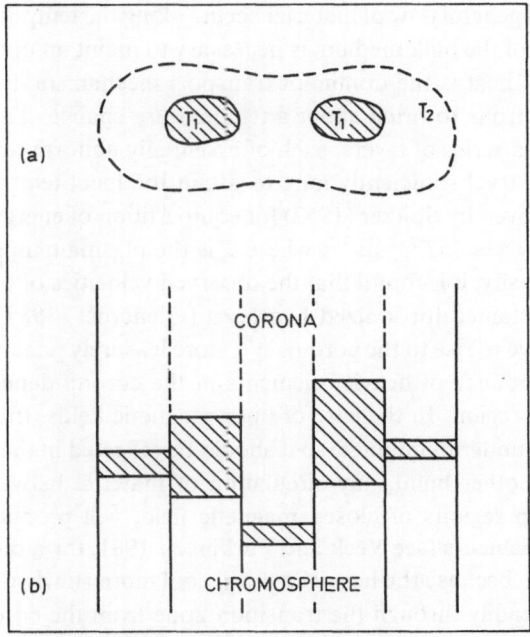

Fig. 4.   This provides an alternative explanation for observations illustrated in Figure 1. The discrete cool features in (a) are assumed to lie where the transition layer (cross-hatched) is thick. On the other hand, the corona is essentially uniformly thick in each column, so they should be indistinguishable when viewed in coronal emission. Some of the thicker regions of transition zone should coincide with intense regions of magnetic field, where downflow of material is likely.

encased in a reasonably cylindrical local field. Figure 4 represents a region of vertical magnetic field across which there is considerable variation in thickness of the transition zone. This means that the brighter features within the appropriate waveband are located at those parts of the disk where the zone is thick. On the other hand, the variation in the location of the base of the corona makes little difference to the net emission from such an extensive region, so all the columns represented in the diagram can be equally bright when viewed in 'hotter' spectral lines. Assuming that magnetic arches link the network to 'quieter' surrounding areas, any pressure difference between the feet (Meyer and Schmidt, 1968) will tend to produce downflow of material in the network (Lites et al., 1976), which results in a thick transition zone there.

4. Steep Temperature Gradients

Once realistic modelling of the transition zone had indicated how thin the layer really was, it became apparent that the local ionization equilibrium was barely attained in the case of some of the heavier ions. That is, transfer of energy to an ion at any level in the atmosphere involves multiple collisions with electrons, and some of the heavy particles manage to random-walk through the layer before reaching the electron temperature, even in a 'static' atmosphere (Zirin, 1963). Local equilibrium is an even more necessary requirement where a general flow of material occurs along the temperature gradient since heating (or cooling) of the bulk medium is necessary to maintain upflow (or downflow). Since conduction of heat is the dominant transport mechanism, ions have to interact with the carrier electrons to bring about a temperature change. Thus, if the transition zone is divided into a series of layers, each of essentially uniform temperature, material must reside at each level sufficiently long to attain the local temperature. Taking the characteristic time given by Spitzer (1962) for equipartition of energy between electrons and protons, namely $\tau \simeq 25T^{3/2}$ ns^{-1}, where T is the plasma temperature and n is the electron number density, it is found that the observed velocities of order tens of km s^{-1} can only just be sustained for ionized hydrogen (Glencross, 1981). Where this is so, heavier elements have to rise to the corona at a more leasurely pace: whether this shows up as an underabundance of heavier elements in the corona depends on the specific configuration of the region. In the case of open magnetic fields, there is no process to compensate for this underabundance, so it should be reflected in the composition of the solar wind. On the other hand, any circulation of material between the corona and chromosphere, as in regions of closed magnetic field, will produce a 'more-normal' distribution of the elements (see Veck and Parkinson 1981, for a discussion of coronal abundances). This is because the heavy ions will cool more slowly than protons, so that they descend less readily through the transition zone from the corona to the chromoshere: this provides some compensation for the previous slow ascent. A detailed examination of ionization equilibrium is given by Joselyn et al. (1979).

It has to be added that the values taken from Spitzer (1962) for conductivity and equipartition times for sharing of energy between electrons and ions are only appropriate where Maxwellian distributions of particle energies apply. This will not be so where a steep temperature gradient exists, so care has to be taken in carrying out detailed calculations (Roussel-Dupré, 1980).

5. Coronal Holes

The earlier discussion concentrated mainly on the characteristics of the transition zone at regions of closed magnetic field, though coronal holes provide an opportunity for observing open-field structures. The difference in appearance of the two regions is most marked when emission from plasma at 10^6 K and higher is examined, while the intensity of emission characteristic of most of the transition zone changes little between the areas (typically 25% variation). The techniques used to interpret the observations are those mentioned earlier, which indicate that the temperature gradient at holes is relatively low (Monroe and Withbroe, 1972; Withbroe and Gurman, 1973; Huber et al., 1975). There are many possible reasons for this being so, and it is unclear which cause is dominant. On the one hand, it is possible that input of heat to the lower corona above holes is far less than to material within closed loops; this would be the case if heating involved resistive processes in sheets of material (see Kuperus et al., 1981) which are more readily maintained when both ends of the magnetic flux tube are captive. Alternatively, the heat input to all regions might be similar but the heat-loss mechanisms differ. Noci (1973) and Pneuman (1973) pointed out that the solar wind represents a significant term in the heat balance for coronal holes, though it must be recognised that the maintenance of a large-scale flow of material along closed loops also plays a dominant role where the upflow occurs, even if the circulation ensures there is no net loss of material and energy from the whole arch. Another explanation for the difference in appearance between coronal holes and other quiet regions on the Sun, proposed by Hearn (1977), requires the transverse components of the transition-region magnetic-fields to differ such that thermal conduction towards the chromosphere is greater in the holes.

With these and other uncertainties to be included, modelling of the transition zone at coronal holes, as elsewhere, can at best provide only some general impressions of conditions in the region. Among a number of attempts to explain details are papers by Kopp and Holzer (1976), Elzner and Elwert (1980) and others.

6. Transient Effects

Although attempts to deal with steady conditions in the transition zone are fraught with uncertainties, observations, are now showing up even more complications such as the existence of short-lived transient effects. These variable features have become detectable through the development of instruments providing greatly improved spatial resolution (about 1 arc second) and excellent time resolution (typically seconds) in addition to spectral resolution of about 0.1 Å in the extreme ultraviolet, which is sufficient to show Doppler motions. With such an instrument on a rocket, Bruckner et al. (1977) detected spectral lines having very broad non-gaussian profiles from small areas of network and plage. Blue-shifted jets with velocities of order 200 km s^{-1} appeared to originate at these regions, and the authors point out that such energetic events could well contribute to the solar wind. More recently, the same group (Dere et al., 1981) has examined dramatic changes in the profile of C IV emission produced in small regions. They associate these variations on a time scale of tens of seconds with plasma cooling rapidly within the transition zone. These rapid changes are observed in addition to flows at about 10 km s^{-1} (Brueckner et al., 1977) though it is unclear how long these more-general flow patterns last from the brief rocket flights.

It is not altogether surprising that rapid changes are seen during these direct observations of the transition zone since ground-based studies show activity associated with spicules developing at the relevant level. The relationship between these different features is unclear however, and the source of energy is unknown. There are clear attractions is supposing that some 'flare-like' mechanism can occur to eject material at high velocities, since this would provide plasma for the solar wind. On the other hand, if circulation of material along magnetic arches is a common feature (Glencross, 1980, 1981; Cargill and Priest, 1980; Noci, 1981), some transient changes could be due to relatively cool plasma becoming embedded within this pre-existing flow. The conclusion to be reached therefore is that numerous ideas are available to explain features observed in the transition zone, but there is still great need for more-comprehensive sets of observations to restrict speculation.

Acknowledgements

The author wishes to thank the organizers of the IXth Lindau Workshop: 'The Source Region of the Solar Wind' and the participants for the opportunity to discuss these problems of mutual interest.

Athay, R. G.: 1966, *Astrophys. J.* **145**, 784.
Athay, R. G.: 1981, *Astrophys. J.* **249**, 340.
Brueckner, G. E., Bartoe, J.-D. F., and Van Hoosier, M. E.: 1977, *Proc. OSO*-8 *Workshop.*
Cargill, P. J. and Priest, E. R.: 1980, *Solar Phys.* **65**, 251.
Delache, P.: 1969, in *Chromosphere-Corona Transition Region*, NCAR publication, p. 183.
Dere, K. P., Bartoe, J.-D. F., Brueckner, G. E., Dykton, M. D., and Van Hoosier, M. E.: 1981, *Astrophys. J.* **249**, 333.
Ellison, M. A.: 1944, *Monthly Notices Roy. Astron. Soc.* **104**, 22.
Elzner, L. R. and Elwert, G.: 1980, *Astron. Astrophys.* **86**, 188.
Foukal, P.: 1975, *Solar Phys.* **43**, 327.
Foukal, P.: 1976, *Astrophys. J.* **210**, 575.
Frisch, H.: 1972, *Space Sci. Rev.* **13**, 455.
Gabriel, A. H.: 1976, *Phil. Trans. Roy. Soc.* London **A281**, 339.
Glencross, W. M.: 1980, *Astron. Astrophys.* **83**, 65.
Glencross, W. M.: 1981, *Solar Phys.* **73**, 67.
Hearn, A. G.: 1977, *Solar Phys.* **51**, 159.
Huber, M. C. E., Foukal, P. V., Noyes, R. W., Reeves, E. M., Schmahl, E. J., Timothy, J. G., Vernazza, J. E., and Withbroe, G. L.: 1975, *Astrophys. J.* **194**, L115.
Jordan, C.: 1977, *IAU Transactions* **XVIA** (Reports 1976), Part 2, p. 66.
Jordan, C.: 1980, *Astron. Astrophys.* **86**, 355.
Joselyn, J., Munro, R. H., and Holzer, T. E.: 1979, *Solar Phys.* **64**, 57.
Kopp, R. A. and Holzer, T. E.: 1976, *Solar Phys.* **49**, 43.
Kuperus, M.: 1969, *Space Sci. Rev.* **9**, 713.
Kuperus, M., Ionson, J. A., and Spicer, D. S.: 1981 *Ann. Rev. Astron. Astrophys.* **19**, 7.
Lites, B. W., Bruner, E. C., Chipman, E. G., Shine, R. A., Rottman, G. J., White, O. R., and Athay, R. G.: 1976, *Astrophys. J.* **210**, L111.
Meyer, F. and Schmidt, H. U.: 1968, *Z. Angew. Math. Mech.* **48**, T218.
Monroe, R. H. and Withbroe, G. L.: 1972, *Astrophys. J.* **176**, 511.
Noci, G.: 1973, *Solar Phys.* **28**, 403.
Noci, G.: 1981, *Solar Phys.* **69**, 63.
Pneuman, G. W.: 1973, *Solar Phys.* **28**, 247.
Pneuman, G. W. and Kopp, R. A.: 1977, *Astron. Astrophys.* **55**, 305.
Pneuman, G. W. and Kopp, R. A.: 1978, *Solar Phys.* **57**, 49.
Pottasch, S.: 1963, *Astrophys. J.* **137**, 945.
Roussel-Dupré, R.: 1980, *Solar Phys.* **68**, 243.
Spitzer, L.: 1962, *The Physics of Fully Ionized Gases*, 2nd ed., Interscience, New York.
Veck, N. J. and Parkinson, J. H.: 1981, *Monthly Notices Roy. Astron. Soc.* **197**, 41.
Withbroe, G. L. and Gurman, J. B.: 1973, *Astrophys. J.* **183**, 279.
Withbroe, G. L. and Noyes, R. W.: 1977, *Ann Rev. Astron. Astrophys.* **15**, 363.
Zirin, H.: 1968, *Astrophys. J.* **154**, 799.

ACCELERATION OF THE SOLAR WIND†

EGIL LEER*, THOMAS E. HOLZER, and TOR FLÅ*

*High Altitude Observatory, National Center for Atmospheric Research**, Boulder, CO 80307, U.S.A.*

Abstract. In this review, we discuss critically recent research on the acceleration of the solar wind, giving emphasis to high-speed solar wind streams emanating from solar coronal holes. We first explain why thermally driven wind models constrained by solar and interplanetary observations encounter substantial difficulties in explaining high speed streams. Then, through a general discussion of energy addition to the solar wind above the coronal base, we indicate a possible resolution of these difficulties. Finally, we consider the question of what role MHD waves might play in transporting energy through the solar atmosphere and depositing it in the solar wind, and we conclude by examining, in a simple way, the specific mechanism of solar wind acceleration by Alfvén waves and the related problem of accelerating massive stellar winds with Alfvén waves.

1. Introduction

On the basis of observations of continuous auroral activity (Birkeland, 1908, 1913), 27-day recurrent geomagnetic activity (e.g., Chapman and Bartels, 1940), and the anti-solar alignment of ionized comet tails (Biermann, 1951), it has long been suspected that the Sun emits ionized particles continuously. In 1958, Parker suggested a physical mechanism whereby the predominantly hydrogen plasma of the Sun's outer atmosphere could be accelerated to supersonic speeds and thus flow continuously away from the Sun, through interplanetary space, as the solar wind. Parker (1958) argued that, given the high temperature (and the consequent high thermal conductivity (Chapman, 1957)) of the solar corona and the low pressure of the interstellar medium, the only possible steady state of the outer solar atmosphere is a supersonic expansion driven by the thermal pressure gradient force. The existence of the solar wind was subsequently confirmed by *in situ* spacecraft observations outside the terrestrial magnetosphere (Gringauz *et al.*, 1960, 1961, 1967; Bonetti *et al.*, 1963; Scherb, 1964; Snyder and Neugebauer, 1964). These and later observations (e.g., reviews by Axford, 1968; Hundhausen, 1972; Feldman *et al.*, 1977) indicated a high degree of variability of most solar wind parameters in lower-speed wind and a relative uniformity of most parameters in high-speed wind (so-called high-speed streams). The high-speed solar wind streams (Neugebauer and Snyder, 1966) were found frequently to recur with approximately a 27-day period (the solar rotation period as viewed from the Earth), and these recurrent streams were associated with the 27-day recurrent geomagnetic activity. By implication, the solar sources of high-speed streams were taken to be the, as yet unidentified, solar M-regions, which had been invoked as the solar sources of recurrent geomagnetic activity (e.g., Chapman and Bartels, 1940). Solar and interplanetary observations during and after the 1973-4 Skylab period led to the identification of these M-regions as solar

† Paper presented at the IX-th Lindau Workshop 'The Source Region of the Solar Wind'.
* On leave from the Auroral Observatory, Institute of Mathematical and Physical Sciences, University of Tromsø, N-9001 Tromsø, Norway.
** The National Center for Atmospheric Research is sponsored by the National Science Foundation.

coronal holes (Krieger *et al.*, 1973; Neupert and Pizzo, 1974; Bell and Noci, 1976; Hansen *et al.*, 1976; Hundhausen *et al.*, 1978; Nolte *et al.*, 1976; Sheeley *et al.*, 1976; Wagner, 1976; Hundhausen, 1977). The relationship among coronal holes, high-speed streams, and geomagnetic activity is illustrated in Figure 1.

It was realized by Parker (1965) that, while a purely thermally driven wind could readily produce the observed low to moderate solar wind speeds, the high-speed wind might require the addition of energy to the wind above the coronal base. In fact, it is

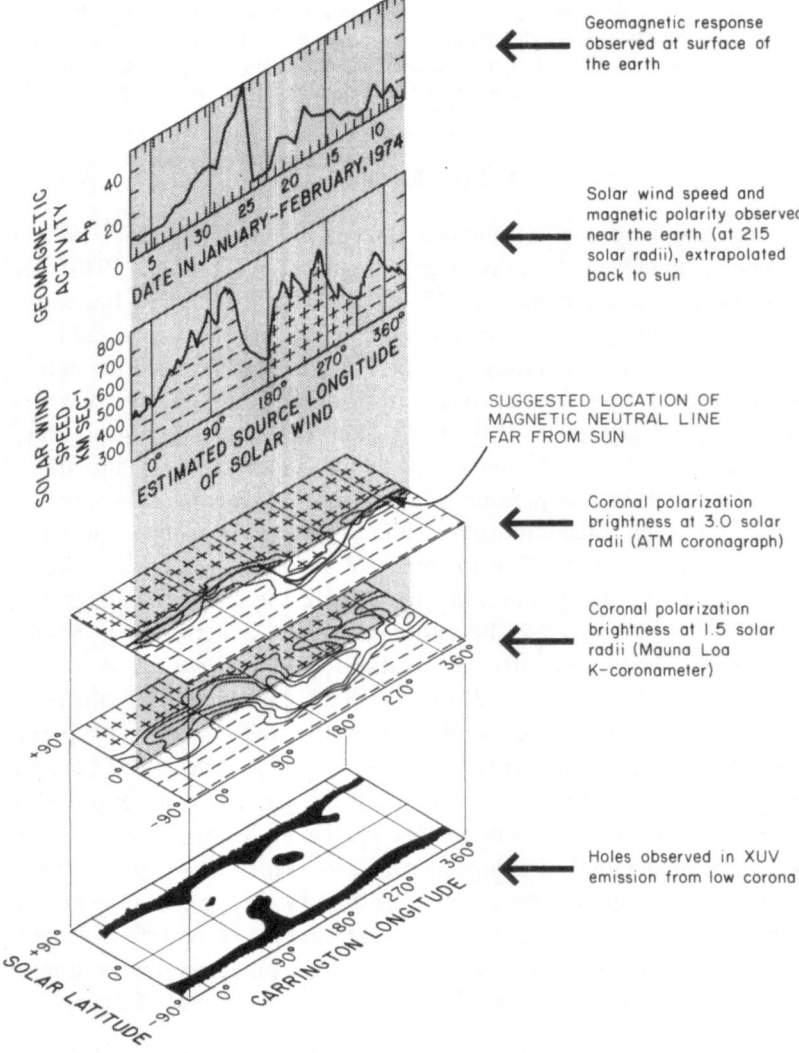

Fig. 1. The three-dimensional relations among coronal hole observations at different heights in the solar atmosphere, the solar wind structure observed in the ecliptic plane, and the resulting geomagnetic activity for Carrington solar rotation 1610, in early 1974. The magnetic polarity is indicated by + signs for magnetic fields pointing out of the Sun and − signs for fields pointing into the Sun (from Hundhausen and Holzer, 1980).

readily shown that if the solar wind temperature monotonically decreases from the coronal base to the orbit of Earth, the observational constraints of modest solar wind mass-flux densities in high-speed streams and moderate plasma pressures at the base of coronal holes require such extended energy addition in the region of supersonic flow (Holzer and Leer, 1980; Leer and Holzer, 1980). Hence, observations of high-speed streams and coronal holes place the most stringent requirements on theoretical descriptions of solar wind acceleration; these requirements are the most complete, as well, because of the unambiguous identification of coronal holes as the solar source of high speed streams and the extensive observations of coronal holes currently available. It seems appropriate, therefore, that in the present review of solar wind acceleration we concentrate primarily on coronal holes and high-speed solar wind streams.

We shall begin by considering models of thermally driven winds with no energy added above the coronal base (Section 2) and then discuss the general effects of energy addition (in the forms of heat and momentum addition) to the subsonic and supersonic regions of the wind (Section 3). Having indicated a possible need for outward energy transport from the coronal base by some means other than advection or thermal conduction, we shall consider the propagation of hydromagnetic waves in the solar atmosphere and solar wind and the role such waves might play in transporting the additional energy (Section 4). Finally, we shall discuss the specific acceleration mechanism involving Alfvén waves interacting with the solar wind – this being both the most thoroughly studied and conceptually the simplest energy addition mechanism (Section 5). The extension of this mechanism to the description of massive winds from cool, low-gravity stars will be briefly outlined in Section 6, and a few closing remarks will be presented in Section 7.

2. Thermally Driven Solar Wind

Our present understanding of the solar wind is based on the work of Parker (1958, 1960, 1963, 1964a, b, 1965). His first solar wind paper (Parker, 1958) dealt with the spherically symmetric expansion of an isothermal corona and provided a sound physical description of the dynamical effects important in the accelaration of the solar wind. Parker (1963, 1964a, b, 1965) later extended this work to include the effects of non-spherical flow and of energy balance in the coronal expansion. The latter work concentrated on classical thermal conduction in a spherical expansion, but the effects of non-classical conduction and energy addition above the coronal base were also considered. In light of observations that have become available over the past 15-y, several other workers have expanded on this theoretical basis and clarified the roles that the physical effects discussed by Parker play in the acceleration of the solar wind. In reviewing our present understanding of this subject, we shall begin with the case of a thermally driven solar wind: that is, a wind in which the driving force is the thermal pressure gradient force and the retarding force results from the solar gravitational field.

There have been several approaches taken toward describing a thermally driven solar wind, ranging from the original isothermal model to conductive, viscous, one-fluid and two-fluid models (for references to this work see reviews by Holzer and Axford (1970), Hundhausen (1972), Barnes (1975), Hollweg (1978a), Holzer (1979)). To illustrate the physical effects that are important in a thermally driven wind, we shall discuss results from inviscid, one-fluid models including various descriptions of thermal conduction and both spherical and more rapidly expanding flow geometries. We use inviscid, one-fluid models because viscosity is negligible, except in the vicinity of steep velocity gradients (Parker, 1963), and because two-fluid models do not provide additional information that would be of interest in the following discussion. Rapidly expanding flow geometries must be considered because they seem to be characteristic of coronal holes, in which the magnetic field diverges rapidly (e.g. Altschuler *et al.*, 1972; Munro and Jackson, 1977), and the flow is expected to follow the field near the Sun. The equations for mass, momentum, and energy balance which we shall use to describe the steady, radial flow of this thermally driven electron-proton solar wind are thus

$$nmuA = \mathscr{F} = \text{const.} \tag{1}$$

$$u \frac{du}{dr} = -\frac{1}{nm} \frac{dp}{dr} - \frac{GM}{r^2}, \tag{2}$$

$$3nuk \frac{dT}{dr} = 2ukT \frac{dn}{dr} - \frac{1}{A} \frac{d}{dr}(qA), \tag{3}$$

where A is the cross-sectional area of an infinitesimal, radial flow tube, n is the electron (proton) number density, m is the proton mass, u is the radial flow speed, T is half the sum of the electron and proton temperatures, $p = 2nkT$, q is the radial heat flux density, and k, G, and M are the Boltzmann constant, gravitational constant, and solar mass. An alternative form for the energy balance equation is

$$\mathscr{F}\left(\tfrac{1}{2}u^2 + 5 \frac{kT}{m} - \frac{GM}{r} + \frac{A}{\mathscr{F}} q\right) = F = \text{const.} \tag{4}$$

If the magnetic field is radial, and the solar wind plasma is assumed to be collision-dominated, the heat flux density can be described classically (Spitzer, 1962) by

$$q = q_a = -\kappa_0 T^{5/2} \frac{dT}{dr}, \tag{5}$$

where $\kappa_0 = 7.8 \times 10^{-7} \text{ erg cm}^{-1} \text{ s}^{-1} \text{ K}^{-7/2}$. If account is taken of the spiral shape of the interplanetary magnetic field (Parker, 1958) and of the strong inhibition of thermal conduction across the magnetic field (e.g., Chapman and Cowling, 1939), then the heat flux density takes the form (Parker, 1964b).

$$q = q_b = q_a \cos^2 \theta, \tag{6}$$

where θ is the angle between the radial direction and the local magnetic field. Of course, the solar wind plasma is not collision-dominated, except, perhaps, relatively near the Sun, and the classical description of thermal conduction must be modified (e.g., Parker, 1964b; Perkins, 1973). Hollweg (1976) has suggested that in a collisionless solar wind plasma the heat flux density can be represented by

$$q = q_c = \tfrac{3}{2}\alpha nukT , \tag{7}$$

where α is an arbitrary parameter, which will be taken here to be 4. Hollweg (1976) represents q by (6) when the mean free path of a thermal electron is less than half the radial distance and by (7) otherwise. Recently, Scudder and Olbert (1979a, b) have derived a kinetic description (including Coulomb collisions) of solar wind electrons, and their results indicate that the solar wind heat flux may not be adequately described by any of the above models. Although detailed calculations relevant to acceleration of the solar wind are not yet available, we shall make use of existing information (Olbert, 1981) to discuss (later in this section) the possible implications of this kinetic description for a thermally driven wind.

Durney (1972) obtained numerical solutions to (1)–(3) and (5) for spherical symmetry $(A \sim r^2)$, and some of these are displayed in Figure 2a to illustrate the dependence of particle flux density and flow speed at 1 AU $(n_E u_E$ and $u_E)$ on the coronal base density (n_0) and temperature (T_0). We see that the particle flux density $(n_E u_E)$ increases rapidly with increasing T_0, if n_0 is fixed. Similarly, for $n_0 \lesssim 10^7$ cm^{-3}, the solar wind flow speed (u_E) increases as T_0 increases, but for $n_0 \gtrsim 10^8$ cm^{-3}, u_E decreases with increasing T_0. It is worth spending a little time trying to understand this behavior of $n_E u_E$ and u_E, for by doing this we can lay a basis for understanding all the important physical effects (associated with accelerating the solar wind) that we shall consider in the remainder of the paper.

Let us first consider the dependence of the solar wind mass flux on coronal temperature (cf. Parker, 1958, 1964a). For an isothermal corona, (2) is readily integrated from the coronal base $(r = r_0)$ to the critical point $(r = r_c = GMm/4\beta_c kT$, where $\beta = (r/2A)(dA/dr))$ to yield the flow speed at the coronal base (Leer and Holzer, 1979)

$$u_0 = \left(\frac{2kT}{m}\right)^{1/2} \cdot \left(\frac{r_c}{r_0}\right)^2 f_c \exp\left\{-\left[\frac{GMm}{2kTr_0} + \tfrac{1}{2} - 2\beta_c\right]\right\} \tag{8}$$

where $f = r_0^2 A/r^2 A_0$, and we assume $u_0^2 \ll 2kT/m$. Because $n_E u_E \propto u_0$, it is clear that for a strongly gravitationally bound corona $(GM/r_0 \gg 2kT/m)$ the solar wind mass flux $(\propto n_E u_E)$ is dominated by the exponential coronal temperature dependence shown in (8). Hence, a small increase in coronal temperature produces a large increase in solar wind mass flux. (The isothermal assumption is not especially restrictive in the above analysis, for T can be taken to represent an average coronal temperature in the region of subsonic solar wind flow (Leer and Holzer, 1979)).

If our physical intuition were based entirely on polytropic (including isothermal) models of the solar wind, we would immediately conclude that this strong dependence

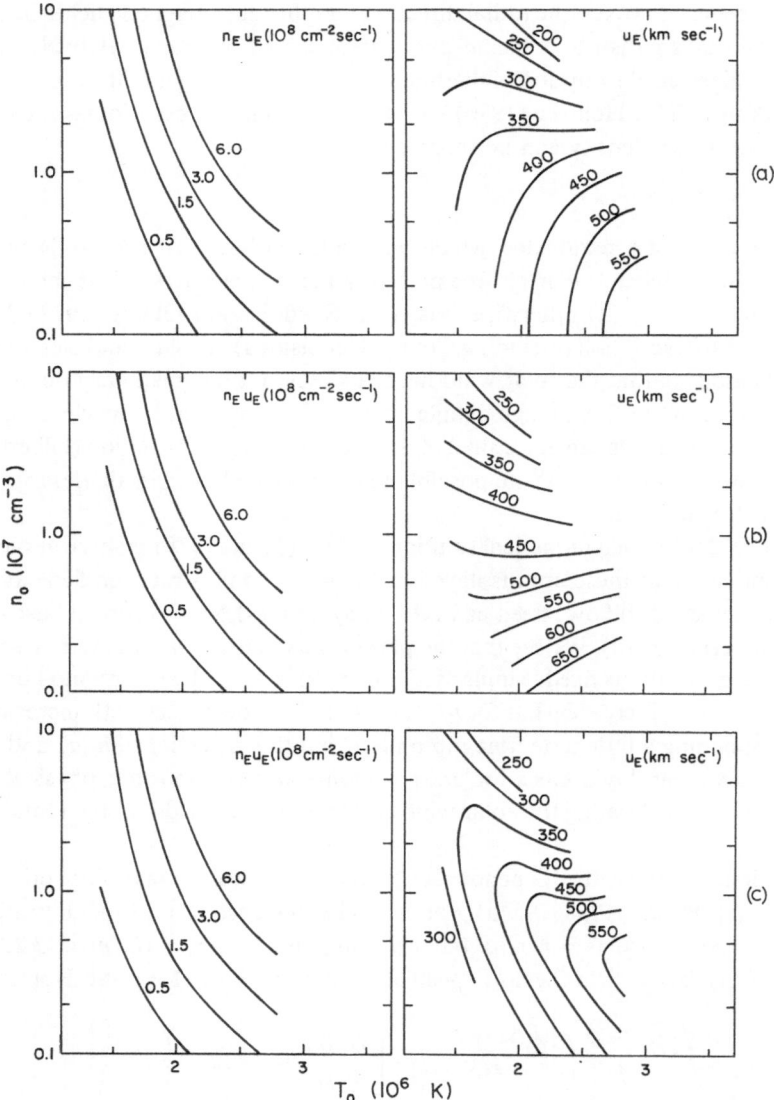

Fig. 2.   Contours of constant particle flux density at 1 AU ($n_E u_E$) and flow speed at 1 AU (u_E) shown in the $n_0 - T_0$ plane (where n_0 and T_0 are the coronal base density and temperature) for the spherically symmetric flow of a conductive, thermally driven wind: (a) Durney's (1972) solutions for a classical (collision-dominated) conduction law; (b) solutions of Durney and Hundhausen (1974) for magnetically inhibited thermal conduction; solutions for the collisionlessly inhibited thermal conduction description used by Hollweg (1976) (from Holzer and Leer, 1980).

of mass flux on coronal temperature has no effect on the solar wind flow speed (u_E), because the mass flux scales out of polytropic representations (e.g. Parker, 1964a; Hundhausen, 1972). In fact, the same sort of scaling occurs when T is specified as a function of r ($T = T_0 \Theta(r)$): in either case, u_E depends only on T_0. If, however, a proper energy equation (e.g., (3)) is considered, it is clear that the magnitude of the mass flux determines the relative importance of thermal conduction and, say, expansive cooling. In Durney's (1972) model, when the mass flux is small (viz., for $n_0 \lesssim 10^7 \text{cm}^{-3}$) conduction dominates everywhere, and the relevant solution of (3) is $T = T_0(r_0/r)^{2/7}$ (Parker, 1964a; Chapman, 1957). As noted above, this case, like the polytropic case, leads to a monotonic increase of u_E with increasing T_0, owing to the increasing strength of the pressure gradient force with increasing T_0. Yet, for larger mass fluxes, conduction ceases to dominate in certain domains, and expansive cooling leads to a more rapid temperature decline and a consequent decrease in efficiency of acceleration of the flow by the pressure gradient force. If the base density and temperature are large enough (e.g., $T_0 \gtrsim 1.5 \times 10^6$ at $n_0 \approx 10^8 \text{cm}^{-3}$ or $T_0 \gtrsim 2 \times 10^8$ K at $n_0 \approx 2 \times 10^7 \text{cm}^{-3}$), the expansive cooling is so effective that conduction is unimportant at $r = r_E = 1$ AU, and virtually all of the solar wind energy flux at 1 AU is carried by the flow. In this case, (4) can be evaluated at the coronal base and at 1 AU to yield

$$F/\mathscr{F} \approx u_E^2/2 \approx q_0 A_0/\mathscr{F} + 5kT_0/m - GM/r_0 . \tag{9}$$

Now we can understand why Durney's results (Figure 2a) indicate that u_E decreases as T_0 increases, when n_0 and T_0 are large enough: the heat flux density at the coronal base (q_0) does not increase as rapidly with T_0 ($q_0 \propto T_0^{7/2}$) as does the mass flux ($\mathscr{F} \propto u_0$; cf. (8)), so the conductive energy per unit mass ($q_0 A_0/\mathscr{F}$) decreases more rapidly than the enthalpy per unit mass ($5kT_0/m$) increases, and the total energy per unit mass supplied to the solar wind ($F/\mathscr{F} \approx u_E^2/2$) decreases.

This line of argument is especially important for what follows, because it is clear from solar wind observations (e.g. Hundhausen, 1972) that $F/\mathscr{F} \approx u_E^2/2$ (i.e., at 1 AU the solar wind energy flux is transported almost entirely by the flow). Hence, if we can determine the outward energy flux in the corona (F) and the resulting mass flux ($\mathscr{F}$), we then know (for realistic models) what the flow speed at 1 AU must be ($u_E \approx \sqrt{2F/\mathscr{F}}$).

Let us turn next to the model of Durney and Hundhausen (1974), in which inhibition of the radial heat flux by the spiral interplanetary magnetic field is considered (i.e., (1)–(3) and (6) are solved). Some results of this model are shown in Figure 2b, and comparison with Figure 2a indicates that the solar wind mass flux is only slightly affected by the conduction inhibition, but the flow speed at 1 AU is increased significantly over a broad range of values of n_0 and T_0. The principal effect of inhibition of thermal conduction in a *strongly conduction-dominated* solar wind flow is the establishment of a temperature plateau, a region of elevated temperature extending from the coronal base out to the distance (in this case, several tens of solar radii) where the inhibition becomes effective (e.g. Parker, 1964b; Durney and Hundhausen, 1974). This temperature plateau has little effect on the average temperature in the region of subsonic flow and thus little effect on the mass flux, but it significantly enhances the pressure gradient in the supersonic region and thus leads to a higher-speed solar wind. In terms of energy balance, one can say that this inhibition reduces the conduction flux at the coronal base, but substantially increases the efficiency with which the conduction flux is transformed to flow energy (through the pressure gradient force), resulting in a net increase in the energy flux supplied to the flow at 1 AU. Together with only a modest change in the mass flux, this leads to an increase in the flow energy per unit mass ($u_E^2/2$). A notable feature of Figure 2b (see also Figure 2a) is the monotonic decrease of u_E with increasing n_0. This results from the fact that the mass flux depends more strongly on base density than do either the conduction flux density at the base or the efficiency of conversion from conductive to flow energy, so that the energy per unit mass supplied to the flow is decreased when n_0 is increased (Holzer and Leer, 1980).

The only substantive difference between the magnetic inhibition of conduction discussed by Durney and Hundhausen (1974) and the collisionless inhibition discussed by Hollweg (1976) (see Figure 2c) is that the latter inhibition can take place much nearer the Sun and can more strongly inhibit the conduction. For very low base densities, electrons in the coronal plasma are collisionless everywhere above the coronal base, and in Hollweg's (1976) model $q = q_c$ everywhere, so that the flow is polytropic (cf. (7)). For low to moderate densities, however, the electrons are assumed collision-dominated near the Sun (i.e., $q = q_b$), and the point where the conduction is inhibited (i.e., where the plasma is assumed to become collisionless and q becomes q_c) moves rapidly away from the Sun with increasing n_0. The resulting increase in conduction flux with increasing n_0 is more rapid than the increase in mass flux, and given the uniformly high efficiency of conversion from conduction energy to flow energy associated with collisionless inhibition, the energy per unit mass supplied to the flow ($u_E^2/2$) increases (Holzer and Leer, 1980). This effect is seen in the lower left portion of the n_0–T_0 plane shown in Figure 2c, where u_E increases with increasing n_0.

It is clear from Figure 2 that high speed solar wind streams, in which $u_E > 600$ km s^{-1} and $n_E u_E \approx 3 \times 10^8$ cm^{-2} s^{-1} (e.g., Feldman et al., 1977), cannot be produced in these three conductive models unless n_0 is taken to be unrealistically low. (Withbroe (1977) indicates that $n_0 T_0 \gtrsim 10^{14}$ cm^{-3} K, which implies $n_0 \gtrsim 5 \times 10^7$ cm^{-3}.) We then must ask whether the rapidly diverging flow expected in solar coronal holes can lead to a substantial increase in u_E for the same three types of conductive model. This question has been addressed by Holzer and Leer (1980), who have extended the models of Durney (1972), Durney and Hundhausen (1974), and Hollweg (1976) to include the effects of rapidly diverging flow geometries. Figure 3 shows their results for a flow geometry described by (Kopp and Holzer, 1976):

$$A = A_0(r/r_0)^2 f, \tag{10}$$

$$f = (f_{\max} e^{(r - r_1)/\sigma} + f_1)/(e^{(r - r_1)/\sigma} + 1), \tag{11}$$

$$f_1 = 1 - (f_{\max} - 1) e^{(r_0 - r_1)/\sigma}, \tag{12}$$

with $\sigma = 0.1 R_\odot$, $r_1 = 2 R_\odot$, and $f_{\max} = 7$. Here, σ measures the radial scale over which a flow tube expands faster than it would in a spherical geometry, r_1 indicates the radial location where the rapid expansion occurs, and $f_{\max}$ is the ratio as $r \to \infty$ of the area of a rapidly expanding flow tube to that of a spherically expanding flow tube, when their areas are equal at $r = r_0$. In the chosen flow geometry, three critical points may exist, and in some cases the only acceptable solution exhibits a shock transition between the

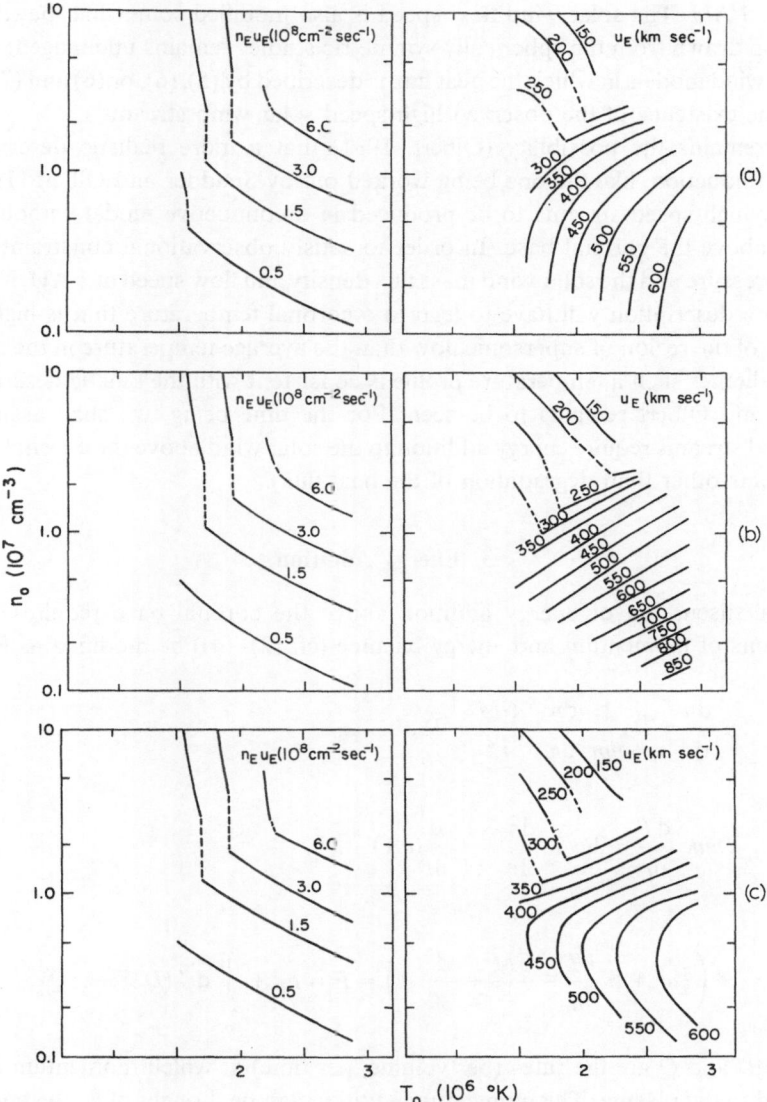

Fig. 3. Same as Figure 2, except for a rapidly diverging flow instead of a spherically symmetric flow. The flow geometry is characterized by $\sigma = 0.1 R_\odot$, $r_1 = 2 R_\odot$, and $f_{max} = 7$, which means that the flow tube area increases by a factor of 7 more than in the spherical case, and most of the rapid area increase takes place in $1.9 R_\odot \lesssim r \lesssim 2.1 R_\odot$. The dashed lines indicate regions where a shock transition exists somewhere between the inner and outer critical points (cf. Figure 5a and Holzer (1977)) (from Holzer and Leer, 1980).

inner and outer critical points (Holzer, 1977; Habbal and Tsinganos, 1982). Such solutions were not calculated in this study, but their region of existence is indicated by the dashed lines in Figure 3. It is found that a rapidly expanding flow geometry leads to a slightly higher mass flux density at the coronal base, but a slightly lower mass flux density at 1 AU. The solar wind flow speed is also modified somewhat, but the basic conclusion drawn from the spherically symmetric studies remains unchanged: conductive solar wind models in which the heat flux is described by (5), (6), or (6) and (7) cannot explain the existence of the observed high-speed solar wind streams.

There remains the possibility (Olbert, 1981) that a more realistic description of thermal conduction, like the one being worked out by Scudder and Olbert (1979a, b), will allow high-speed streams to be produced in a conductive model without energy addition above the coronal base. In order to satisfy observational constraints on the coronal pressure and the solar wind mass flux density and flow speed at 1 AU, it appears that such a description will have to lead to a coronal temperature that is higher in at least part of the region of supersonic flow than the average temperature in the subsonic region. Whether such a temperature profile is consistent with the kinetic description of Scudder and Olbert remains to be seen. For the time being, we shall assume that high-speed streams require energy addition to the solar wind above the coronal base (by some means other than degradation of the heat flux).

3. Energy Addition

A general discussion of energy addition above the coronal base requires that our descriptions of momentum and energy balance (cf. (2)–(4)) be modified as follows:

$$u \frac{du}{dr} = -\frac{1}{nm} \frac{dp}{dr} - \frac{GM}{r^2} + D, \tag{13}$$

$$3nuk \frac{dT}{dr} = 2ukT \frac{dn}{dr} - \frac{1}{A} \frac{d}{dr}(qA) + Q, \tag{14}$$

$$\mathscr{F}\left(\tfrac{1}{2}u^2 + 5 \frac{kT}{m} - \frac{GM}{r} + \frac{A}{\mathscr{F}} q\right) = F = F_0 + \int_{r_0}^{r} dr' (D\mathscr{F} + AQ), \tag{15}$$

where nmD and Q are the rates (per volume, per time) at which momentum and heat are added to the plasma. The momentum addition can be thought of as the application of an outward-directed body fore per unit mass, D. An energy flux associated with both the heat and momentum addition can be defined by

$$\Delta F = \int_{0}^{\infty} dr(D\mathscr{F} + AQ), \tag{16}$$

and this will prove a useful parameter in the following discussion of the effects of energy addition on the solar wind particle flux density, flow speed, and temperature at 1 AU ($n_E u_E$, u_E, and T_E). In this discussion, we shall find it instructive to consider heat addition and momentum addition separately.

Let us begin with heat addition, setting $D = 0$ and writting

$$Q = Q_{i\,\exp}\left[-100\left(\frac{r}{r_i} - 1\right)^2 \right]. \tag{17}$$

(17) describes the heating of the solar wind in a narrow region centered at $r = r_i$ and allows us to study the effects of adding heat at various locations in the flow by solving (1), (13), (14), and (17) for several values of r_i ranging from $r_i = r_0$ to $r_i \gtrsim 1$ AU. The magnitude of the heating can be specified in such a study by requiring that the energy flux added to the wind be independent of the location where the energy is added: i.e., $(\partial/\partial r_i)\,(\Delta F) = 0$. Because of our desire to match as closely as possible the observation (e.g. Hundhausen, 1972) that $F_E/\mathscr{F} \approx u_E^2/2$, we shall use Hollweg's (1976) representation of thermal conduction (cf. Section 2), in which the plasma is assumed collision-dominated ($q = q_b$: see (6)) wherever the mean free path of thermal electrons is less than half the radial distance, and it is assumed collisionless ($q = q_c$: see (7)) otherwise. The difficulty with the classical description of thermal conduction ($q = q_a$ everywhere) is that frequently $q_E A_E/\mathscr{F} \gtrsim u_E^2/2$, and a significant fraction of the energy flux added by heating above the coronal base is carried past 1 AU by an enhanced thermal conduction flux, causing the predicted flow speed (u_E) to be unrealistically low (see the results of Pneuman (1980) for an example of this difficulty).

The above model (with collisionless inhibition of conduction) has been studied by Leer and Holzer (1980) for spherically symmetric flow, and their results are shown in Figure 4a. Coronal base parameters are taken to be $n_0 = 10^8$ cm^{-3} and $T_0 = 1.4 \times 10^6$ K, and the magnitude of the heat addition is prescribed by $\Delta F/F_R = 0$, 0.3, and 1 in the three cases shown. The reference energy flux, F_R, is the solar wind energy flux ($F = F_0$) in the absence of energy addition above the coronal base ($\Delta F = 0$), and it generally differs from the coronal base energy flux (F_0) in the presence of energy addition ($\Delta F \neq 0$). The critical (sonic) point of the flow occurs near $10R_\odot$ in the reference case ($\Delta F = 0$), but moves inward (as far as $4R_\odot$ for $\Delta F = 1$) when heat is added to the subsonic flow. The addition of heat in the subsonic region increases the local temperature and thus increases the mass flux ($\mathscr{F} \propto n_E u_E$). This mass flux increase very nearly balances the increase in energy flux (ΔF), so the flow speed at 1 AU ($u_E \approx \sqrt{2F_E/\mathscr{F}}$) is not significantly changed. When, however, heat is added in the supersonic region, the temperature in the subsonic region is generally unaffected, and the mass flux ($\mathscr{F}$) is unchanged; the increase in energy flux (ΔF) thus leads to an increase in the flow speed at 1 AU ($u_E \approx \sqrt{2F_E/\mathscr{F}}$). As the location of the heat addition moves farther out in the solar wind, there is less time before the wind reaches 1 AU for the conductive energy and internal energy to be converted into flow energy through the action of the pressure gradient force. As a result, T_E increases and u_E decreases as r_i increases (in the supersonic region); evidently, if heat is added too far out in the wind, the temperature at 1 AU can become unrealistically large. (The decrease in u_E below its reference value when $r_i \approx 1$ AU results from the localized inward pressure gradient force associated with the spatially narrow heating function used here.) Quite similar

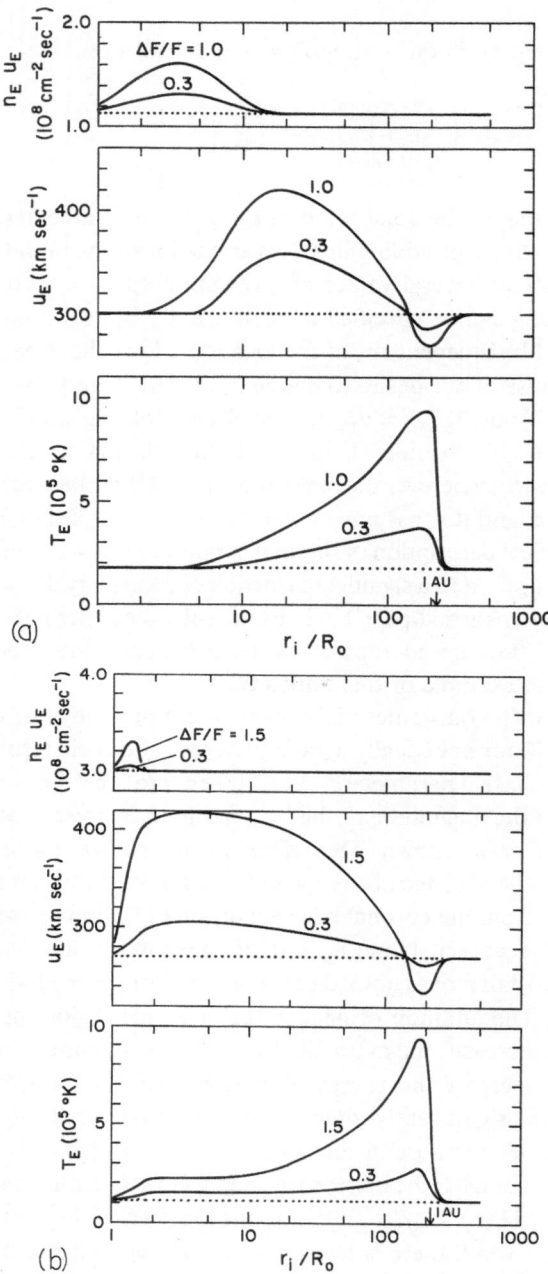

Fig. 4.   Values at 1 AU of the proton flux density ($n_E u_E$), flow speed (u_E), and temperature (T_E) for a range of locations (r_i) of a spatially narrow heat addition function: (a) spherically symmetric flow, with $n_0 = 10^8$ cm^{-3} and $T_0 = 1.4 \times 10^6$ K; (b) rapidly diverging flow ($f_{max} = 7$, $\sigma = 0.1 R_\odot$, $r_1 = 2 R_\odot$), with $n_0 = 1.6 \times 10^7$ cm^{-3} and $T_0 = 2.5 \times 10^8$ K. The dotted lines indicate the reference values of parameters (i.e., values for $\Delta F = 0$) and the solid lines correspond to (a) $\Delta F/F_R = 0.3$, 1.0 and (b) $\Delta F/F_R = 0.3$, 1.5. The sonic (critical)point in the reference models ($\Delta F = 0$) is near $10 R_\odot$ in (a) and near $1.8 R_\odot$ in (b) (from Leer and Holzer, 1980).

results are found for a rapidly expanding flow geometry (Figure 4b): heat addition in the subsonic region increases the solar wind mass flux, but has little effect on the flow speed at 1 AU; heat addition in the supersonic region does not affect the mass flux, but increases the flow speed at 1 AU. Of course, in the rapidly expanding flow geometry, the sonic point lies much nearer the Sun, so u_E can be increased with heat addition much nearer the Sun than in the spherically symmetric models.

Now let us consider the effects of momentum addition, setting $Q = 0$ and writing (cf. (17))

$$D = D_i(A/A_i) \exp\left[-50\left(\frac{r}{r_i} - 1\right)^2\right]. \tag{18}$$

Solutions to (1), (13), (14), and (18) are shown in Figure 5a and b for spherical and rapidly expanding flow geometries and the same model parameters as used in the heat addition study (Figure 4) (Leer and Holzer, 1980). The dashed lines in Figure 5a indicate solutions (not calculated) in which shocks occur between the inner and outer critical points (cf. Figure 3). As in the case of heat addition, energy addition by direct acceleration (momentum addition) leads to an enhanced mass flux when the energy is added in the region of subsonic flow, but not when it is added in the supersonic region. The mass flux enhancement, however, is somewhat larger when a given amount of energy is added as momentum (Figure 5) than when the same amount of energy is added as heat (Figure 4). When all the energy is added between the coronal base and the (innermost) critical point (at $r = r_c$), the increase in mass flux ($\mathscr{F} - \mathscr{F}_R$) over the reference model mass flux ($\mathscr{F}_R$) is represented approximately by

$$\mathscr{F} \approx \mathscr{F}_R \exp(mI/2k \langle T \rangle) \tag{19}$$

where $I = \int_{r_0}^{r_c} dr\, D$, and $\langle T \rangle$ is the average subsonic-region temperature (Leer and Holzer, 1979). The increase in mass flux resulting from momentum addition in the subsonic region is due to the increased velocity scale height and the inward motion of the critical point, which are characteristic consequences of momentum addition, heat addition, or increased coronal base temperature.

Because the addition of energy by direct acceleration (momentum addition) in the subsonic region leads to an enhancement of the mass flux that is significantly larger than that associated with the addition of the same amount of energy by heating, it is not surprising that momentum addition in the subsonic region leads to a substantial decrease in flow speed at 1 AU (see u_E in Figures 5a and b). As with heat addition, the addition of momentum in the supersonic region increases u_E (again, because $\mathscr{F}$ is unchanged and F_E is increased), but the momentum addition is more uniformly efficient in increasing u_E (cf. Figures 4 and 5), because adding momentum has little effect on the temperature profile, and virtually all of the energy flux at 1 AU is carried as flow energy, regardless of how near 1 AU the momentum is added. In fact T_E is generally decreased by momentum addition, either because of an increased expansive cooling (for addition in the supersonic region) or because of an inward movement of the point at which inhibition of thermal conduction occurs (for addition nearer the Sun).

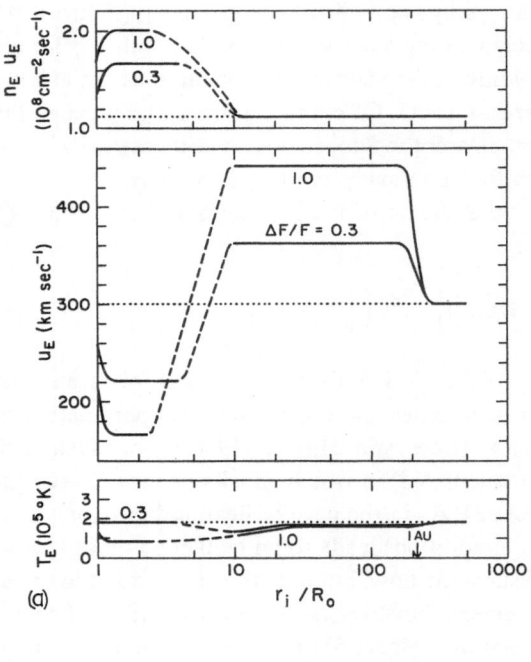

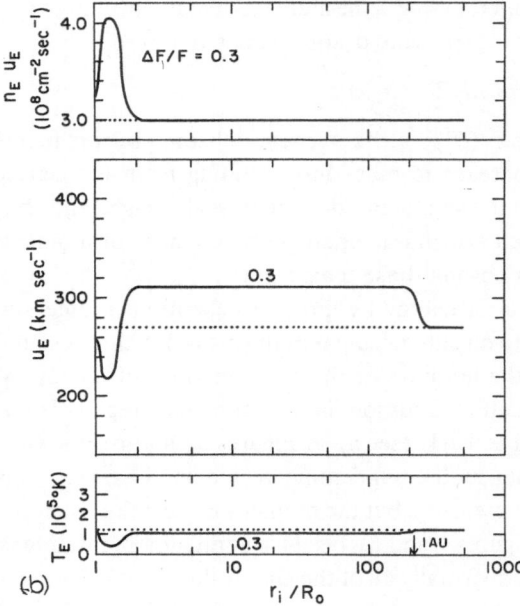

Fig. 5.   Same as Figure 4, but for momentum addition instead of heat addition. The dashed portions of the
curves in Figure 5a indicate solutions passing through an inner critical point, shocking to subsonic flow, and
then passing through an outer critical point, beyond which the flow remains supersonic (cf. Figure 3 and
Holzer (1977)) (from Leer and Holzer, 1980).

Two other aspects of the general energy addition problem that are of interest here are momentum loss in the subsonic region and energy exchange between the subsonic and supersonic regions (see Leer and Holzer (1980) for a detailed discussion). As can be inferred from Figure 5, momentum loss (i.e., an additional inward-directed force that decelerates the flow) should decrease the solar wind mass flux and increase the flow speed at 1 AU. This effect has been demonstrated for the frictional force associated with a substantial, but not unreasonable, abundance of He^{++} in the corona (Leer and Holzer, 1979), and it is conceivable that some other type of inward force, perhaps electromagnetic in nature, might be effective in the subsonic region. One can also infer from Figures 4 and 5 that any process whereby energy is removed from the subsonic region and deposited in the supersonic region will lead to a decrease in the solar wind mass flux and an increase in the flow speed at 1 AU. Such a process might involve the generation of, say, hydromagnetic waves in the subsonic region and their propagation into and dissipation in the supersonic region.

It is clear from the above discussion that an energy flux (other than advective or conductive) emanating from the coronal base will be most effective in producing high-speed solar wind streams if it is deposited in the solar wind in the region of supersonic flow. There may, however, be no need for such an energy flux if there exists a process which removes momentum from (decelerates the flow in) the subsonic region or which removes energy from the subsonic region and redeposits it in the supersonic region.

4. Alfvén Waves in the Solar Atmosphere

We have indicated that it may be necessary to add energy to the solar wind above the coronal base to accelerate high-speed streams. Hydromagnetic waves represent one means whereby this energy might be transported from the lower solar atmosphere to the region in the solar wind where it is deposited, and Alfvénic fluctuations observed at 1 AU (Belcher and Davis, 1971) could well represent remnants of such a transport process. Fast-mode and slow-mode hydromagnetic waves have a compressional component and generally suffer stronger damping than Alfvén waves (e.g. Barnes, 1979), so even if a substantial energy flux in all three modes were present in the corona, one would expect to see the remnant energy flux only in the Alfvén mode at 1 AU. In this section and the following section, we shall concentrate on the transport of energy by Alfvén waves in the solar atmosphere and the solar wind, not because Alfvén waves are more likely to be important than fast-mode and slow-mode waves, nor because energy transport by means other than hydromagnetic waves is unlikely, but because Alfvén waves are the most thoroughly studied energy transport mechanism, and we can gain a good bit of general understanding of energy transport through a careful consideration of Alfvén waves.

Hydromagnetic waves may be generated in the corona (e.g., Barnes, 1969), or they may be generated in the lower solar atmosphere and propagate into the corona. In the latter case, propagation through the transition region (between the chromosphere and corona) involves certain difficulties (e.g. Osterbrock, 1961): fast-mode waves are refracted strongly and can carry no significant energy flux through the transition region, into the corona; slow-mode waves steepen rapidly in the transition region (if they are not already shocks in the upper chromosphere) and the resulting shock waves dissipate their energy below or very near the coronal base. In addition, observations of solar oscillations in the middle chromosphere (Athay and White, 1979a) place a lower limit on the upward energy flux density carried by slow-mode waves of about 2×10^4 erg cm^{-2} s^{-1} and an upper limit of about 10^5 erg cm^{-2} s^{-1} (Athay and White, 1979b), which is much less than the energy flux density required to heat the corona and drive the solar-wind in coronal holes ($\approx 5 \times 10^5$ erg cm^{-2} s^{-1}), to say nothing of that required to heat the upper chromosphere. These observations (Athay and White, 1979a, b) can also be used to place limits on the energy flux carried by Alfvén waves, but a careful analysis of the behavior of Alfvénic disturbances in the lower solar atmosphere must first be carried out, and this is done below.

Because of the very small density scale-height in the lower solar atmosphere, a moderate wave period (min to h) of an Alfvénic disturbance is comparable to or longer than an Alfvénic travel-time across a characteristic scale-length of the medium (i.e., $\omega \lesssim dv_A/ds$, where ω is the disturbance frequency and v_A the Alfvén speed), so the WKB approximation, normally used to describe Alfvén-wave propagation, breaks down. Analysis of Alfvénic disturbances, therefore, requires a reconsideration of the conservation laws and Maxwell's equations. For slow oscillations of the plasma in the solar atmosphere, the mass and momentum conservation laws, Ampère law, and Faraday's law can be written

$$\frac{\partial \rho}{\partial t} + \nabla \cdot \rho \mathbf{u} = 0 , \tag{20}$$

$$\rho \left(\frac{\partial}{\partial t} + \mathbf{u} \cdot \nabla \right) \mathbf{u} = -\nabla p - \rho \, \frac{GM}{r^2} \, \hat{e}_r + \mathbf{j} \times \mathbf{B}/c , \tag{21}$$

$$\nabla \times (\mathbf{E} + \mathbf{u} \times \mathbf{B}/c) = 0 , \tag{22}$$

$$\nabla \times \mathbf{B} = \frac{4\pi}{c} \, \mathbf{j} , \tag{23}$$

$$\nabla \times \mathbf{E} = -\frac{1}{c} \frac{\partial \mathbf{B}}{\partial t} , \tag{24}$$

where $\rho = nm$. Let us consider toroidal Alfvénic disturbances in an axisymmetric background plasma. First, (20)–(24) yield two equations for the ϕ components (in spherical coordinates) of $\mathbf{u}$ and $\mathbf{B}$:

$$\rho\left(\frac{\partial}{\partial t} + \mathbf{u} \cdot \nabla\right) u_\phi + \frac{\rho u_\phi}{r}(u_r + u_\theta \tan \theta) + \frac{1}{r \sin \theta} \frac{\partial p}{\partial \phi} =$$

$$= -\frac{1}{4\pi}\left[\frac{1}{2r \sin \theta}\frac{\partial B^2}{\partial \phi} - \right.$$

$$\left. - (\mathbf{B} \cdot \nabla)B_\phi - \frac{B_\phi}{r}(B_r + B_\theta \cot \theta)\right], \tag{25}$$

$$\left(\frac{\partial}{\partial t} + \mathbf{u} \cdot \nabla\right) B_\phi + \frac{u_\phi}{r}(B_r + B_\theta \cot \theta) =$$

$$= (\mathbf{B} \cdot \nabla)u_\phi + \frac{B_\phi}{r}(u_r + u_\theta \cot \theta) -$$

$$- B_\phi\left(\frac{\partial}{\partial t} + \mathbf{u} \cdot \nabla\right) \ln \rho. \tag{26}$$

Now, if we assume axial symmetry ($\partial/\partial\phi = 0$), take the background flow velocity and magnetic field to be radial ($u_\theta = B_\theta = 0$), and assume that terms of second order in the disturbance quantities ($\delta v = u_\phi$, $\delta B = B_\phi$) can be neglected (allowing the neglect of $\partial\rho/\partial t$), we find, following Heinemann and Olbert (1980), that in the limit $\sqrt{u} \ll \sqrt{v_A}$

$$\left(\frac{\partial}{\partial t} - v_A \frac{\partial}{\partial r}\right) f = -\tfrac{1}{2}g \frac{dv_A}{dr}, \tag{27}$$

$$\left(\frac{\partial}{\partial t} + v_A \frac{\partial}{\partial r}\right) g = \tfrac{1}{2}f \frac{dv_A}{dr}, \tag{28}$$

where $\delta v = \delta v_0(\rho_0/\rho)^{1/4}(g-f)/(g_0-f_0)$, $\delta B = \delta B_0(\rho/\rho_0)^{1/4}(g+f)/(g_0+f_0)$, and $v_A^2 = B_r^2/4\pi\rho$.

If $f, g \propto \exp(-i\omega t)$, then (27) and (28) have simple solutions in the two limits $\omega \ll v_A'$ and $\omega \gg v_A'$ (where $v_A' = dv_A/dr$). In the low-frequency limit ($\omega \ll v_A'$) the first term on the left side in both (27) and (28) is negligible, and we have the two solutions

$$f_\pm = \pm g_\pm = f_0\left(\frac{v_A}{v_{A0}}\right)^{\pm 1/2} e^{-i\omega t}, \tag{29}$$

$$\delta v_- = \delta v_0\left(\frac{r}{r_0}\right) e^{-i\omega t}, \tag{30}$$

$$\delta B_- = \delta B_0 \left(\frac{r_0}{r}\right)\left(\frac{v_{A0} v'_{A0}}{v_A v'_A}\right) e^{-i\omega t} =$$

$$= -i\delta v_- \left(\frac{B_r}{2 v_A}\right)\left(\frac{\omega}{v'_A}\right) = 0\left(\frac{\omega}{v'_A}\right), \tag{31}$$

$$\delta v_+ = \delta v_0 \left(\frac{r}{r_0}\right) \frac{\ln (\omega/v'_A)}{\ln (\omega/v'_{A0})} e^{-i\omega t} =$$

$$= i\delta B_+ \left(\frac{v_A}{B_r}\right)\left(\frac{\omega}{v'_A}\right) \ln\left(\frac{\omega}{v'_A}\right) = 0\left(\frac{\omega}{v'_A}\right) \tag{32}$$

$$\delta B_+ = \delta B_0 \left(\frac{r_0}{r}\right)\left(\frac{v'_A v_{A0}}{v_A v'_{A0}}\right) e^{-i\omega t}. \tag{33}$$

These solutions have a straightforward physical interpretation: if a toroidal oscillation is imposed on the atmosphere at $r = r_1$, the first $(-)$ solution describes the rigid-body oscillation of the lower-density (higher v_A) region of the atmosphere (in $r > r_1$), and the second $(+)$ solution describes the oscillation of the higher-density (lower v_A) region $(r < r_1)$. In the outer region, only a negligible twist of the magnetic field is required to maintain rigid-body behavior (i.e., $\delta B/B_r \ll \delta v/v_A$), but in the inner region, the oscillation can only be maintained by a relatively large twist (i.e., $\delta B/B_r \gg \delta v/v_A$), because of the substantial inertia of that region. If the Alfvén speed continues to decrease with decreasing r in the inner region, eventually $\omega/v'_A > 1$, and rigid-body oscillation will give way to the propagation of a torsional Alfvén wave, as is described below. In the high-frequency limit ($\omega \gg v'_A$, the WKB limit), the coupling terms on the right sides of (27) and (28) are negligible, and these equations have solutions describing inward (f) and outward (g) propagating waves:

$$f = f_0 e^{-i(\omega t + \int k \, dr)} \tag{34}$$

$$g = g_0 e^{-i(\omega t - \int k \, dr)} \tag{35}$$

$$\delta v = \left(\frac{\rho_0}{\rho}\right)^{1/4}\left[a_1 e^{-i(\omega t + \int k \, dr)} - a_2 e^{-i(\omega t - \int k \, dr)}\right] \tag{36}$$

$$\delta B = \frac{B_{r0}}{v_{A0}}\left(\frac{\rho}{\rho_0}\right)^{1/4}\left[a_1 e^{-i(\omega t + \int k \, dr)} - a_2 e^{-i(\omega t - \int k \, dr)}\right] \tag{37}$$

where a_1 and a_2 are constants. In general, of course, all the terms in (27) and (28) must be included. For the special case of an atmosphere in which the Alfvén speed varies exponentially with height (i.e., $v_A = v_{A0} \exp[(r - r_0)/h_A]$, where h_A is a constant), (27) and (28) have the general solution (Ferraro and Plumpton, 1958)

$$\delta v = i \left(\frac{r}{r_0}\right) \left[a_3 J_0 \left(\frac{\omega}{v_A'}\right) + a_< Y_0 \left(\frac{\omega}{v_A'}\right) \right] e^{-i\omega t}, \tag{38}$$

$$\delta B = -\left(\frac{r}{r_0}\right) \frac{B_r}{v_A} \left[a_3 J_1 \left(\frac{\omega}{v_A'}\right) + a_4 Y_1 \left(\frac{\omega}{v_A'}\right) \right] e^{-i\omega t}, \tag{39}$$

where J and Y are Bessel functions of the first and second kinds, a_3 and a_4 are constants, and $\omega/v_A' = \omega h_A/v_A$.

Using an exponentially varying Alfvén speed, one can construct a very simple model of the solar atmosphere by taking $h_A = 200$ km in $r \le r_0$, and $h_A = \infty$ in $r > r_0$, where r_0 represents the coronal base (i.e., the top of the transition region). In such a model, we have an analytic description of the behavior of toroidal Alfvénic disturbances: δv and δB are described by (38) and (39) in the lower solar atmosphere ($r \le r_0$) and by (36) and (37) in the corona ($r > r_0$). We can thus gain a clear understanding of all relevant physical effects, while closely reproducing the results of Hollweg (1978b), who solved numerically for δv and δB, using essentially (38) and (39), in a solar atmosphere described by 16 exponential layers. If we require that there be no inward propagating wave in $r > r_0$ (i.e., $a_1 = 0$ and $a_2 = -\delta v_0$), and that the Poynting flux and the velocity amplitude be continuous across r_0 (i.e., δv and δB are continuous across r_0), then (36)–(39) yield the real parts of δv and δB in $r < r_0$:

$$\mathrm{Re}\,\delta v = \frac{\delta v_0}{J_0 Y_1 - J_1 Y_0} \left(\frac{r}{r_0}\right) \left\{ \left[Y_1 J_0 \left(\frac{\omega}{v_A'}\right) - J_1 Y_0 \left(\frac{\omega}{v_A'}\right) \right] \cos(-\omega t + kr_0) - \right.$$

$$\left. + \left[Y_0 J_0 \left(\frac{\omega}{v_A'}\right) - J_0 Y_0 \left(\frac{\omega}{v_A'}\right) \right] \sin(-\omega t + kr_0) \right\}, \tag{40}$$

$$\mathrm{Re}\,\delta B = \left(\frac{B_r}{v_A}\right) \frac{\delta v_0}{J_0 Y_1 - J_1 Y_0} \left(\frac{r}{r_0}\right) \left\{ \left[Y_0 J_1 \left(\frac{\omega}{v_A'}\right) - J_0 Y_1 \left(\frac{\omega}{v_A'}\right) \right] \times \right.$$

$$\times \cos(-\omega t + kr_0) - \left[Y_1 J_1 \left(\frac{\omega}{v_A'}\right) - J_1 Y_1 \left(\frac{\omega}{v_A'}\right) \right] \times$$

$$\left. \times \sin(-\omega t + kr_0) \right\}, \tag{41}$$

where Bessel functions without arguments explicitly stated are evaluated at $\omega/v_{A0}' = \omega h_A/v_{A0}$, and we have set $\int k\,dr = kr_0$ at r_0. As is clearly illustrated in Hollweg's (1978b) model (see his Table 1, which is based on Gingerich et al. (1971) and Vernazza et al. (1973)), the 8 order of magnitude density change from the photosphere to the corona leads to a nearly 4 order of magnitude change in the Alfvén speed; hence, there is a broad range of frequencies over which waves can propagate in the photosphere, but cannot propagate at higher levels and are thus reflected (i.e., $\omega/v'_A \gg 1$ at the photosphere, but $\omega/v'_A < 1$ higher up). For $n = 5 \times 10^{16}$ cm^{-3} and $B = 10$ G, $\omega/v'_A \gg 1$ if $\omega \gg 5 \times 10^{-4}$ s^{-1}, so waves with periods of several minutes or less propagate in the photosphere according to the short-wave-length (WKB) approximation. We can describe these waves by using the complex equivalents of (40) and (41) and taking the asymptotic form of the Bessel functions for large arguments ($\omega/v'_A \gg 1$):

$$\delta v = \frac{\delta v_0}{J_0 Y_1 - J_1 Y_0} \left(\frac{r}{r_0}\right) \left(\frac{v'_A}{2\pi\omega}\right)^{1/2} e^{-i(\omega t - kr_0)} \times$$

$$\times \left\{ [(Y_1 + J_0) - i(Y_0 - J_1)] e^{i(\omega/v'_A - \pi/4)} + \right.$$

$$\left. + [(Y_1 - J_0) - i(Y_0 + J_1)] e^{i(\pi/4 - \omega/v'_A)} \right\} \tag{42}$$

$$\delta B = \left(\frac{B_r}{v_A}\right) \frac{\delta v_0}{J_0 Y_1 - J_1 Y_0} \left(\frac{r}{r_0}\right) \left(\frac{v'_A}{2\pi\omega}\right)^{1/2} e^{-i(\omega t - kr_0)} \times$$

$$\times \left\{ [(Y_1 + J_0) - i(Y_0 - J_1)] e^{i(\omega/v'_A - \pi/4)} - \right.$$

$$\left. - [(Y_1 - J_0) - i(Y_0 + J_1)] e^{i(\pi/4 - \omega/v'_A)} \right\} . \tag{43}$$

(42) and (43) clearly show that at these relatively high frequencies the Alfvénic oscillations in the photosphere are composed of outward and inward propagating waves, the latter having arisen from reflection at higher levels (but below the coronal base). The reflection coefficient (i.e., the ratio of the energy flux density in the downward propagating waves to that in the upward propagating waves) characterizing the atmosphere overlying the photosphere is thus

$$R = \frac{(Y_1 + J_0)^2 + (Y_0 - J_1)^2}{(Y_1 - J_0)^2 + (Y_0 + J_1)^2} . \tag{44}$$

For waves that propagate in the short-wave-length limit from the photosphere to the top of the transition region, there is virtually no reflection:

$$R \approx 0 \qquad (\omega \gg v'_{A0}) . \tag{45}$$

For a wave that must 'tunnel through' to the corona because its wavelength becomes greater than the local scale-hieght well before the corona is reached, most of the energy is reflected:

$$R \approx 1 - 2\pi\omega h_A/v_{A0} \qquad (\omega \ll v'_{A0}) \tag{46}$$

and a standing wave is produced in the lower solar atmosphere, as pointed out by Hollweg (1978b). Hollweg (1972, 1978b, 1982) claims that the standing wave pattern is caused by reflections at discontinuities in the density scale height. In fact, coupling between inward and outward propagating waves is an integrated effect, and a finite jump in the refractive index is required to produce an inward propagating wave of finite amplitude (cf. Alfvén and Fälthammar, 1963; Heinemann and Olbert, 1980). We must agree, therefore, with the interpretation of Ferraro and Plumpton (1958): viz., that the Alfvén waves are reflected due to the continuous variation of the refractive index. The features of such a standing wave pattern are readily determined using (40) and (41), and one example, for $\omega/v'_{A0} = 10^{-3}$, is shown in Figure 6.

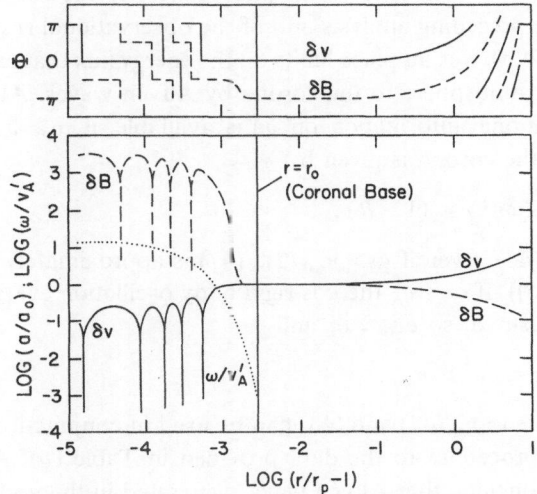

Fig. 6. Amplitudes (a_v and a_B) and phases (Φ_v and Φ_B) of δv and δB, and the dimensionless ratio ω/v'_A plotted as functions of height above the photosphere ($r/r_p - 1$). The amplitudes and phases are defined by $\delta v = a_v \cos(-\omega t + \Phi_v)$ and $\delta B = a_B \cos(-\omega t + \Phi_B)$, where $a_{v0} = \delta v_0$ and $a_{B0} = \delta B_0$. Below the coronal base ($r \le r_0$) the Alfvén speed scale-height, h_A, is 200 km, and above the coronal base ($r > r_0$)$h_A = \infty$. For $\log(r/r_p - 1) > 1$, the a_v and a_B curves approach straight lines. Regions of standing waves ($\log(r/r_p - 1) \lesssim -2.8$), rigid-body oscillation ($-2.75 \lesssim \log(r/r_p - 1) < -2.55$), and outward propagating waves ($\log(r/r_p - 1) > -2.55$, corresponding to $r > r_0$) can be seen.

In Figure 6 we see that the toroidal Alfvénic disturbances propagate outward as simple waves above the coronal base ($r > r_0$), give rise to rigid-body oscillation of the atmosphere just below the coronal base, and produce a standing wave pattern in the lower atmosphere. The minima in the standing wave pattern are imperfect nodes (i.e., δv, $\delta B \ne 0$ at the nodes). The reflection caused by the Alfvén-speed gradient is very efficient at this frequency $\omega = 10^{-3} v'_{A0}$: reference to (46) indicates that only 0.6% of the energy carried by upward propagating waves in the photosphere reaches the corona; the rest is reflected. It has been noted (Hollweg, 1978b) that if the generation of the Alfvén waves takes place at some particular location near the top of the convection zone, then at certain 'resonant' frequencies a node will be located at the generation point, and a modest velocity perturbation (δv) will correspond to a very large magnetic perturbation (δB) and thus to a very large energy flux at that frequency. Of course, to produce such a large energy flux by driving the wave at a node there are three necessary requirements:

(1) the wave driver must be phase-coherent over many wave periods; (2) the wave generation region must be narrow in height and remain at a fixed height over many wave periods; (3) the frequency for which the preceding two conditions are met must correspond to the presence of a node in the wave generation region. It seems unlikely that these three requirements, or their equivalent, will be met in the real solar atmosphere, and we conclude that the large 'resonant' energy fluxes discussed by Hollweg (1978b) are an artefact of the model considered and should not be invoked in discussing the transport of energy from the photosphere to the corona.

On the basis of the preceding analysis and of the observational results presented by Athay and White (1979b), we can place limits on the energy flux that can be transported from the lower solar atmosphere to the corona by Alfvén waves. At the atmospheric levels where observational information on δv is available, if $\omega \gg v'_A$ the energy flux density passing into the corona is given by

$$\phi_A = nm \langle \delta v^2 \rangle v_A (1 - R),$$
(47)

where R is given by (46). Even if $\omega > v'_{A0}/2\pi$, (47) is approximately correct provided we take $R = 0$ (cf. (45)). If $\omega \ll v'_A$ there is rigid body oscillation everywhere above the level where δv is measured, so $\delta v_0 \approx \delta v$ and

$$\phi_A \approx n_0 m < \delta v^2 > v_{A0}.$$
(48)

When ω is comparable to v'_A (47) and (48) can be used in conjunction to estimate ϕ_A. Applying the above procedure to the data provided in Table 1 of Athay and White (1979b) leads us to conclude that Alfvén waves generated in the upper chromosphere or below, with wave periods between about 30 and 3000 s, can provide an energy flux density of no more than 1×10^5 erg cm^{-2} s^{-1} to the corona. For wave periods of longer than one hour, a large-scale Alfvén wave might give rise to a Doppler shift of spectral lines rather than simply to line broadening, which is the effect on which the preceding interpretation is based. For such long periods, however, the atmosphere will be in rigid body motion everywhere from the middle chromosphere to the corona, so that an energy flux entering the corona of more than 10^5 erg cm^{-2} s^{-1} at these periods would correspond to chromospheric velocities of more than 10 km s^{-1}. As such large chromospheric Doppler shifts are not observed, the upper limit given above would seem to apply to these longer periods as well. Finally, one might expect that very short period waves ($\tau \lesssim$ a few seconds) could carry larger energy fluxes to the corona, given the observational constraints on the velocity amplitude in the lower solar atmosphere, but these waves tend to be damped strongly by viscous and frictional effects in the photosphere and the chromosphere (Osterbrock, 1961). Of course, energy flux densities of the order 10^5 erg cm^{-2} s^{-1} in the corona generally imply non-linear waves in the photosphere and lower chromosphere, and such waves should be strongly damped in these lower regions (Parker, 1960; Osterbrock, 1961). We conclude, therefore that an energy flux density of more than 10^5 erg cm^{-2} s^{-1} carried into the region of supersonic solar wind flow by Alfvén waves cannot have been transported upward from the photosphere or chromosphere by Alfvén waves alone and thus must have arisen, at least in part, from the generation of Alfvén waves in the transition region or the lower corona.

In the simple model described above, with $h_A = \infty$ above the coronal base, Alfvén waves of all frequencies propagate as short-wave-length (WKB) waves in the corona and the solar wind. In reality, of course, v_A does vary above the coronal base, and waves with periods longer than a few hours do not propagate so simply in this region. The description of non-WKB effects becomes a bit more complicated in this case, however, because one can no longer assume that $u \ll v_A$, and (27) and (28) must be replaced (Heinemann and Olbert, 1980) by

$$\left[\frac{\partial}{\partial t} + (u - v_A)\frac{\partial}{\partial s}\right] f = \tfrac{1}{2}g(u - v_A)\frac{d\ln v_A}{dr}, \tag{49}$$

$$\left[\frac{\partial}{\partial t} + (u + v_A)\frac{\partial}{\partial s}\right] g = \tfrac{1}{2}f(u + v_A)\frac{d\ln v_A}{dr}, \tag{50}$$

where ds is a length element along the magnetic field ($\mathbf{u} \times \mathbf{B} = 0$), which is no longer assumed radial, and f and g are related to δv and δB through

$$\delta v = \tfrac{1}{2}\eta^{1/4}\left(\frac{f}{1 - \eta^{1/2}} + \frac{g}{1 + \eta^{1/2}}\right), \tag{51}$$

$$\delta B = \frac{B}{v_A}\,\tfrac{1}{2}\eta^{1/4}\left(\frac{f}{1 - \eta^{1/2}} - \frac{g}{1 + \eta^{1/2}}\right), \tag{52}$$

where $\eta = \rho/\rho_a$, and the subscript a refers to the Alfvén point, where $u = v_A$. Well inside the Alfvén point, $u \ll v_A$, and the discussion of simple wave propagation and rigid body oscillation in connection with (27) and (28) applies quite well, but when $u \gtrsim v_A$ the problem is more complicated, and numerical solutions to (49) and (50) are generally required (cf. Figure 7). There is, however, a simple case of some interest in which analytic solutions to these equations exist: i.e., the case in which $\omega \to 0$ ($\partial/\partial t$ terms negligible) for a toroidal Alfvénic disturbance that is symmetric about the solar rotation axis. This is, of course, the case considered by Weber and Davis (1967), in which the Alfvénic disturbance of zero frequency corresponds to solar rotation; the solutions to (49) and (50) are (Heinemann and Olbert, 1980)

$$f \propto \left[\left(\frac{v_A}{v_{Aa}}\right)^{1/2} - \left(\frac{v_{Aa}}{v_A}\right)^{1/2}\right] e^{-i\omega t}, \tag{53}$$

$$g \propto \left[\left(\frac{v_A}{v_{Aa}}\right)^{1/2} + \left(\frac{v_{Aa}}{v_A}\right)^{1/2}\right] e^{-i\omega t}, \tag{54}$$

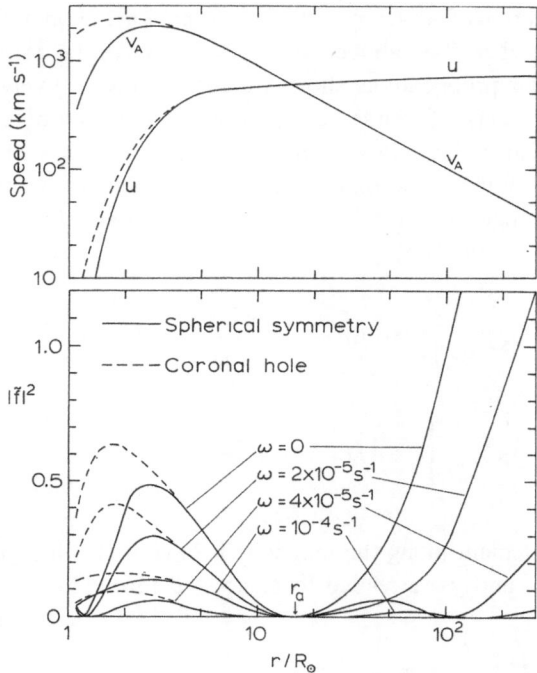

Fig. 7.   The background flow speed and Alfvén speed as functions of radial distance for a spherically symmetric flow geometry (solid lines) and for the average rapidly diverging flow geometry implied by the coronal hole boundary of Munro and Jackson (1977) (dashed lines). The inward-propagating wave amplitudes, $|\tilde{f}|^2$, for different wave frequencies ω, are shown as functions of radial distance for the spherical and rapidly diverging flow geometries (adapted from Heinemann and Olbert (1980)).

and these give velocities and magnetic fields agreeing with those of Weber and Davis (1967). In obtaining (53) and (54), the boundary condition $f_a = 0$ is applied. That this is always a sensible condition is readily seen by considering the facts that f and g correspond to inward and outward propagating waves (though f and g are not proper eigenmodes of the system) and that in $r > r_a$ the inward wave (f) is convected outward by the super-Alfvénic flow.

If we define $\tilde{f} = f e^{i\omega t}$ and $\tilde{g} = g e^{i\omega t}$ and normalize such that $|\tilde{g}|^2 - \tilde{f}^2 = 1$, then the condition $|\tilde{f}_a|^2 = 0$ (and thus $|\tilde{g}_a|^2 = 1$) leads to the results (Heinemann and Olbert, 1980) shown in Figure 7 for toroidal Alfvénic disturbances in a spherically symmetric and a rapidly expanding coronal expansion. When the coupling between outward and inward propagating waves is weak, $|\tilde{f}|^2 \ll |\tilde{g}|^2$, and the disturbance propagates outward as a simple (WKB) wave. Inside the Alfvén radius (i.e., in $r < r_a$), the coupling is very weak for wave periods of a few hours or less, but is significant for periods of a day or longer. Coronal hole geometries, which lead to a decrease in the scale length over which v_A varies, cause only slightly stronger coupling (near the coronal base) than is found for spherical geometry. Outside the Alfvén radius (i.e., in $r > r_a$), the coupling again is very weak for periods of a few hours or less, but strong for periods of more than a day. In all regions the maximum coupling is attained as $\omega \rightarrow 0$ and

$|\tilde{f}|^2 \rightarrow \frac{1}{4}(v_A/v_{Aa} + v_{Aa}/v_A) - \frac{1}{2}$. (For a more detailed discussion of these points, see Heinemann and Olbert (1980).)

The force exerted by the Alfvénic oscillations on the solar wind plasma is composed of a centrifugal force and a Lorentz force (the centrifugal force being important only when the coupling is strong). This force varies on a time scale $\tau/4$, where $\tau = 2\pi/\omega$ is the wave period, and this time scale can be compared with the solar wind expansion time, $\tau_{exp} = |(u/n)(dn/dr)|^{-1}$. If $\tau \ll \tau_{exp}$ the solar wind responds to the average effect of the force over a wave period, but if $\tau \gtrsim \tau_{exp}$ the variation of the force over a wave period must be considered. In the following section we shall assume that $\tau \ll \tau_{exp}$, although for waves with $\tau \gtrsim 1$ hour this assumption will break down, most severely in the vicinity of the sonic point, where τ_{exp} is smallest.

5. Interaction of Alfvén Waves with the Solar Wind

A simple illustration of energy addition to the solar wind is provided by the interaction of Alfvén waves with the solar wind. The simplicity arises, in part, from the fact that Alfvén waves transport energy along the magnetic field, which allows us to continue to describe solar wind flow along an isolated infinitesimal flow tube. Because linear damping of these waves is very weak above the coronal base, their energy is not dissipated as heat in the plasma until the wave amplitude becomes very large (i.e., $\delta B \approx B$), and for reasonable wave energy fluxes this does not occur in the region of subsonic flow. Consequently, Alfvén waves, through the force associated with the gradient in their energy density, supply only momentum (not heat) to the subsonic solar wind. In the supersonic region, both heat and momentum are added, as non-linear dissipation becomes important. This energy addition by Alfvén waves, then, affects both the solar wind mass flux and energy flux, and determination of its effect on the solar wind flow speed at 1 AU requires careful consideration.

The subject of Alfvén waves in the solar wind has been considered by several workers (e.g., Alazraki and Couturier, 1971; Belcher, 1971; Hollweg, 1973, 1978a; Jacques, 1977; see also references given by Hollweg, 1978a). These authors have all made use of the short-wave-length (WKB) approximation, as shall we in the following discussion (cf. Section 4 and Heinemann and Olbert, 1980). In general, little attention has been given to the coronal base boundary conditions: frequently, the base density has been allowed to take on unrealistically low values so as to produce a reasonable solar wind mass flux. We shall reproduce here the essential results of the several studies mentioned above, but we shall fix the coronal base pressure at a realistic value and investigate the effects of Alfvén waves on the mass flux, energy flux, and flow speed at 1 AU, in a manner similar to that employed in discussing the general subject of energy addition in Section 3. Our approach will involve simplifying the model of Hollweg (1978a), so that it retains the basic physics but admits analytic solutions, which should provide us with greater physical insight into the problem (cf. Leer *et al.*, 1980).

Let us begin by reconsidering the conservation laws used to describe energy addition in the solar wind. The mass conservation Equation (1) remains unchanged, and the momentum Equation (13) is specialized by associating the force per unit mass, D, with the Alfvén wave pressure gradient:

$$D = -\frac{1}{\rho} \frac{d}{dr} \left(\frac{\langle \delta B^2 \rangle}{8\pi} \right). \tag{55}$$

We replace the energy Equation (15) by the assumption that the temperature is constant in the subsonic region ($T = T_0$ in $r_0 \leq r \leq r_c$) and that the solar wind energy flux at the critical point is given by

$$F_c = \mathscr{F} \left[\tfrac{1}{2}u_c^2 + (5 + \tfrac{3}{2}\alpha) \frac{kT_0}{m} - \frac{GM}{r_c} \right]. \tag{56}$$

This treatment of energy balance produces results very similar to those that are obtained using Hollweg's (1976) description of thermal conduction (cf. Sections 2 and 3 and Hollweg, 1978a), but allows us to derive an analytic description and thus to illuminate, in a simple way, the important physical effects of Alfvén waves on the solar wind.

The force on the solar wind due to Alfvén waves, given in (55), can be expressed in terms of coronal base parameters and the solar wind density by employing the principle of conservation of wave action (e.g., Dewar, 1970; Bretherton, 1970; Jacques, 1977), which allows us to write the Alfvén-wave energy density, ε, and energy flux, F_W, as follows:

$$\varepsilon = \frac{\langle \delta B^2 \rangle}{4\pi} = \varepsilon_0 \frac{M_{A0}}{M_A} \left(\frac{1 + M_{A0}}{1 + M_A} \right)^2 \tag{57}$$

$$F_W = F_{W0} \frac{1 + \tfrac{3}{2}M_A}{1 + \tfrac{3}{2}M_{A0}} \left(\frac{1 + M_{A0}}{1 + M_A} \right)^2 \tag{58}$$

where $M_A = u/v_A$ and $F_W = \varepsilon(v_A + \tfrac{3}{2}u)A$. Now we can write the equation of motion describing the solar wind expansion in the presence of Alfvén waves in a standard form:

$$\frac{1}{u} \frac{du}{dr} \left(u^2 - \frac{2kT^*}{m} \right) = \frac{2kT^*}{m} \frac{1}{A} \frac{dA}{dr} - \frac{GM}{r^2}, \tag{59}$$

where we have defined the effective temperature, T^*, by

$$\frac{2kT^*}{m} = \frac{2kT_0}{m} + \frac{1}{4} \left(\frac{1 + 3M_A}{1 + M_A} \right) \langle \delta v^2 \rangle, \tag{60}$$

and in the WKB limit $\langle \delta v^2 \rangle = \langle \delta B^2 \rangle / 4\pi\rho$. (59) has the same form as the equation of motion describing a solar wind expansion driven only by the thermal pressure, with

$p = 2nkT^*$; indeed, $(2kT^*/m)^{1/2}$ measures the speed at which a compressive wave propagates parallel to the magnetic field in the subsonic solar wind plasma in the presence of relatively short-wave-length Alfvén waves. Thus the critical point at $r = r_c$, where

$$r_c = \frac{GMm}{4kT_c^* \beta_c},$$ (61)

$$u_c = (2kT_c^*/m)^{1/2},$$ (62)

is appropriately considered the point separating subsonic from supersonic flow. To understand why the presence of Alfvén waves modifies the speed with which a compressive wave propagates, we must remember that the Alfvén-wave energy density depends on the plasma density (cf. (57));

$$\langle \delta B^2 \rangle \propto \rho^{1/2} \zeta(\rho),$$ (63)

where $\zeta(\rho) = [(1 + M_{A0})/(1 + M_A)]^2 \approx 1$ for $M_A \ll 1$. When the plasma is compressed, therefore, both the internal energy density of the plasma ($\propto \rho T$) and the Alfvén-wave energy density ($\propto \langle \delta B^2 \rangle$) are increased, and the restoring force in a compressive wave, which depends on the gradient of the total perturbed energy density, is larger than it would be in the absence of Alfvén waves.

We are now in a position to calculate the effect of Alfvén waves on the solar wind mass flux, energy flux, and flow speed at 1 AU. Making use of definitions of f and β in Section 2, we can integrate (59) from the coronal base ($r = r_0$) to the critical point ($r = r_c$) to obtain the solar wind proton flux density at 1 AU:

$$n_E u_E = n_0 (v_T^2 + \tfrac{1}{4} \langle \delta v_c^2 \rangle)^{1/2} \times$$
$$\times \left(\frac{f_c r_c^2}{f_E r_E^2} \right) \exp \left[-\frac{v_{e0}^2}{2v_T^2} + (2\beta_c + \tfrac{3}{2}) \frac{\delta v_c^2}{4v_T^2} + (2\beta_c - \tfrac{1}{2}) \right],$$ (64)

where $v_e^2 = 2GM/r, v_T^2 = (2kT/m)^{1/2}$, and we have assumed $u_0^2 \ll 2kT_0/m, (n_c/n_0)^{1/2} \ll 1$, and $M_{A0} < M_{AC} \ll 1$. When $\delta v_c \to 0$, (64) reduces to (8), as it should. (64) indicates that the presence of Alfvén waves in the solar wind tends to increase the mass flux, as was anticipated in the discussion of momentum addition in Section 3. The mass flux ($\propto n_E u_E$) can be expressed in terms of coronal base parameters (except for f_c and β_c) if we can relate δv_c to δv_0. From (57) we find $\langle \delta v_0^2 \rangle = \langle \delta v_c^2 \rangle (n_c/n_0)^{1/2}$ and use of (64) enables us to obtain an implicit equation expressing the desired relation:

$$\langle \delta v_0^2 \rangle^2 = \langle \delta v_c^2 \rangle^2 \exp \left[-\frac{v_{e0}^2}{2v_T^2} + (2\beta_c + \tfrac{3}{2}) \frac{\langle \delta v_c^2 \rangle}{4v_T^2} + (2\beta_c - \tfrac{1}{2}) \right].$$ (65)

Since virtually all the solar wind energy flux is carried by the flow at 1 AU in a model like the one we are using, the flow speed at 1 AU is given by

$$\tfrac{1}{2}u_E^2 = (F_c + F_{Wc})/\mathscr{F}, \tag{66}$$

where $\mathscr{F} = n_E m u_E f_E r_E^2$ is given by (64), F_c by (56), and F_{Wc} by (58). (64)–(66) allow

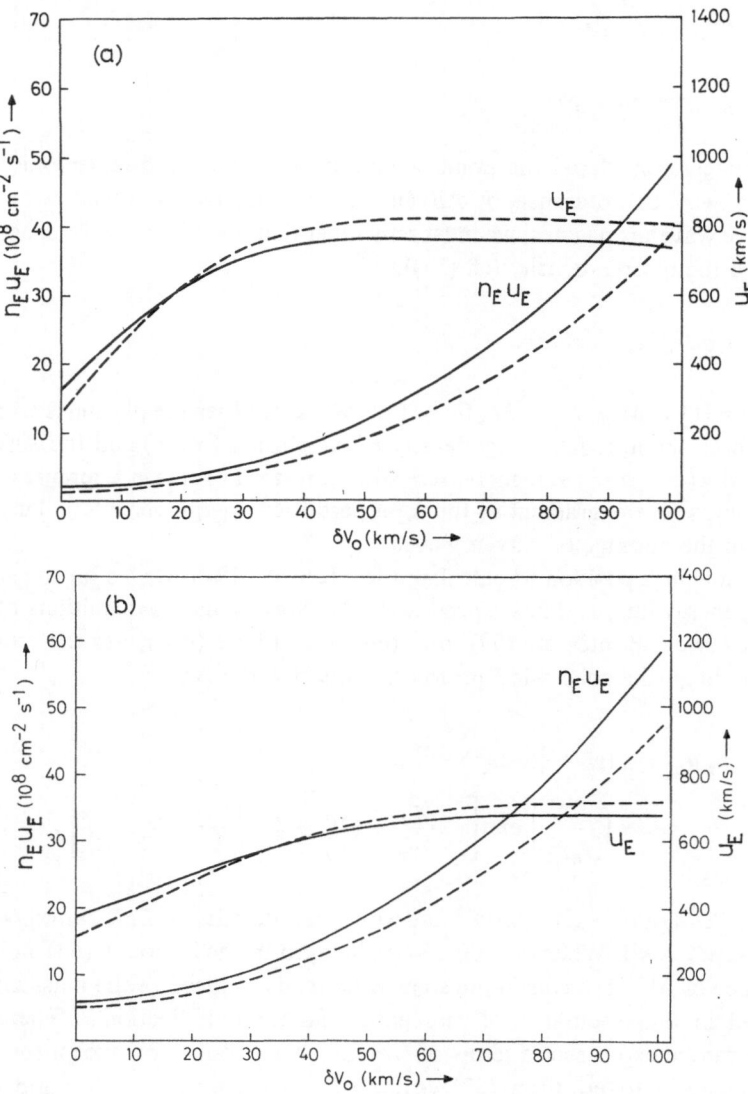

Fig. 8.   Proton flux density $(n_E u_E)$ and flow speed (u_E) at 1 AU as functions of the coronal-base Alfvén-wave amplitude $(\delta v_0 = \sqrt{\langle \delta v_0^2 \rangle})$. The coronal-base pressure is specified by $n_0 T_0 = 2 \times 10^{14}$ cm^{-3} K, the (radial) magnetic field at 1 AU is $B_E = 4 \times 10^{-5}$ G, and the coronal temperature is $T_0 = 1.1 \times 10^6$ K in (a) and $T_0 = 1.3 \times 10^6$ K in (b). Results for both spherically symmetric flow (solid curves) and rapidly expanding flow (dashed curves), with $f_E = 7$, $f_c = 5$, and $\beta_c = 1.5$, are shown.

us to calculate u_E and $n_E u_E$ in terms of coronal base parameters, and some results of such calculations are shown in Figures 8, 9, and 10.

Figure 8 illustrates the effects of varying the Alfvén-wave velocity amplitude at the coronal base (δv_0) on the solar wind mass flux ($\propto n_E u_E$) and flow speed (u_E). When the wave energy density is small in comparison with the thermal energy density (i.e., $\langle \delta v_0^2 \rangle \ll v_T^2$) the mass flux is determined primarily by the temperature, and varying δv_0 affects $n_E u_E$ very little. For a sufficiently large magnetic field, however, the energy flux carried by the waves ($\propto B_0 \langle \delta v_0^2 \rangle \rho_C^{1/2}$) can be large enough that increasing δv_0 can substantially increase u_E when $n_E u_E$ is essentially unchanged (cf. (66)). This behavior is illustrated in Figure 8 for $\delta v_0 \lesssim 30$ km s^{-1}. For larger δv_0, the Alfvén waves begin to play a significant role in determining the mass flux, leading to such a rapid increase in $n_E u_E$ with increasing δv_0 that u_E peaks and actually begins to decrease slowly as δv_0 continues to increase (cf. (66)). For higher coronal temperature ($T_0 = 1.1 \times 10^8$ K in Figure 8a and $T_0 = 1.3 \times 10^8$ K in Figure 8b), the change in u_E produced by a given

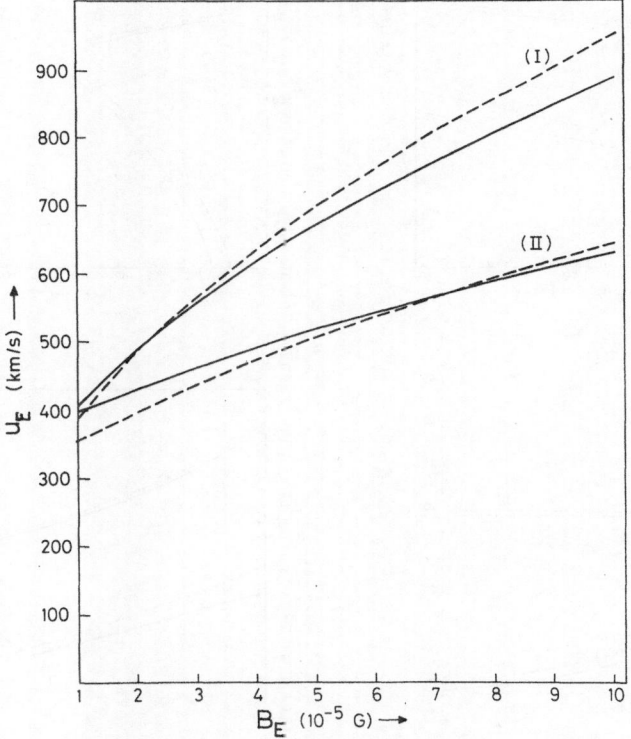

Fig. 9.   Flow speed at 1 AU (u_E) as a function of magnetic field strength at 1 AU (B_E), for $\delta v_0 = 20$ km s^{-1}, $T_0 = 1.1 \times 10^6$ K (I) and 1.3×10^6 K (II), and otherwise the same parameters as Figure 8. The particle flux densities at 1 AU are $n_E u_E = 3.5 \times 10^8$ cm^{-2} s^{-1} (I) and 8.3×10^8 cm^{-2} s^{-1} (II) for spherical symmetry (solid curves) and 2.9×10^8 cm^{-2} s^{-1} (I) and 7.0×10^8 cm^{-2} s^{-1} (II) for rapidly expanding flow (dashed curves).

Alfvén wave energy flux ($\propto B_0 \langle \delta v_0^2 \rangle \rho_0^{1/2}$) is not as large, because the mass flux is larger for the higher temperature and the energy per unit mass supplied by the Alfvén waves is correspondingly lower. Note that rapidly expanding flow geometries (dashed curves) do not lead to results significantly different from those for spherically symmetric flow.

The importance of the magnitude of the magnetic field is illustrated in Figure 9, where δv_0 (= 20 km s^{-1}) is held constant. Because the mass flux depends on $\langle \delta v_0^2 \rangle$, not B_0, whereas the energy flux depends on both, only u_E is plotted as a function of B_0 (note: $B_0 = B_E f_E r_E^2 / r_0^2$). The monotonic increase of the Alfvén-wave energy flux with increasing B_0 gives rise to the increase in u_E with increasing $B_E (\propto B_0)$, and the higher flow speeds associated with lower coronal temperature reflect the effect of the lower mass flux arising from lower T_0 (cf. (64), (66)). Figure 10 shows several of the requirements

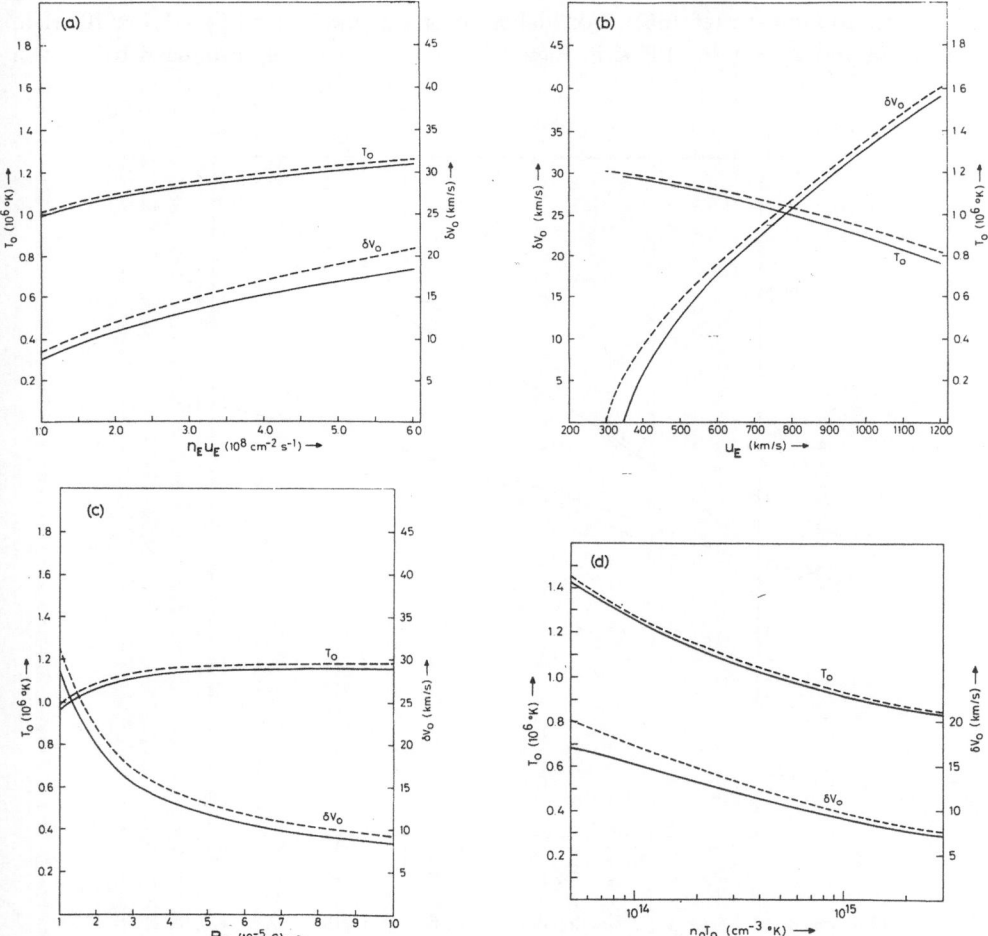

Fig. 10. Parameter variations for the basic model used in Figures 8 and 9, with the additional reference parameters of $B_E = 4 \times 10^{-5}$ G in (a), (b), and (d), $u_E = 500$ km s^{-1} in (a), (c), and (d), and $n_E u_E = 3 \times 10^8$ cm^{-2} s^{-1} in (b), (c), and (d).

on coronal parameters if Alfvén waves are to play a role in driving high-speed solar wind streams, given observational constraints on $n_E u_E$, u_E, and $n_0 T_0$: (b) larger wave amplitudes and lower coronal temperatures lead to higher-speed winds ($n_E u_E$ and $n_0 T_0$ fixed); (a) a narrow range of values of T_0 and δv_0 are possible if a high-speed stream is to have a reasonable mass flux; (d) even a broad range of observationally acceptable values of the coronal base pressure (say, $10^{14} \lesssim n_0 T_0 \lesssim 6 \times 10^{14} \text{ cm}^{-3} \text{ K}$) provide little latitude for variations of δv_0 and T_0; (c) the same is true for acceptable values of $B_0 (= B_E f_E r_E^2 / r_0^2)$.

From the preceding results, one can see that it is not unreasonable to suppose that high-speed streams can be produced in a solar wind with Alfvén waves present, if $\delta v_0 \approx 20-30 \text{ km s}^{-1}$. Yet the observational constraints at 1 AU of $n_E u_E \approx 3 \times 10^8 \text{ cm}^{-2} \text{ s}^{-1}$ and $2 \times 10^{-5} \lesssim B_E \lesssim 7 \times 10^{-5} \text{ G}$ (e.g., Feldman *et al.*, 1977) and at the coronal base of $10^{14} \lesssim n_0 T_0 \lesssim 6 \times 10^{14} \text{ cm}^{-3} \text{ s}^{-1}$ (e.g. Withbroe, 1977) place such tight restrictions on possible values of δv_0 and T_0 that one must be very careful in drawing any definitive conclusions. At present, we can only say that the high-speed stream models invoking Alfvén waves as a source of additional energy can be made consistent with the limited coronal observations available (see Withbroe (1982) for inferences regarding δv_0 and T_0).

6. Stellar Winds Driven by Alfvén Waves

Late-type giant and supergiant stars frequently exhibit cool, massive stellar winds (see reviews by Cassinelli (1979) and Cassinelli and MacGregor (1982) and references therein), and it has been suggested that Alfvén waves might play an important role in driving such winds (Belcher and Olbert, 1975; Hartmann and MacGregor, 1980; Holzer *et al.*, 1982). Because the models proposed to describe these stellar winds differ in important respects from that for the solar wind, discussed in Section 5, it will be instructive to consider briefly a representative stellar wind model. In such a model (e.g. Hartmann and MacGregor, 1980), the temperature of the atmosphere ($\approx 10^4$ K) is so low that the thermal pressure gradient force is relatively unimportant, but the stellar gravity is so weak ($M_* R_\odot^2 / M_\odot R_*^2 \approx 10^{-4}$) that a moderate Alfvén-wave energy flux density ($\approx 10^6 \text{ erg cm}^{-2} \text{ s}^{-1}$ at the stellar surface) can drive a quite massive wind. The winds, however, are so cool and dense that frictional damping of the Alfvén waves (cf. Section 4 and Osterbrock, 1961) and radiative energy loss from the plasma play important roles. We can understand most of the relevant physical effects in such models by taking a simple approach, similar to that of Section 5.

Let us consider a one-fluid, isothermal model: mass conservation is given by (1), momentum conservation by (13) and (55), and energy conservation by $T = T_0 = $ constant, in place of (14) or (15). The isothermal assumption is a reasonable first approximation, because energy balance in such an atmosphere consists basically of a balance between wave heating and radiative cooling, and the latter depends so strongly on temperature that the variation of temperature through the region of subsonic flow is generally quite modest, even if the wave-heating rate varies considerably. An important implication of this energy balance is that a substantial fraction of the Alfvén-wave energy flux can be lost to the stellar radiation field, whereas in the solar case virtually the entire wave flux is converted into the flow energy of the wind.

If the local damping length of the Alfvén waves is L, then (57) and (58) are modified by the same exponential factor: viz., (57) becomes

$$\varepsilon = \frac{\langle \delta B^2 \rangle}{4\pi} = \varepsilon_0 \frac{M_{A0}}{M_A} \left(\frac{1 + M_{A0}}{1 + M_A} \right)^2 \exp\left(-\int_{r_0}^{r} dr' L^{-1} \right), \tag{67}$$

where the subscript 0 refers to the reference level $r = r_0$. Combining (1), (3), (13), (55), and (67), the stellar wind equation of motion for spherical symmetry is obtained:

$$\frac{1}{u} \frac{du}{dr} \left[u^2 - v_T^2 - \frac{1}{4} \left(\frac{1 + 3M_A}{1 + M_A} \right) \langle \delta v^2 \rangle \right] =$$

$$= \frac{2}{r} \left[v_T^2 + \frac{1}{4} \left(\frac{1 + 3M_A}{1 + M_A} \right) \langle \delta v^2 \rangle + \frac{1}{4} \frac{r}{L} \langle \delta v^2 \rangle - \frac{1}{4} v_e^2 \right]. \tag{68}$$

(Note that (68) differs from (59) only in the damping term, $(r/4L) \langle \delta v^2 \rangle$.) Before discussing solutions of (68), let us consider the limit of weak damping ($L \to \infty$), low temperature ($v_T^2 \ll \langle \delta v^2 \rangle$), strong radial magnetic field ($M_{AC} \ll 1$), and moderate wave amplitude ($\delta v_0^2 \ll v_{e0}^2$). It is then readily shown that

$$r_c = 1.75 r_0, \tag{69}$$

$$\rho_c = \rho_0 (1.75 \langle \delta v_0^2 \rangle / v_{e0}^2)^2, \tag{70}$$

$$\rho_0 u_0 = \frac{1}{2} (1.75)^{7/2} \rho_0 v_{e0}^{-3} \langle \delta v_0^2 \rangle^2. \tag{71}$$

If we take r_0 to be the stellar radius, B_* to be the stellar surface magnetic field in Gauss, and f_{w*} to be the Alfvén wave energy flux density in units of 10^6 erg cm^{-2} s^{-1} at $r = r_0$, and if we define $R_* = r_0/R_\odot$ and $M_* = M/M_\odot$, the mass loss rate driven by Alfvén waves, in units of solar masses per year, is

$$-\dot{M} = 1.8 \times 10^{-13} \left(\frac{f_{w*} R_*}{B_*} \right)^2 \left(\frac{R_*}{M_*} \right)^{3/2}. \tag{72}$$

If all the Alfvén wave energy flux is eventually converted into flow energy, the asymptotic flow speed, u_∞, can be calculated from the energy conservation requirement that $u^2 - v_e^2 + 2 \langle \delta v^2 \rangle / M_A = $ const:

$$u_\infty^2 = v_{e0}^2 \left[2.8 \frac{B_*^2}{f_{w*}} \left(\frac{M_*}{R_*} \right)^{1/2} - 1 \right] \tag{73a}$$

$$= v_{e0}^2 \left(\frac{8}{7 M_{AC}} - \frac{3}{7} \right). \tag{73b}$$

For $R_* \gg M_* \gg 1$, as is the case for the cool giants and supergiants under consideration, (72) indicates that undamped Alfvén waves can drive a quite massive wind, when their energy flux density is of the order 10^6 erg cm^{-2} s^{-1} and the stellar magnetic field is a few Gauss or less. If the requirement $M_{AC} \ll 1$ (which is necessary for (69)–(73) to be valid) is met, then (73b) indicates that the asymptotic flow speed is much larger than the gravitational escape speed at the stellar surface. As we shall see, this large asymptotic flow speed seems to raise difficulties for the models of Alfvén-wave-driven stellar winds.

The problem we encountered in the solar wind (cf. Section 2, 3, and 5) was to explain the relatively low mass flux and high asymptotic flow speed. For winds from cool giants and supergiants the problem is just the reverse: it is necessary to explain very large mass fluxes and very low asymptotic flow speeds (i.e., $u_\infty^2 \ll v_{e0}^2$). (73b) illustrates the difficulty one encounters in trying to drive such winds with undamped Alfvén waves: viz., too high an asymptotic flow speed is produced. As we shall see later, this difficulty continues to plague us even when the effects of wave damping are included. Before discussing damping, however, let us consider the transition from thermally dominated winds (cf. Sections 2 and 5) to wave dominated winds (cf. (69)–(73)), in the context of undamped Alfvén waves.

For this purpose, we can use (56) and (64)–(66), for which it was assumed that $M_{A0} < M_{AC} \ll 1$, $u_0^2 \ll 2kT_0/m$, and $(a_c/n_0)^{1/2} \ll 1$. We can readily eliminate the latter two assumptions and include the generally small terms that become important as $r_c \to r_0$. Solutions to these equations are shown in Figure 11. In a thermally dominated wind, the critical point is seen to move inward with increasing temperature, eventually reaching the coronal base (cf. Figure 11a in the region $v_T^2/v_{e0}^2 > 5 \times 10^{-2}$ and (61) in the limit $\langle \delta v^2 \rangle \to 0$). As the temperature decreases, however, a maximum critical point radius is reached (near $v_T^2/v_{e0}^2 = 5 \times 10^{-2}$ for $10^{-4} \lesssim \langle \delta v_0^2 \rangle / v_{e0}^2 \lesssim 10^{-2}$), and as the temperature continues to decrease the critical point moves inward, approaching $r_c = 1.75 r_0$, as long as $\delta v_0^2/v_{e0}^2 \lesssim 10^{-2}$ (cf. (69)). This behavior of the critical point is an indication of the transition to the dominance of wave effects over the thermal effects, and this transition to wave-dominance occurs when $\langle \delta v_0^2 \rangle \ll v_T^2$, because $\langle \delta v^2 \rangle (\propto \rho^{-1/2})$ increases rapidly with radial distance as the density scale-height decreases (cf. (65)). In the limit $v_T^2/v_{e0}^2 \to 0$, the Alfvén waves completely control the radial atmospheric structure, and the density scale-height is determined by $\langle \delta v_0^2 \rangle / v_{e0}^2$ (cf. (70)). As the velocity amplitude of the waves becomes large ($0.1 \lesssim \langle \delta v_0^2 \rangle / v_{e0}^2 < 1$), the critical point, of course, moves inward (from $r_c = 1.75 r_0$) toward the atmospheric base ($r_c = r_0$). The mass flux ($\propto u_0$) is seen (Figure 11b) to increase with increasing temperature in the region of thermal dominance and to increase with increasing wave-amplitude in the region of wave dominance, as one would expect. Figure 11 illustrates the fact that the solar wind is thermally dominated; the wind models for the cool-giants and supergiants we are considering fall in the region of wave-dominance. The division between the thermally driven and wave-driven winds can be represented approximately by the line (in Figure 11b)

$$\left(\frac{v_T}{v_{e0}}\right)^4 \approx \frac{1}{100}\left(\frac{\delta v_0}{v_{e0}}\right), \tag{74}$$

with both wave pressure and thermal pressure playing a significant role in determining the mass flux in a narrow region around this line. As shown in Section 5, even when Alfvén waves are unimportant in determining the mass flux ($\delta v_0 \ll 100 v_T^4 / v_{e0}^3$), they may (for a large enough magnetic field) play an important role in determining the asymptotic flow speed, u_∞.

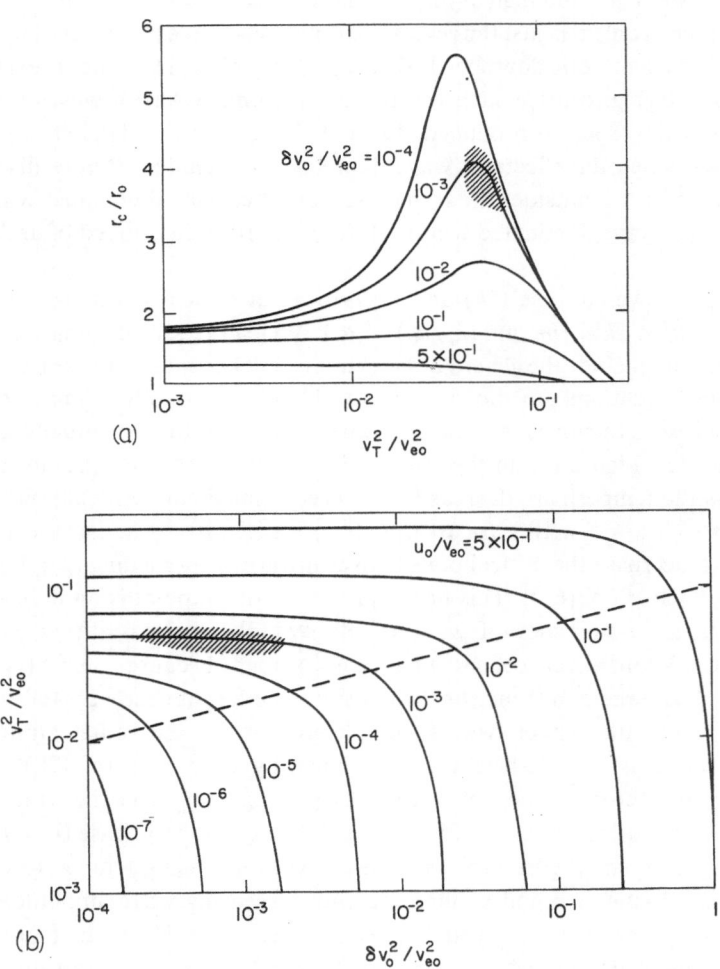

Fig. 11.   Alfvén-wave driven stellar winds. (a) Dimensionless critical radius (r_c/r_0) as a function of the dimensionless thermal energy per unit mass (v_T^2/v_{e0}^2) and of the dimensionless wave energy per unit mass ($\langle \delta v_0^2 \rangle / v_{e0}^2$) at the atmospheric base. (b) Dimensionless measure of the stellar mass loss rate (u_0/v_{e0}), as a function of v_T^2/v_{e0}^2 and $\langle \delta v_0^2 \rangle / v_{e0}^2$. The dashed line shows the approximate boundary between thermally dominated mass loss (to the left and above) and wave-dominated mass loss (to the right and below). The hatched areas represent the range of solar wind parameters.

Now let us turn to the subject of the damping of Alfvén waves and its effect on the mass flux and asymptotic flow speed of a wave-driven wind. One can infer from Section 3 that to maintain the large mass flux produced by undamped Alfvén waves (cf. (72)) and to decrease substantially the corresponding asymptotic flow speed (cf. (73)), it is necessary to maintain the work done on the flow by waves in the region of subsonic flow and to minimize the work done in the supersonic region. To do this the waves must be damped just beyond the critical point, and as $r_c \approx 1.75r_0$ for the wave-driven winds, one would expect the required damping length to be $L \approx r_0$. Hartman and MacGregor (1980) have constructed Alfvén-wave-driven wind models in which the damping length, L, is assumed constant and taken to be $L = R_*$ (where $r_0 = R_*$ and (68) is used). These authors did, indeed, find that this particular damping

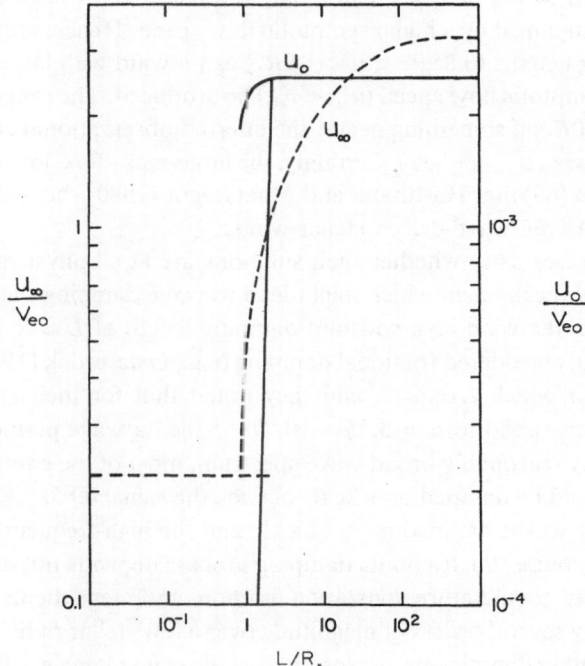

Fig. 12. Asymptotic flow speed in units of gravitational escape speed at the atmospheric base (u_∞/v_{e0}) and a dimensionless measure of the mass loss rate (u_0/v_{e0}) as functions of the Alfvén-wave damping length in units of stellar radii (L/R_*). For strong damping ($L/R_* \ll 1$), the mass loss is negligible and u_∞ corresponds to the solution passing through the outermost (thermal) critical point. For weak damping ($L/R_* \gg 1$), both u_0 and u_∞ correspond to the solution passing through the inner (wave-driven) critical point. For moderate damping ($L/R_* \approx 1$) both critical solutions exist and u_0 and u_∞ for both are shown, the lower values of u_0 and u_∞ corresponding to the solution passing through the outer critical point. Parameters corresponding to model 6 of Hartmann and MacGregor (1980) have been used: $M_* = 16M_\odot$, $R_* = 400R_\odot$, $T = 10^4$ K, $n_0 = 10^{11}$ cm^{-3}, $B_0 = 10$ G, $f_{w0} = 3.36 \times 10^6$ erg cm^{-2} s^{-1}, $\mu = 0.667\, m_H$.

length leads to a large mass flux and an asymptotic flow speed less than the surface escape speed. One might ask, however, just how sensitive these results are to the choice of the damping length and what physical mechanism(s) can produce the required damping length. The first of these questions is answered in Figure 12, where the mass flux ($\propto u_0$) and asymptotic flow speed are shown over a range of damping lengths for the stellar parameters used by Hartmann and MacGregor (1980) in their model 6. For $L < 0.85R_*$ the waves are damped so rapidly that they cannot drive a wind, but the thermal pressure gradient in the (assumed) isothermal atmosphere drives a wind that is subsonic out to $r_c = 31.4R_*$ and has a small mass flux and low asymptotic flow speed ($u_\infty = 0.22v_{e0}$). For $0.85R_* < L < 2.25R_*$ both thermally driven ($r_c = 31.4R_*$) and wave-driven ($r_c \approx 1.75R_*$) winds are possible, the former exhibiting a mass flux that increases very rapidly with increasing L and the latter exhibiting a uniformly large mass flux. The thermally driven wind again has a low asymptotic flow speed, but the wave-driven wind has a flow speed that increases rapidly with increasing L, such that $u_\infty > v_{e0}$ for $L > 1.9R_*$. For $L > 2.25R_*$, only a wave-driven wind exists, but its large mass flux is accompanied by a high asymptotic flow speed. Hence, only for a very small range of damping lengths ($0.85R_* < L < 1.9R_*$) can a wind with large mass flux and a relatively low asymptotic flow speed ($u_\infty < v_{e0}$) be produced. The range is much smaller ($0.85R_* < L < 1.0R_*$) if something nearer the inferred observational constraint on flow speed is applied (viz., $u_\infty < v_{e0}/2$). Certainly, the large-mass-flux, low-asymptotic-flow-speed solutions to (69) that Hartmann and MacGregor (1980) showed would not seem to be typical of Alfvén-wave-driven stellar winds.

The question arises as to whether such solutions are at all physically relevant: viz., is there a physical mechanism which might lead to wave damping that would produce the same effect on the wind as a constant damping length of $L \approx R_*$? Hartmann and MacGregor (1980) considered frictional damping (e.g., Osterbrock (1961), Kulsrud and Pierce (1969)), for which $L \propto \omega^{-2}$, and they noted that for their stellar parameters $L = R_*$ would correspond to $\omega = 3.55 \times 10^{-4}\,\text{s}^{-1}$ (i.e., a wave period of nearly 5 hr). Obviously, for any reasonably broad wave spectrum, most of the energy flux carried by Alfvén waves would be damped on a scale outside the range $0.85R_* < L < 1.0R_*$, with the low-frequency waves damped more quickly and the high-frequency waves damped more slowly. (Of course, the frictional damping process depends not only on frequency, but also on density, temperature, ionization fraction, and magnetic field intensity, so L generally varies by several orders of magnitude over a few stellar radii outward from the stellar surface.) Other damping processes, such as those involving nonlinear interactions or mode-conversion, will also lead to an effective damping length of $L \approx R_*$ only for a small range of stellar parameters, if at all (e.g., Holzer et al., 1982).

Although we are forced to conclude that a low asymptotic flow speed ($u_\infty^2 \ll v_{e0}^2$) will probably not be produced in an Alfvén-wave driven stellar wind, we need not necessarily conclude that the massive winds from cool giants and supergiants are not driven by Alfvén waves. This is because in these wave-driven winds the asymptotic flow speed may not be approached until $u \gtrsim v_A$, which may occur farther from the star than the point at which observers normally infer asymptotic flow speed. In other words, stellar wind observers must make sure they are measuring u_∞ and not the flow speed in, say, the region beyond which little spectral information about the atmosphere is available, because of its low density. In any case, if $u_\infty^2 \ll v_{e0}^2$, then almost any mechanism for driving the stellar wind will face problems similar to those faced by the Alfvén-wave mechanism.

7. Closing Remarks

In this review, we have discussed critically recent research on the acceleration of the solar wind (and of related stellar winds), giving emphasis to high-speed solar wind streams emanating from solar coronal holes. We have seen that existing conductive solar wind models cannot produce high-speed streams with a reasonable mass flux (given a reasonable coronal base pressure), although more realistic descriptions of the thermal conduction could modify this result (Olbert, 1981). It seems, therefore, that the addition of energy to the solar wind above the coronal base is required to accelerate the wind to speeds at 1 AU of more than 600 km s^{-1}. A significant fraction of this energy must be added in the region of supersonic flow in order to increase the flow speed, because subsonic energy addition tends to increase the solar wind mass flux as much as or more than the solar wind energy flux (thus producing little change in or a decrease of the wind energy-per-unit mass, which determines the flow speed at 1 AU).

The energy that must be added to the solar wind to produce high speed streams may be supplied in any number of ways, but we have chosen to concentrate, for illustrative purposes, on Alfvén waves, because they are relatively easy to describe and have been quite well studied. One should not, however, conclude that the Alfvén wave is likely to be the most important type of hydromagnetic wave in accelerating the solar wind or other stellar winds. In a fluctuating, magnetized atmosphere, all hydromagnetic wave modes will be present and will be exchanging energy with each other, and any or all or none may play a significant role in accelerating the wind. For example, it is possible that the refractive properties of fast-mode waves may enable them to drive high-speed solar wind streams in coronal holes, despite a modest upward wave energy flux density at the coronal base (Habbal *et al.*, 1982). In any case, certain aspects of the physical effects involved in the interaction of Alfvén waves with the solar (a stellar) atmosphere and wind are common to other hydromagnetic modes and an understanding of these effects should provide a useful basis for studying the other modes and even other processes not involving hydromagnetic waves.

There have recently been suggestions of alternative mechanisms for the acceleration of the solar wind which we have not touched on in this review. R. N. Thomas and his colleagues (see Heidman and Thomas (1980) and references therein) have suggested that a stellar chromosphere, corona, and wind have a common origin in an intrinsic mass flux from within a star, and that the stellar corona plays no role in the determination of the mass flux, but that the conversion of flow energy to internal energy plays a significant role in heating the chromosphere and corona. This suggestion, which contrasts sharply with Parker's theory (cf. Sections 2 and 3), has not been worked out in detail by Thomas and colleagues, and the only quantitative analyses of it that have been made have failed to provide any support for it (Parker, 1981; Wolfson and Holzer, 1982). Another suggestion of a solar wind acceleration mechanism involves the revival by Pneuman (1982) of Schlüter's 'melon seed' model for ejecting a bubble of magnetized plasma from the solar atmosphere (Schlüter, 1957; Parker, 1957). The driving force expelling such a bubble is the Lorentz force, and the excess energy necessary for the bubble to escape the solar gravitational field is provided by the injection of the bubble into the atmosphere, which serves to distort (thus adding energy to) the ambient magnetic field. No quantitative analyses have yet been presented to describe the formation and injection of the bubble and to indicate the magnitude of the effect such bubbles might have on the solar wind, so we cannot yet evaluate the possible relevance of such an acceleration mechanism.

Acknowledgements

We are grateful to B.C. Low for his careful reading of and useful comments on the manuscript, to R. G. Athay and A. J. Hundhausen for several valuable discussions, and to A. K. Lynch for typing the manuscript. Two of us (E. L. and T. F.) express our gratitude for the hospitality of R. MacQueen and the High Altitude Observatory during our visit.

References

Alazraki, G. and Couturier, P.: 1971, *Astron. Astrophys.* **13**, 380.
Alfvén, H. and Fälthammar, C.-G.: 1963, *Cosmical Electrodynamics*, Oxford.
Altschuler, M. D., Trotter, D. E., and Orrall, F. Q.: 1972, *Solar Phys.* **26**, 354.
Athay, R. G. and White, O. R.: 1979a, *Astrophys. J. Suppl.* **39**, 333.
Athay, R. G. and White, O. R.: 1979b, *Astrophys. J.* **229**, 1147.
Axford, W. I.: 1968, *Space Sci. Rev.* **8**, 331.
Barnes, A.: 1969, *Astrophys. J.* **155**, 311.
Barnes, A.: 1975, *Rev. Geophys. Space Phys.* **13**, 1049.
Barnes, A.: 1979, in *Solar System Plasma Physics*, Vol. I, North-Holland, Amsterdam.
Beicher, J. W.: 1971, *Astrophys. J.* **168**, 509.
Belcher, J. W. and Davis, Jr., L.: 1971, *J. Geophys. Res.* **76**, 3534.
Belcher, J. W. and Olbert, S.: 1975, *Astrophys. J.* **200**, 369.
Bell, B. and Noci, G.: 1976, *J. Geophys. Res.* **81**, 4508.
Biermann, L.: 1951, *L. Astrophys.* **29**, 274.
Birkeland, K.: 1908, *The Norwegian Aurora Polaris Expedition 1902-1903*, Vol. I, H. Aschehoug and Co., Christiania.
Birkeland, K.: 1913, *The Norwegian Aurora Polaris Expedition 1902-1903*, Vol. II, H. Aschehoug and Co., Christiania.
Bonetti, A., Bridge, H. S., Lazarus, A. J., Lyon, E. J., Rossi, R., and Scherb, F.: 1963, *J. Geophys. Res.* **68**, 1963.
Bretherton, F. P.: 1970, in *Mathematical Problems in the Geophysical Sciences*, American Mathematical Society, Providence.
Cassinelli, J. P.: 1979, *Ann. Rev. Astron. Astrophys.* **17**, 275.
Cassinelli, J. P. and MacGregor, K. B.: 1982, To appear in Sturrock, P. A., Holzer, T. E., Mihalas, D., and Ulrich, R. (eds.), *Physics of the Sun*, The National Academy of Sciences, Washington, D. C.
Chapman, S.: 1957, *Smithsonian Contrib. Astrophys.* **2**, 1.

Chapman, S. and Bartels, J.: 1940, *Geomagnetism*, Clarendon, Oxford.
Chapman, S. and Cowling, T. G.: 1939, *Mathematical Theory of Non-uniform Gases*, Cambridge University Press.
Dewar, R.: 1970, *Phys. Fluids* **13**, 2710.
Durney, B. R.: 1972, *J. Geophys. Res.* **77**, 4042
Durney, B. R. and Hundhausen, A. J.: 1974, *J. Geophys. Res.* **79**, 3711.
Feldman, W. C., Asbridge, J. R., Bame, S. J. and Gosling, J. T.: 1977, in O. R. White (ed.), *The Solar Output and Its Variation*, Colorado Associated University Press.
Ferraro, V. C. A. and Plumpton, C: 1958, *Astrophys. J.* **127**, 459.
Gingerich, O., Noyes, R. W., Kalkofen, W., and Cuny, Y.: 1971, *Solar Phys.* **18**, 347.
Gringauz, K. T., Bezrukikh, V. V., Ozerov, V. D., and Rybchinskiy, R. E.: 1960, *Soviet Phys. 'Doklady'* (English Transl.) **5**, 361.
Gringauz, K. T., Bezrukikh, V. V., Ozerov, V. D., and Rybchinskiy, R. E.: 1961, *Space Res.* **2**, 539.
Gringauz, K. T., Bezrukikh, V. V., Ozerov, V. D., and Rybchinskiy, R. E.: 1967, *Cosmic Research* **5**, 216.
Habbal, S. R., Flå, T., Holzer, T. E., and Leer, E.: 1982, submitted to *Astrophys. J.*
Habbal, S. R., and Tsinganos, K. C.: 1982, submitted to *J. Geophys. Res.*
Hansen, R. T., Hansen, S. F., and Sawyer, C.: 1976, *Planet. Space Sci.* **24**, 381.
Hartmann, L. and MacGregor, K. B.: 1980, *Astrophys. J.* **242**, 260.
Heidmann, N. and Thomas, R. N.: 1980, *Astron. Astrophys.* **87**, 36.
Heinemann, M. and Olbert, S.: 1980, *J. Geophys. Res.* **85**, 1311.
Hollweg, J. V.: 1972, *Cosmic Electrodyn.* **2**, 423.
Hollweg, J. V.: 1973, *Astrophys. J.* **181**, 547.
Hollweg, J. V.: 1976, *J. Geophys. Res.* **81**, 1649.
Hollweg, J. V.: 1978a, *Rev. Geophys. Space Phys.* **16**, 689.
Hollweg, J. V.: 1978b, *Solar Phys.* **56**, 305.
Hollweg, J. V.: 1982, personal communication
Holzer, T. E.: 1979, in *Solar System Plasma Physics*, Vol. I, North-Holland, Amsterdam.
Holzer, T. E. and Axford, W. I.: 1970, *Ann. Rev. Astron. Astrophys.* **8**, 30.
Holzer, T. E., Flå, T., and Leer, E.: 1982, To be submitted to *Astrophys. J.*
Holzer, T. E. and Leer, E.: 1980, *J. Geophys. Res.* **85**, 4665.
Hundhausen, A. J.: 1972, *Coronal Expansion and Solar Wind*, Springer, New York.
Hundhausen, A. J.: 1977, in J. B. Zirker (ed.), *Coronal Holes and High Speed Wind Streams*, Colorado Associated University Press.
Hundhausen, A. J., Hansen, R. T., Hansen, S. F., Feldman, W. C., Asbridge, J. R., and Bame, S. J.: 1976, in *Physics of Solar Planetary Environments*, American Geophysical Union, Washington, D. C.
Hundhausen, A. J. and Holzer, T. E.: 1980, *Phil. Trans. Roy. Soc. London* **A297**, 521.
Jacques, S. A.: 1977, *Astrophys. J.* **215**, 942.
Kopp, R. A. and Holzer, T. E.: 1976, *Solar Phys.* **49**, 43.
Krieger, A. S., Timothy, A. F., and Roelof, E. C.: 1973, *Solar Phys.* **29**, 505.
Kulsrud, R. and Pierce, W. P.: 1969, *Astrophys. J.* **156**, 445.
Leer, E., Flå, T., and Holzer, T. E.: 1980, *Il Nuovo Cimento* **36**, 114.
Leer, E. and Holzer, T. E.: 1979, *Solar Phys.* **63**, 143.
Leer, E. and Holzer, T. E.: 1980, *J. Geophys. Res.* **85**, 4681.
Munro, R. H. and Jackson, B. V.: 1977, *Astrophys. J.* **213**, 874.
Neupert, W. M. and Pizzo, V. A.: 1974, *J. Geophys. Res.* **79**, 3701.
Neugebauer, M. and Snyder, C. W.: 1966, *J. Geophys. Res.* **71**, 4469.
Nolte, J. T., Krieger, A. S., Timothy, A. F., Gold, R. E., Roelof, E. C., Vaiana, G., Lazarus, A. J., and Sullivan, J. D.: 1976, *Solar Phys.* **46**, 303.
Olbert, S.: 1981, in *Proceedings of an International School and Workshop on Plasma Astrophysics* ESA SP-161.
Osterbrock, D. E.: 1961, *Astrophys. J.* **134**, 347.
Parker, E. N.: 1957, *Astrophys. J. Suppl.* **3**, 51.
Parker, E. N.: 1958, *Astrophys. J.* **128**, 664.
Parker, E. N.: 1960, *Astrophys. J.* **132**, 821.
Parker, E. N.: 1963, *Interplanetary Dynamical Processes*, Interscience, New York.
Parker, E. N.: 1964a, *Astrophys. J.* **139**, 72.
Parker, E. N.: 1964b, *Astrophys. J.* **139**, 93.

Parker, E. N.: 1965, *Space Sci. Rev.* **4**, 666.
Parker, E. N.: 1981, *Astrophys. J.* **251**, 266.
Perkins, F.: 1973, *Astrophys. J.* **179**, 637.
Pneuman, G. W.: 1980, *Astron. Astrophys.* **81**, 161.
Pneuman, G. W.: 1982, submitted to *Astrophys. J.*
Scherb, F.: 1964, *Space Res.* **4**, 797.
Schlüter, A.: 1957, in H. C. van de Hulst (ed.), *Radio Astronomy, IAU Symp.* **4**, 356.
Scudder, J. D. and Olbert, S.: 1979a, *J. Geophys. Res.* **84**, 2755.
Scudder, J. D. and Olbert, S.: 1979b, *J. Geophys. Res.* **84**, 6603.
Sheeley, Jr., N. R., Harvey, J. W., and Feldman, W. C.: 1976, *Solar Phys.* **49**, 271.
Snyder, C. W. and Neugebauer, M.: 1964, *Space Res.* **4**, 89.
Spitzer Jr., L.: 1962, *Physics of Fully Ionized Gases*, Interscience, New York.
Vernazza, J. E., Avrett, E. H., and Loeser, R.: 1973, *Astrophys. J.* **184**, 605.
Wagner, W. J.: 1976, *Astrophys. J.* **206**, 583.
Weber, E. J. and Davis Jr., L.: 1967, *Astrophys. J.* **148**, 217.
Withbroe, G. L.: 1977, in *Proceedings of the November 7–10, 1977 OSO-8 Workshop*, LASP, University of Colorado.
Withbroe, G. L.: 1982, *Space Sci. Rev.*
Wolfson, R. L. T. and Holzer, T. E.: 1982, *Astrophys. J.* **255**, 610.

PROCESSES AFFECTING ABUNDANCES
IN THE SOLAR WIND*

JOHANNES GEISS

Physikalisches Institut, University of Bern, Sidlerstrasse 5, 3012 Bern, Switzerland

Abstract. Data on composition in the solar wind are summarized and compared with best estimates of abundances in the outer convective zone of the Sun. Several mechanisms of element and isotope fractionation are discussed in relation to observed abundances and their variations.

The evidence available so far indicates that in addition to ion fractionation in the corona there is a separation mechanism operating at low solar altitude that affects solar wind composition. It is suggested that the systematic depletion of helium observed in the solar wind is in part caused by ion-neutral separation in the chromosphere-transition zone. Conditions for this mechanism to be effective are discussed. It is shown that ion-neutral separation is much more pronounced than ion-ion separation under these conditions. Therefore, this mechanism should fractionate elements according to the rate at which first ionization occurs. This implies that isotope fractionation by this mechanism is minor.

Ion-neutral separation may be responsible for the general depletion that is observed in the slow interstream solar wind as well as in the fast streams coming out of coronal holes. However, the occurrences of very low He/H ratios are probably caused in the corona.

1. Introduction

Models of solar wind acceleration indicate that Coulomb friction can be adequate to pull helium and heavier elements out of the gravitational field of the Sun (Geiss *et al.*, 1970; Joselyn and Holzer, 1978; Borrini and Noci, 1979; McKenzie *et al.*, 1979). Thus, these models offer an explanation for the main trends in the observation of solar wind composition:

(1) Helium is usually present, but the frequent occurrences of low He/H ratios indicate that coupling between helium and the proton-electron gas is often marginal.

(2) The occurrence of heavier elements (i.e. O, Ne, Si, Ar, and Fe) with roughly solar abundances (Bame *et al.*, 1975; Geiss *et al.*, 1972) can basically be understood in terms of models that include Coulomb friction. In the corona, these elements rapidly attain high ionic charges, giving them a ratio of Coulomb collision cross-section to weight ($\propto Z_i^2/A_i$) that is higher than this ratio is for helium.

The qualitative agreement between some of the general observations and the predictions of models of solar wind acceleration that include Coulomb collisions does not preclude that other factors affect the abundances in the solar wind. One such factor, the energy or rate of first ionization has been mentioned occasionally (Arrhenius and Alfvén, 1971; Geiss, 1972). In this paper, we review solar wind abundances and discuss whether they are different from the abundances in the outer convective zone of the Sun. We show that it is difficult to explain the observed systematic depletion in the He/H ratio by coronal separation processes alone, and we explore whether retardation in the ionization of helium relative to hydrogen contributes to the general depletion of this element in the solar wind.

* Paper presented at the IX-th Lindau Workshop 'The Source Region of the Solar Wind'.

Space Science Reviews **33** (1982) 201–217.   0038–6308/82/0332–0201$02.55.

2. Abundances of the Helium Isotopes in the Sun and the Solar Wind

In the solar wind, the abundance ratios most extensively investigated are ^{4}He/H and ^{3}He/^{4}He. Neugebauer (1981) has recently published a thorough account of the ^{4}He in the solar wind, its time variation, and the relation of the ^{4}He/H ratio to other solar wind parameters and to observations in the corona. In the low speed solar wind, ^{4}He/H is highly variable with average values of ≤ 0.04 (Robbins et al., 1970; Ogilvie, 1972; Feldman et al., 1977; Neugebauer, 1981a). Since the low speed solar wind probably comes from regions in the lower corona with complicated magnetic field structures, it is very difficult to model the geometric and thermodynamic situation in the source region.

The flux geometry in coronal holes appears to be less complicated. In the high speed wind from these holes, Bame et al. (1977) reported

$$(^4\text{He/H})_{\text{SW, CH}} = 0.048 \pm 0.005 . \tag{1}$$

The remarkable constancy in time of the ^{4}He/H ratio and its reproducibility in different coronal holes suggested to Bame et al. (1977) that, after all, 0.05 might be the true solar helium/hydrogen abundance ratio.

Estimates of the solar helium abundance, usually giving He/H ~ 0.1, are certainly not precise enough to exclude a ratio of 0.05. However, in recent years, helium abundances have been determined with relatively high precision in a variety of galactic and extragalactic objects. The He/H abundance ratio is found to be remarkably constant, supporting the generally held view that helium is essentially primordial and that the He/H ratio in the interstellar gas of the galaxy has been changed only little by stellar nucleosynthesis. Audouze (1981) has recently reviewed the helium determinations and gives

$$(\text{He/H})_{\text{Primordial}} = 0.076 \pm 0.005 . \tag{2}$$

Since there is no evidence for a helium/hydrogen fractionation in the formation process of the Sun, we adopt for the outer convective zone (OCZ)

$$\text{He/H}_{\text{OCZ}} = 0.08 \pm 0.01 . \tag{3}$$

A possible small contribution by stellar production has been included in this value and the error estimate. Comparing (1) and (3), we conclude that ^{4}He is definitely depleted even in the high speed solar wind from coronal holes.

Geiss et al. (1972) have obtained for a total collection time of about five days during 1969–1972 an average solar wind ratio

$$(^3\text{He/}^4\text{He})_{\text{SW}} = (4.3 \pm 0.3) \times 10^{-4} . \tag{4}$$

More recently, Ogilvie et al. (1980) evaluated 4334 mass spectra taken in 1978/79. From a broad distribution of individual ratios, they obtained a remarkably similar average of 4.7×10^{-4}.

Spectrographic estimates of the solar ^{3}He/^{4}He ratio are subject to large uncertainty. Thus, again we have to turn to non-solar observations for obtaining a solar value, bearing in mind that ^{3}He in the outer convective zone of the Sun is the sum of originally present D and ^{3}He (cf. Geiss and Reeves, 1972; Bochsler and Geiss, 1973). D/H has been determined in the dilute interstellar gas through Lyman absorption spectra. The most recent analyses by Laurent *et al.* (1979) and Bruston *et al.* (1981) give

$$(D/H)_{\text{Interstellar}} = (1 \text{ to } 2.5) \times 10^{-5} . \tag{5}$$

If one allows for D destruction between the time of formation of the solar system and now, a somewhat higher ratio than (5) is indicated for the protosolar gas. Thus, we adopt

$$(D/H)_{\text{Protosolar}} = (2 \pm 1) \times 10^{-5} . \tag{6}$$

Direct determinations of the ^{3}He/^{4}He ratio in the interstellar gas are subject to large errors (cf. Rood *et al.*, 1979; Audouze, 1981). Thus, one usually resorts to measurements of the ^{3}He/^{4}He ratio in the planetary component of meteoritic helium (Jeffery and Anders, 1970; Frick and Moniot, 1977; Eberhardt, 1978) giving

$$(^3\text{He}/^4\text{He})_{\text{Meteoritic}} = (1.4 \pm 0.2) \times 10^{-4} . \tag{7}$$

Combining (2), (6), and (7), we obtain for the outer convective zone

$$(^3\text{He}/^4\text{He})_{\text{OCZ}} = (3.9^{+2.5}_{-1.5}) \times 10^{-4} . \tag{8}$$

The larger upper limit allows for a possible admixture of ^{3}He from the solar interior (cf. Schatzman and Maeder, 1981; Geiss and Bochsler, 1981). The derived value for the outer convective zone of the Sun is very close to the ratio (4) measured in the solar wind. The relatively large error of the OCZ estimate precludes any precise conclusion concerning helium isotope fractionation in the solar wind source region. Nevertheless, the similarity of the OCZ and SW ^{3}He/^{4}He ratios indicates that in the solar wind the two He isotopes might often be depleted by a similar factor. Such a similar depletion could hardly be expected to occur at the high temperature prevailing in the SW acceleration region. If, on the other hand, ionization rates affect solar wind abundances, a similar depletion of ^{4}He and ^{3}He would be a natural consequence.

3. Comparison of Elemental Abundances in the Solar Wind and the Outer Convective Zone

The evidence on the abundances in the solar wind of elements heavier than helium is still rather limited. Averages for elemental and isotopic abundances were summarized by Bochsler and Geiss (1976). In the present paper, we discuss primarily element ratios and ^{3}He/^{4}He, and in Table I we give averages for these ratios. For O, Si, and Fe only low speed data have been published so far. The averages given for Ne and Ar represent the results of 5 and 2 foil collection experiments with total collection times of 5 and 3 days respectively. Although the foil collections include some relatively short periods of solar wind at higher velocities, it is difficult to extract from these data information on a possible velocity dependence of the abundances. Thus, determinations of rare ions in high speed streams are still not available.

TABLE I

Comparison of measured solar wind abundances with best estimates of solar abundances

	Solar wind			Refs.	Outer convective zone	Refs.
He/H	0.04 ± 0.01	AV	1962–75	[1]	0.08 ± 0.01	[7]
He/H	0.048 ± 0.005	CH		[2]		
O/H	$(5 \pm 2) \times 10^{-4}$	LS	1969–72	[3]	$(6.9 \pm 2) \times 10^{-4}$	[8]
Si/H	$(7.6 \pm 3) \times 10^{-5}$	LS	1969–72	[3]	$3.9 \times 10^{-5}(1.7)$	[9]
Fe/H	$(5 \pm 3) \times 10^{-5}$	LS	1969–72	[3]	$3.4 \times 10^{-5}(1.7)$	[9]
Ne/He	530 ± 70	AV	1969–72	[4]		
O/Ne					7.5 ± 2	[10]
Ne/H	$(7.5 \pm 2.5) \times 10^{-5}$	AV	1969–72		$1.5 \times 10^{-4}(2)$	
Ne/Ar	41 ± 10	AV	1971–72	[5]		
Ar/H	$1.8 \times 10^{-6}(1.5)$	AV	1971–72		$4.3 \times 10^{-6}(2)$	[11]
^{3}He/^{4}He	$(4.3 \pm 0.3) \times 10^{-4}$	AV	1969–72	[4]	$\left(3.9 {+2.5 \atop -1.5}\right) \times 10^{-4}$	[12]
^{3}He/^{4}He	$(4.7 \pm 1.2) \times 10^{-4}$	AV	1978–79	[6]		

AV: Average of data, as available during the indicated period.
CH: Average observed in high speed streams from coronal holes.
LS: Low speed (interstream) solar wind data.
Numbers in parenthesis are uncertainty factors.

References:
[1] Neugebauer and Snyder (1966); Robbins *et al.* (1970); Ogilvie and Hirshberg (1974); Neugebauer (1981).
[2] Bame *et al.* (1977).
[3] Bame *et al.* (1975); Grünwaldt (1976).
[4] Geiss *et al.* (1970a, 1972).
[5] Cerutti (1974).
[6] Ogilvie *et al.* (1980).
[7] See text.
[8] Photosphere, Ross and Aller (1976).
[9] Based on abundance in Cl chondrites, Cameron (1980).
[10] H II regions, Meyer (1979).
[11] Interpolated by semi-equilibrium abundance method, Cameron (1980).
[12] See text.

Information on the variability of the abundances of elements heavier than He is still scarce. Geiss *et al.* (1972) found small but significant variations in the Ne/He ratio, Zastenker and Yermolaev (1981) reported large variations, particularly in the Si abundance, and Bame *et al.* (1979) showed that O/He and Fe/He in the interstream wind and in flare-expelled plasma are different. We shall return to these observed variations in Section 6.

For discussing possible differences between solar and solar wind composition, we give in Table I our best estimates of the abundances in the outer convective zone. For this purpose, we had, of course, to disregard determinations in the corona, solar wind or solar cosmic rays – data that are often included in the figures given in abundance compilations.

The elemental abundances and the ^{3}He/^{4}He ratio are plotted in Figure 1. So far, a significant difference between solar and average solar wind data is only established for the ^{4}He/H ratio.

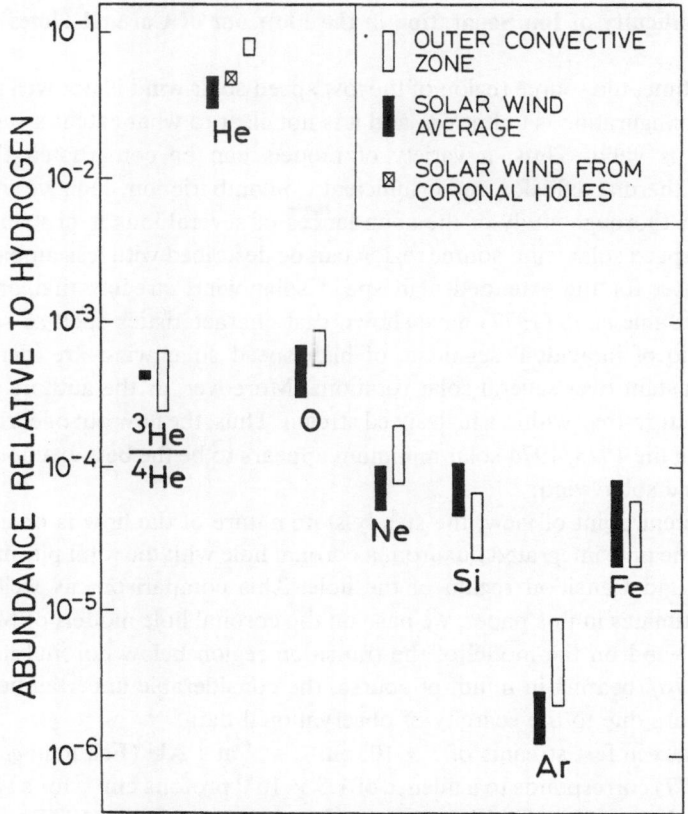

Fig. 1. Elemental abundances and the isotopic ratio ^{3}He/^{4}He as measured in the solar wind are compared to the abundances estimated for the outer convective zone of the Sun. Details on these estimates and references are given in Table I and in the text.

Meyer (1981) and Veck and Parkinson (1981) have derived from X-ray data an inverse relation between ionization potential and coronal abundance and the latter authors suggest that this is due to ambipolar diffusion. There is some indication for a similar trend in the solar wind data (cf. Figure 1), but if one considers the uncertainties in the abundances of *both*, solar wind and outer convective zone, it becomes clear that improved data are needed for firmly establishing such a relation.

Table I shows that, partly for technical reasons, some abundance ratios such as He/Ne, ^{3}He/^{4}He, or Ne/Ar are relatively well determined in the solar wind, whereas the best estimates for the outer convective zone exist for other ratios. An example of the latter is the O/Ne ratio which is remarkably constant in H II regions (Meyer, 1979), and thus apparently quite reliable. Determination of the elemental O/Ne ratio in the solar wind should be feasible by M/Q spectrometers because in a wide range of freezing-in temperatures the dominant fraction of these elements exists in the form of only three ions, O^{6+}, O^{7+}, and Ne^{8+}. These three ions are separable but still similar enough in M/Q and in abundance to allow a good comparison.

4. Difficulty of Ion Separation in the Flow out of Coronal Holes

At the present time, the source region of the low speed solar wind is not well identified, its geometric configuration is unknown, and it is not clear to what extent a steady-state approximation is valid. Thus, a variety of models can be constructed that, by a combination of thermal diffusion and insufficient Coulomb friction, can give any desired ^{4}He/H ratio. A thorough study of the abundances of several ions is probably needed before the low speed solar wind source region can be described with less ambiguity. The situation is better for the extended high speed solar wind streams that originate in coronal holes. Bame *et al.* (1977) have shown that characteristics such as speed, flux and He/H ratio of individual segments of high speed solar wind are identical and remarkably constant over several solar rotations. Moreover, as the authors point out, the flow is 'structure-free' within a high speed stream. Thus, the flow out of coronal holes observed during the 1973/1974 solar minimum appears to be the best example we have of a steady-state solar wind.

From a different point of view, the steady-state nature of the flow is demonstrated by comparing the time integrated flux from a coronal hole with the total plasma content in the corona and transition region of the hole. This comparison, as well as other quantitative estimates in this paper, we base on the coronal hole models of Munro and Jackson (1977) and on the model of the transition region below coronal holes given by Gabriel (1976), bearing in mind, of course, the considerable uncertainties in these models which are due to the scarcity of observational data.

The proton flux in fast streams of 3×10^8 cm^{-2} s^{-1} at 1 AU (Feldman *et al.*, 1976; Bame *et al.*, 1977) corresponds to a fluence of 1.5×10^{15} protons cm^{-2} for a two-month period. Taking the non-radial hole geometry of Munro and Jackson (1977) and Gabriel (1976) into account, we estimate that in a flux tube having a cross section of 1 cm^2 at 1 AU the proton content between the altitudes 3000 km (as given by Gabriel) and $5R_\odot$ is $\sim 1 \times 10^{13}$ protons, i.e. the reservoir between these two altitudes is emptied by the solar wind every 10 h. Thus, in a steady-state outward flow of protons, any pile-up of helium in this reservoir should manifest itself in an increase of the He/H ratio with a time constant of ~ 10 h. Since such an increase is not observed and since inside the fast streams there is very little spatial structure in the speed and in the He/H ratio (cf. Bame *et al.*, 1977), the deficit in helium in the solar wind would have to be compensated by a helium return mechanism that is very smooth in space and time, if the gas supplied to the corona has the solar He/H ratio. These return conditions are difficult to fulfill once the helium is fully ionized: At higher altitudes, any return mechanism would have to overcome the outflow velocity, and it is difficult to see how this could take place without significant structure in space and/or time. At lower altitude (in the low corona and transition zone), thermal diffusion is quite effective in lifting He^{2+} ions upwards, as will be discussed below.

In the transition zone and low corona, the momentum equation can be approximated by the diffusion equation. We shall use here the diffusion equation for minor ions (A_i, Z_i) in a proton-electron plasma (cf. Burgers, 1969)

$$\Delta v_i = D_i \left\{ -\frac{d \ln c_i}{dz} + \alpha_i \frac{d \ln T}{dz} - A_i' \frac{\Gamma}{kT} + \frac{\Delta F_w}{kT} \right\}. \tag{9}$$

Vertical flow geometry as well as $T_i = T_e$ were assumed for simplicity. Δv_i is the difference between the velocities of the rare ions and the protons, and $c_i = n_i/n_p$ is the ion/proton number density ratio. D_i is the diffusion constant and α_i the thermal diffusion coefficient. $\Gamma = m_0 GM_\odot/r^2$ is the local weight per amu, and $A_i' = A_i - (Z_i + 1)/2$ is the 'effective mass number' which takes into account the weight reduction due to the charge separation E-field. ΔF_w stands for the difference between the forces on the ions and the particles of the main gas by other external fields and waves.

Under the assumption that ion-ion collisions are governed by the Coulomb potential $(1/r)$, the constants D_i and α_i can be derived from the formulae given by Burgers (1969) or Schunk and Walker (1969), giving approximately

$$D_i = \frac{6.6 \times 10^7 \, T^{5/2}}{\mu_A^{1/2} \, Z_i^2 n_p}, \tag{10}$$

where μ_A is the reduced mass in atomic mass units. For $Z_i \geq 2$, a useful approximation is

$$\alpha_i = \frac{15}{8} (2\mu_A)^{1/2} \left(1 - \frac{1}{A_i} \right) Z_i^2 + \frac{4}{5} Z_i (Z_i - 1) \qquad (Z_i \geq 2). \tag{11}$$

Both (10) and (11) are valid for rare ions in a p-e gas. However, for low ionic charge and high temperature deviations from the Coulomb potential at close encounters become important. In the temperature range considered here, $T \leq 10^6$ K, we estimate that this effect is small for He^+, He^{2+} and for the charge states $Z \geq 3$ of C, O, and Ne. Our estimates indicate that the Coulomb potential is still a fair approximation for $T \lesssim 7 \times 10^5$ K in the case of doubly charged and for $T \lesssim 2 \times 10^5$ K in the case of singly charged C, O, and Ne.

The diffusion equation (9) can be applied also to neutrals. Below, we shall be concerned with diffusion of helium atoms in a proton-electron gas. Classical collision theory should give a satisfactory approximation in the temperature range $T \leq 5 \times 10^4$ K that we shall consider (Massey, 1982). At long distance He° and H^+ are attracted by the induced dipol potential $(1/r^4)$. However, for energies of a few eV, the deviations from this potential at short range cannot be disregarded. Helbig et al. (1970) and Rich et al. (1971) give a potential for $He^\circ H^+$ which is based on scattering data at 4 eV and on theory. Using this potential, we have numerically calculated momentum transfer cross sections and obtained approximately

$$q_D(\text{cm}^2) \approx 4.2 \times 10^{-16} \, T_4^{-1} \tag{12}$$

(T_4 is the temperature in 10^4 K).

This corresponds to a diffusion constant of

$$D_n = 1.14 \times 10^{21} \frac{T_4^{3/2}}{n_p}. \tag{13}$$

The magnitude of the thermal diffusion coefficient α_n for He$^\circ$ in a p-e gas should be much smaller than α_i for ions (cf. Chapman and Cowling, 1958; Burgers, 1969). We shall disregard here thermal diffusion of He$^\circ$ (cf. Shine et al., 1975). However, for the steepest temperature gradients, this effect is probably not completely negligible.

For obtaining an explicit solution of the motion and concentration of a nuclear species, the diffusion equations (9) of its charge states have to be completed by a set of continuity equations. Since we want to discuss possibilities for element and isotope fractionation in a general way, we shall not aim at deriving solutions for particular models. Instead, we shall discuss types of separation processes which follow from the properties of Equation (9).

Equation (9) is applicable not only to a particular charge state of a nuclear species, but also to the sum of all its charge states, provided ionization and recombination are fast in relation to the rates of change in T and n_e experienced by the moving ions, i.e. provided there is local charge state equilibrium. Δv_i is then the bulk speed of the species, and D_i, α_i, and A_i' are suitable averages that are functions of the altitude z.

As has been shown by Delache (1965, 1967), Jokipii (1965), and Nakada (1969), steady-state solutions of Equation (9) with $\Delta F_w = 0$ lead to strong enrichment of heavier elements in the corona for the estimated steep temperature gradient in the transition zone of the quiet Sun (cf. Dupree, 1972). It has to be noted, however, that the temperature gradient in the transition zone below coronal holes is smaller than the quiet-Sun gradient (cf. Gabriel, 1976).

Abundance observations strongly suggest that diffusive equilibrium normally is not reached. This is not surprising since irregular motions appear to be faster than diffusion velocities. Still, a downward directed diffusion velocity of an element would lead to its depletion in the corona by a degree that would depend on the relative magnitude of diffusive and convective motions (cf. Nakada, 1969). Since a condition for such a net downward motion of a rare ion is $\Delta v_i < 0$ for d ln $c_i/dz = 0$, we consider in the following the equation

$$\Delta v_i = D_i \left\{ \alpha_i \frac{d \ln T}{dz} - A_i' \frac{\Gamma}{kT} \right\} \tag{14}$$

which gives the drift velocity under the influence of temperature gradient and gravitation only. If the gas composition is initially homogenous (c_i = const), Equation (14) gives the initial diffusion velocity. Thus, it gives the direction of diffusive flow, showing whether the gas will move downward or upward relative to hydrogen.

The condition $\Delta v_i < 0$ in Equation (14) defines a 'critical temperature gradient'

$$T' < T_c' = \frac{A_i'}{\alpha_i} \frac{\Gamma}{k}. \tag{15}$$

TABLE II

The thermal diffusion coefficient α_i and the critical temperature gradient T'_c (Equation 15) at $r = R_\odot$ for ions considered in this paper. α_i was calculated after Schunk and Walker (1969).

Ion	α_i	T'_c (K km^{-1})	Ion	α_i	T'_c (K km^{-1})
^{3}He$^+$	1.4	47.0	O$^+$	2.4	206.0
^{3}He^{2+}	7.8	6.3	O^{2+}	11.2	43.0
^{4}He$^+$	1.7	58.0	O^{3+}	26.5	17.0
^{4}He^{2+}	8.7	9.5	O^{4+}	48.2	9.2
C$^+$	2.3	158.0	Ne$^+$	2.4	261.0
C^{2+}	10.9	32.0	Ne^{2+}	11.4	54.0
C^{3+}	25.8	13.0	Ne^{3+}	26.9	22.0
C^{4+}	47.0	6.7	Ne^{4+}	49.0	12.0
			Ne^{5+}	77.5	7.2

In Table II, we give α_i (Coulomb potential approximation) and the critical temperature gradient for ions of several elements (at $r = R_\odot$). According to Gabriel's (1976) model, the temperature gradient varies from > 250 to 10 K km^{-1} in the transition region ($z \leq 30\,000$ km) of coronal holes, i.e. it is everywhere higher than the critical value given by (15) for ^{4}He^{2+}. Thus, ^{4}He^{2+} tends to move upwards relative to hydrogen throughout this transition region. If we assume that the 'effective temperature' profiles given by Munro and Jackson (1977) in their Figure 7b correspond to ion temperatures, we obtain positive Δv_i for ^{4}He^{2+}, even in the case of their lowest profile. We think that these examples indicate how difficult it is to construct a steady-state model which preferentially returns He^{2+} to the Sun.

In Figure 2 we have plotted the Δv_i for various ions as a function of altitude in Gabriel's (1976) transition zone model for coronal holes. Figure 2 allows two important observations which are rather independent of the details of Gabriel's (1976) model: (1) Δv_i is positive for He^{2+} over the whole altitude range considered. He$^+$ moves downward above 6000 km. However, for $z > 6000$ km, the ionization time for He$^+$ with Gabriel's (1976) T and n_e coronal hole parameters is less than 1 s. Thus, helium and for the same reason C, O, and Ne are not preferentially returned to the Sun by bulk downward motion in the temperature-density profile above 4000 km given by Gabriel (1976) for coronal holes. (2) For any temperature profile, O and Ne must be doubly charged to move with He$^+$, and they must be four times charged to behave like He^{2+}.

5. Incomplete Ionization, a Possible Cause for Element Depletion in the Solar Wind

It is likely that mass to the transition zone and corona is supplied inhomogenously, spicules and/or macrospicules being prime candidates (cf. Pneuman and Kopp, 1977). In Table III, we give some typical characteristics of spicules and macro-spicules as they have been summarized by Beckers (1972), Bohlin et al. (1975), Bohlin (1977) and Withbroe (1981).

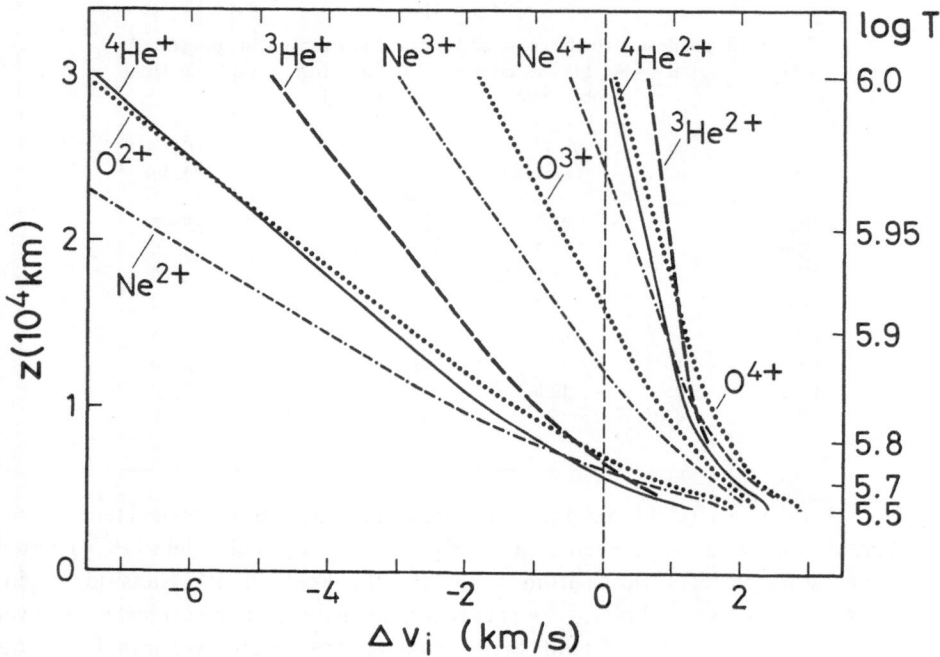

Fig. 2. Diffusion velocities of various ions relative to the main component H^+ in the transition zone below coronal holes (upward velocities are positive). The temperature and density profiles were taken from Gabriel's (1976) model.

TABLE III

Typical characteristics of spicules and macrospicules (cf. Beckers, 1972; Bohlin *et al.*, 1975; Bohlin, 1977)

	Spicules	Macrospicules
Length (km)	10 000	20 000
Width (km)	1 000	7 000
Number density (cm^{-3})	$10^{10} - 10^{11}$	
Apparent velocity (km s^{-1})	25	$\lesssim 90$
Lifetime (s)	400	800
Time of free fall from top (s)	300	400

In the following, we shall discuss a general mechanism of helium depletion in the corona, using the picture of mass supply by spicules whenever specifics are required: The gas moving upwards from the solar surface is dense and cold relative to the surrounding plasma (cf. the spicule models of Beckers, 1972), and at the outset it will be mainly in the neutral state. The degree of ionization of this gas is gradually increased due to UV irradiation from the surrounding and overlying plasma (cf. Beckers, 1972; Zirin, 1975), and due to collisions with the warming electron gas. In both processes, helium will be more slowly ionized than hydrogen. Later, e.g. when the spicule collapses at the end of its lifetime, remaining gas returns to the Sun whereas at least a fraction of the ionized gas is supplied to the transition zone and corona. This ionized gas will have a reduced He/H ratio because the returned neutral gas is enriched in helium. It is clear that this mechanism requires return of some material at a fairly low temperature, i.e. at $T \lesssim 3 \times 10^4$ K.

The models of physical conditions in spicules given by Beckers (1972) that include ionization by UV and electrons give He^+/H^+ ratios of $\lesssim 60\%$ of the element ratio, as long as the temperature remains $\lesssim 20\,000$ K. Beckers' (1972) models are equilibrium models, but since the lifetimes of spicules are only $< 10^3$ s and conditions are changing rapidly, we should also consider ionization rates. In Table IV, we give rates of ionization by UV for zero optical depth. We have adopted the figures given by Banks and Kockarts (1973) for H, He, O, and O^+ for quiet solar conditions at 1 AU. The rates for Ne were calculated from cross section data (Marr and West, 1976) and the quiet solar UV flux spectrum of Banks and Kockarts (1973). Quiet-Sun rates near the solar surface were calculated by multiplying the 1 AU rates with $(r/R_\odot)^2$. In coronal holes, X-ray and UV fluxes are reduced and the spectrum is modified in comparison to quiet Sun conditions (Huber et al., 1974; Reeves, 1976; Bohlin, 1977; Withbroe, 1981). Coronal lines from ions that exist at temperatures $\gtrsim 10^6$ K are strongly suppressed, whereas the intensity of chromospheric and transition zone lines are reduced by only 25–30%, with the exception of the He II emission which is decreased by about 50% (cf. Bohlin, 1977; Withbroe, 1981). We have used the available information for adapting the ionization rates at $1R_\odot$ to the UV spectrum in coronal holes. The results are given in the third line of Table IV, the corresponding ionization times in the last line.

TABLE IV

UV ionization rates for zero optical depth. The last line gives ionization times for coronal hole conditions at $1R_\odot$

Condition	Distance	Units	H	He	O	O^+	Ne
Quiet Sun	1 AU	$10^{-8}\,s^{-1}$	5.5	4.2	17	8.5	13
Quiet Sun	$1R_\odot$	$10^{-3}\,s^{-1}$	2.5	1.9	7.9	3.9	6.0
Coronal Hole	$1R_\odot$	$10^{-3}\,s^{-1}$	2.1	1.1	5.3	1.8	3.2
Ionization time		s	480	910	190	560	310

Beckers (1972) has reasoned that UV radiation below 504 Å, the ionization limit of He, can penetrate to the center of spicules. But for longer wave lengths, we cannot exclude appreciable attenuation. However, in the material near the sides and the top of the spicules, the rates for zero optical depth may still serve as a guide to relative ionization rates, especially when species with similar ionization potential are compared.

Table III and IV allow us to make the following observations: (1) Ionization by UV may well be significant (cf. Zirin, 1975): the number of photons capable of ionizing helium that are leaving coronal holes is $\sim 10^2$ times larger than the number of leaving helium ions. (2) Ionization by UV of spicular material should be significant because UV ionization times are comparable to spicule lifetimes (Table III). (3) Among all the elements, helium is most slowly ionized by UV (for zero optical depth). At any depth, ionization by UV of He is slower than of Ne, and ionization by UV of O is faster than of H. (4) UV is insignificant for producing He^{2+} and Ne^{2+}. Production of O^{2+} by UV could only be significant (cf. Table IV) if the gas is maintained for some time at $(1-2) \times 10^4$ K.

If the UV radiation field is negligible, and if collisional ionization and recombination are in equilibrium, H and He are ionized at $\sim 15\,000$ and $\sim 25\,000$ K, respectively, i.e. there is a relatively small temperature difference. However, the appropriate parameter to consider here is the enthalpy, i.e. the heat that must be supplied (at constant pressure) to get partial ionization of an element. If we assume equilibrium between collisional ionization and recombination, the enthalpy per atom of a gas of solar composition with an initial temperature of 5000 K must be increased by 3.5 eV for 10% of H to be ionized, and by 22 eV for obtaining 10% ionization of He. Thus, when the gas is heated up, there is a large interval in which hydrogen is partly ionized but the He^+/H^+ ratio remains low.

We still need to discuss whether there are processes that can sufficiently separate neutrals from ions. Several authors have recognized the potential importance of diffusion on EUV line intensities (Tworkowski, 1975; Shine *et al.*, 1975; Meyer and Nussbaumer, 1979; Roussel-Dupré, 1980). Shine *et al.* (1975) have studied helium diffusion in the quiet-Sun transition zone by solving the time dependent one-dimensional diffusion equation with collisional ionization and recombination as source and sink terms. They find appreciable abundance changes at altitudes of a few hundred km after only 100 s. However, outflow is not considered in their work.

We shall discuss here the problem of separation between neutrals and ions in a general way, and try to derive some relations that are not restricted in their applicability to a particular configuration. A significant fraction of helium will remain neutral only for $T \lesssim 3 \times 10^4$ K. At such low temperatures, ion diffusion velocities are small even for large temperature gradients. If we consider the extreme case in which diffusion is essentially driven by the thermal gradient, we observe (cf. Equation (9)) that in a proton-electron gas the Δv_i's for all heavier ions ($A \geq 3$) are the same within a factor of about 2:

$$\Delta v_i \approx 1.5 \times 10^{-2} \frac{T^{5/2}}{n_p} \frac{\mathrm{d}\ln T}{\mathrm{d}s} \tag{16}$$

(Δv_i in km s^{-1}, ds in km, n_p in cm^{-3}).

This results from the fact that the diffusion constant is proportional to $1/Z_i^2$ and the thermal diffusion coefficient roughly proportional to Z_i^2 (cf. Equation (10) and (11)). The similarities of Δv_i's are also apparent at the lowest altitudes in Figure 2.

For the downward drift velocity of neutral helium, under the influence of gravity, we obtain from (9) and (13)

$$\Delta v_n (\text{km s}^{-1}) \approx 1.5 \times 10^9 \frac{T_4^{1/2}}{n_p(\text{cm}^{-3})} . \tag{17}$$

In order to compare the magnitudes of the downward velocity of He$^\circ$ with the diffusion velocities of the ions in the direction of the positive temperature gradient, we take $T \lesssim 3 \times 10^4$ K and assume that the scale height of T is not smaller than 50 km. Then we obtain from Equation (16)

$$\Delta v_i (\text{km s}^{-1}) \lesssim \frac{5 \times 10^7}{n_p(\text{cm}^{-3})} . \tag{18}$$

Equations (17) and (18) allow some general inferences: (1) Parallel to **B**, the separation velocity between neutral helium and the plasma is nearly two orders of magnitude larger than the relative velocities between ions of different mass and/or charge. Thus, ion-neutral separation will dominate over ion-ion separation under the circumstances discussed here. Perpendicular to **B**, the dominance of neutral over ion diffusion is even more pronounced. (2) If the time available for separation is of the order of the lifetime of spicules (~ 500 s), we obtain for $T_4 = 3$ a neutral-ion displacement s (km) $= 130 \times 10^{10}/n$ (cm^{-3}). Parallel to **B**, a displacement of a few hundred km is probably needed for having a significant separation of ions and neutrals. Thus, if we have diffusive separation only under the influence of gravity and the thermal gradient, the proton number density should be $< 10^{10}$ cm^{-3}. Perpendicular to **B**, diffusion of neutrals over shorter distances could lead to an effective separation, i.e. the limit on the density would be higher. (3) Separation could be helped by fields and waves that act on the charged particles and not on the neutrals. However, for making a significant contribution, and thus for increasing the upper limit on the density, such additional forces need to impart an acceleration to the plasma that is at least of the order of magnitude of the gravitational acceleration (cf. Equation (9)). In this case, neutral-ion separation will be much stronger again than the separation between different ions, because even parallel to **B**, the diffusion constant at $T \lesssim 3 \times 10^4$ K for He$^\circ$ is larger than for all ions by a factor of $> 10^2$ (cf. Equations (10) and (13)).

6. Conclusions

Observations of differences between the composition in the Sun and the solar wind, and of time variations of solar wind abundances indicate that element and isotope fractionation is not the result of one process, but that probably several mechanisms have to be distinguished.

6.1. QUASI-STEADY-STATE FRACTIONATION IN THE CORONA

The strong variability of He/H in the low speed solar wind, and the observation that average element depletion does not seem to be a function of atomic mass (cf. Figure 1), can be explained by models in which coupling between the ions is accomplished by Coulomb interaction. Geiss *et al.* (1970) have concluded that the relevant coupling parameter in the solar wind acceleration region is Z_i^2/A_i'. Furthermore, they showed that this parameter has to be modified when thermal diffusion (again a Coulomb collision effect) is taken into account; and that this effect, however, is not very significant in the flat temperature profiles that are inferred for the solar wind acceleration region.

If for an ion species Coulomb friction is insufficient, it will accumulate in the lower part of the acceleration region. Such a *dynamic accumulation* should be most pronounced for ^{4}He, because among the ions of the elements He to Fe existing at $T > 10^6$ K, ^{4}He^{2+} has the lowest Z_i^2/A_i'.

We may consider *dynamic accumulation* and *static stratification* as limiting cases of ion separation mechanisms in the corona. *Static stratification* obtains when the local bulk speed is low enough for diffusive equilibrium to be reached. If the temperature gradient is smaller than the critical gradient T_c' (Equation (15)), ions are stratified with concentration scale heights proportional $1/A_i'$ (from Equation (9) with $\Delta v_i = 0$), i.e. protons are enriched at high altitudes relative to all other elements. The small negative temperature gradient that exists above the temperature maximum in the corona tends to accentuate the proton enrichment.

Recently, Borrini *et al.* (1981) have demonstrated that minimal He/H ratios coincide with sector boundaries. They suggest that these low helium abundances could be caused if the velocities of H and He in streamers begin to be *relatively* similar only at rather high altitude, i.e. if solar wind acceleration in streamers effectively starts from the upper levels of a stratified corona. If this is *static stratification* in the sense described above, we would expect an *overabundance of H* relative to the other elements in the escaping gas.

6.2. NON-STEADY-STATE EFFECTS

Occasionally, solar wind plasmas with very high helium abundances (He/H $\gtrsim 0.15$) are observed (Robbins *et al.*, 1970; Hirshberg *et al.*, 1970; Bame *et al.*, 1979). The latter authors have identified some of these solar wind samples as flare expelled plasma. The anomalously high helium abundance can be explained (Hirshberg *et al.*, 1970; Hundhausen, 1972) by the high local He/H density ratio that resulted from *dynamic accumulation* or *static stratification* prior to the flare. Bame *et al.* (1979) found that O and Fe are much less enriched than He in the flare-expelled plasma, relative to the interstream solar wind. This may indicate that the high He abundance in this plasma is – at least partly – due to *dynamic accumulation*.

6.3. ANOMALOUS ^{3}He ENHANCEMENTS

Bame *et al.* (1968) and Grünwaldt (1976) observed some occurrences of high ^{3}He abundance with ^{3}He/^{4}He ratios of $\sim 2 \times 10^{-3}$, i.e. a 5-fold increase above the average (cf. Table I). Ogilvie *et al.* (1980) have confirmed these observations with a mass spectrometer and found among their evaluated spectra 6% with ^{4}He/^{3}He ≤ 600. The event studied by Grünwaldt (1976) is particularly relevant. For a period of two days, ^{3}He/^{4}He was $\sim 2 \times 10^{-3}$ while H : He : O was quite normal. Obviously it is hard to explain these observations with a steady-state Coulomb friction model, even if one includes thermal diffusion. On the other hand, in view of its duration, this event cannot be dismissed as a strange transient. Possibly, ^{3}He enhancements in the solar wind are caused by a mechanism which is similar to the one causing the ^{3}He enhancements in flares (cf. Fisk, 1978), only that the solar wind mechanism is much weaker.

6.4. INCOMPLETE IONIZATION

There are three observations that point towards ion-neutral separation as a possible mechanism affecting solar wind composition: (a) Studies of Meyer (1981) and Veck and Parkinson (1981) indicate that corona abundances are related to the first ionization potential ϕ_I. They suggest that elements with $\phi_I \gtrsim 9$ eV have low abundances relative to hydrogen. (b) The steady, systematic helium depletion in the solar wind from coronal holes is difficult to explain by ion separation in the acceleration region (Section 4). (c) The average ^{3}He/^{4}He ratio in the solar wind is very close to the ratio inferred for the outer convective zone; the latter is, however, subject to a rather large uncertainty (cf. Table I and Figure 1). A similar reduction of ^{3}He and ^{4}He is not expected from separation processes in the corona, but it would naturally result from separation processes that depend on incomplete ionization.

In Section 4, we have discussed the circumstances under which ion-neutral separation could take place. We give limits on temperature and density above which the separation mechanism becomes ineffective. These limits are valid for various geometries in which the mechanism might operate, e.g. in or around spicule-like features or in a flat surface. We have shown that ion-neutral separation is more effective than ion-ion separation at temperatures low enough for the partial survival of neutral atoms. Thus, this mechanism would primarily separate according to the fraction of the species that is ionized at moderate temperature. Separation of ions with different mass or charge states would be less important. Of all the elements, helium would be most affected in this process, whether ionization is primarily by electron collision or by UV. Other elements might also be depleted in relation to the rate at which they become ionized. If ionization is primarily by electron collisions, depletion would most likely occur for He, Ne, and Ar. Abundance *enhancements* relative to hydrogen could result for elements with low ionization potential, e.g. Mg, Si, Fe; and also C, if ionization is by electron collisions. Isotopic fractionation as a result of this process should be small.

The ion-neutral separation mechanism could operate inside and outside coronal holes, and it is difficult to estimate where He/H separation would be more effective. We suggest, however, that the occurrences of very low He/H ratios in the solar wind are not primarily due to this mechanism, but that they are caused in the corona.

Acknowledgements

The author obtained valuable advice from Sir Harrie Massey and Martin Huber. He has benefitted from discussions with P. Bochsler, F. Bühler, A. Bürgi, and D. T. Young, and he thanks G. Troxler for her help in preparing the manuscript. This work was in part supported by the Swiss National Science Foundation.

References

Arrhenius, G. and Alfvén, H.: 1971, *Earth Planet. Sci. Letters* **10**, 253.

Audouze, J.: 1981, Paper presented at *Study Week on Cosmology and Fundamental Physics*, Città del Vatican.

Bame, S. J., Hundhausen, A. J., Asbridge, J. R., and Strong, I. B.: 1968, *Phys. Rev. Letters* **20**, 393.

Bame, S. J., Asbridge, J. R., Feldman, W. C., Montgomery, M. D., and Kearney, P. D.: 1975, *Solar Phys.* **43**, 463.

Bame, S. J., Asbridge, J. R., Feldman, W. C., and Gosling, J. T.: 1977, *J. Geophys. Res.* **82**, 1487.

Bame, S. J., Asbridge, J. R., Feldman, W. C., Fenimore, E. E., and Gosling, J. T.: 1979, *Solar Phys.* **62**, 179.

Banks, P. M. and Kockarts, G.: 1973, *Aeronomy*, Academic, New York.

Beckers, J. M.: 1972, *Ann. Rev. Astron. Astrophys.* **10**, 73.

Bochsler, P. and Geiss, J.: 1973, *Solar Phys.* **32**, 3.

Bochsler, P. and Geiss, J.: 1967, *Transactions IAU* **XVI B**, 120.

Bohlin, J. D.: 1977, in J. B. Zirker (ed.), *Coronal Holes and High Speed Wind Streams*, Colorado Assoc. University Press, Boulder, p. 27.

Bohlin, J. D., Vogel, S. N., Purcell, J. D., Sheeley, N. R. Jr., Tousey, R., and Van Hoosier, M. E.: 1975, *Astrophys. J.* **197**, L133.

Borrini, G. and Noci, G.: 1979, *Solar Phys.* **64**, 367.

Borrini, G., Gosling, J. T., Bame, S., J., Feldman, W. C., and Wilcox, J. M.: 1981, *J. Geophys. Res.* **86**, 4565.

Bruston, P., Audouze, J., Vidal-Madjar, A., and Laurent, C.: 1981, *Astrophys. J.* **243**, 161.

Burgers, J. M.: 1969, *Flow Equations for Composite Gases*, Academic, New York.

Cameron, A.G.W.: 1980, to appear in *A Festschrift in Honor of Willy Fowler's 70th Birthday*.

Cerutti, H.: 1974, PhD Thesis, University of Bern.

Chapman, S. C. and Cowling, T. G.: 1958, *The Mathematical Thory of Non-Uniform Gases*, Cambridge University Press.

Delache, P.: 1965, *Comp. Rend. Acad. Sci. (France)* **261**, 643.

Delache, P.: 1967, *Ann. Astron.* **30**, 827.

Dupree, A. K.: 1972, *Astrophys. J.* **178**, 527.

Eberhardt, P.: 1978, *Proc. Lunar Planet. Sci. Conf. 9th*, p. 1027.

Feldman, W. C., Asbridge, J. R., Bame, S. J., and Gosling, J. T.: 1976, *J. Geophys. Res.* **81**, 5054.

Feldman, W. C., Asbridge, J. R., Bame, S. J., and Gosling, J. T.: 1977, in O. R. White (ed.), *The Solar Output and its Variation*, Colorado Associated Press, Bouler, p. 351.

Fisk, L. A.: 1978, *Astrophys. J.* **224**, 1048.

Frick, U. and Moniot, R. K.: 1977, *Proc. Lunar Sci. Conf. 8th*, p. 229.

Gabriel, A. H.: 1976, in R. M. Bonnet and Ph. Delache (eds.), 'The Energy Balance and Hydrodynamics of the Solar Chromosphere and Corona', *IAU Colloq.* **36**, 375.

Geiss, J.: 1972, in C. P. Sonett, P. J. Coleman Jr., and J. M. Wilcox (eds.), *Solar Wind*, NASA SP-308, p. 559.

Geiss, J. and Reeves, H.: 1972, *Astron. Astrophys.* **18**, 126.

Geiss, J. and Bochsler, P.: 1981, in H. Rosenbauer (ed.), *Solar Wind Four*, MPAE, 3411 Lindau, F.R.G., p. 403.

Geiss, J., Hirt, P., and Leutwyler, H.: 1970, *Solar Phys.* **12**, 458.

Geiss, J., Eberhardt, P., Bühler, F., Meister, J., and Signer, P.: 1970a, *J. Geophys. Res.* **75**, 5972.

Geiss, J., Bühler, F., Cerutti, H., Eberhardt, P., and Filleux, Ch.: 1972, *Apollo 16 Prel. Sci. Rep.*, NASA SP-315.

Grünwaldt, H.: 1976, *Space Res.* **16**, 681.

Helbig, H. F., Millis, D. B., and Todd, L. W.: 1970, *Phys. Rev.* **A2**, 771.

Hirshberg, J., Alksne, A., Colburn, D. S., Bame, S. J., and Hundhausen, A. J.: 1970, *J. Geophys. Res.* **75**, 1.

Huber, M. C. E., Foukal, P. V., Noyes, R. W., Reeves, E. M., Schmahl, E. J., Timothy, J. G., Vernazza, J. E., and Withbroe, G. L.: 1974, *Astrophys. J.* **194**, L115.

Hundhausen, A. J.: 1972, *Coronal Expansion and Solar Wind*, Springer, Berlin Heidelberg New York.

Jeffery, P. M. and Anders, E.: 1970, *Geochim. Cosmochim. Acta* **34**, 1175.

Jokipii, J. R.: 1965, California Institute of Technology Thesis; 1966, in R. J. Mackin and M. Neugebauer (eds.), *The Solar Wind*, Pergamon Press, New York, p. 215.

Joselyn, J. and Holzer, T. E.: 1978, *J. Geophys. Res.* **83**, 1019.

Laurent, C., Vidal-Madjar, A., and York, D. G.: 1979, *Astrophys. J.* **229**, 923.

Marr, G. V. and West, J. B.: 1976, *Atomic Data and Nuclear Data Tables* **18**, 497.

Massey, H. M. S.: 1982, personal communication.

McKenzie, J. F., Ip, W.-H., and Axford, W. I.: 1979, *Astrophys. and Space Sci.* **64**, 183.

Meyer, J.-P.: 1979, in *Les Elements et leurs Isotopes dans l'Univers*, 22nd Liège Internat. Astrophys. Symp., University of Liège Press, p. 489.

Meyer, J.-P.: 1981, *17th Int. Cosmic Ray Conf., Paris* **3**, 149.

Meyer, A. and Nussbaumer, H.: 1979, *Astron. Astrophys.* **78**, 33.

Munro, R. H. and Jackson, B. V.: 1977, *Astrophys. J.* **213**, 874.

Nakada, M. P.: 1969, *Solar Phys.* **7**, 302.

Neugebauer, M.: 1981, *Fundamentals of Cosmic Physics* **7**, 131.

Neugebauer, M.: 1981a, in H. Rosenbauer (ed.), *Solar Wind Four*, MPAE, 3411 Lindau, F.R.G., p. 425.

Neugebauer, M. and Snyder, C. W.: 1966, *J. Geophys. Res.* **71**, 4469.

Ogilvie, K. W.: 1972, *J. Geophys. Res.* **77**, 4227.

Ogilvie, K. W. and Hirshberg, J.: 1974, *J. Geophys. Res.* **79**, 4595.

Ogilvie, K. W., Coplan, M. A., Bochsler, P., and Geiss, J.: 1980, *J. Geophys. Res.* **85**, 6021.

Pneuman, G. W. and Kopp, R. A.: 1977, *Astron. Astrophys.* **55**, 305.

Reeves, E. M.: 1976, *Solar Phys.* **46**, 53.

Rich, W. G., Bobbio, S. M., Champion, R. L., and Doverspike, L. D.: 1971, *Phys. Rev. A* **4**, 2253.

Robbins, D. E., Hundhausen, A. J., and Bame, S. J.: 1970, *J. Geophys. Res.* **75**, 1178.

Rood, R. T., Wilson, T. L., and Steigman, G.: 1979, *Astrophys. J. Letters* **227**, L97.

Ross, J. E. and Aller, L. H.: 1976, *Science* **191**, 1223.

Roussel-Dupré, R.: 1980, *Astrophys. J.* **241**, 402.

Schatzman, E. and Maeder, A.: 1981, *Astron. Astrophys.* **96**, 1.

Schunk, R. W. and Walker, J. C. G.: 1969, *Planet. Space Sci.* **17**, 853.

Shine, R., Gerola, H., and Linsky, J. L.: 1975, *Astrophys. J.* **202**, L101.

Tworkowski, A. S.: 1975, *Astrophys. Letters* **17**, 27.

Veck, N. J. and Parkinson, J. H.: 1981, *Monthly Notices Roy. Astron. Soc.* **197**, 41.

Withbroe, G. L.: 1981, in A. O. Benz, Y. Chmielewski, M. C. E. Huber, and H. Nussbaumer (eds.), *Activity and Outer Atmospheres of the Sun and Stars*, Swiss Society of Astronomy and Astrophysics, Saas-Fee, p. 1.

Zastenker, G. N. and Yermolaev, Yu. I.: 1981, *Planet. Space Sci.* **29**, 1235.

Zirin, H.: 1975, *Astrophys. J.* **199**, L63.

OBSERVATIONS OF CORONAL STRUCTURE DURING SUNSPOT MAXIMUM†

N. R. SHEELEY, Jr., R. A. HOWARD, M. J. KOOMEN, and D. J. MICHELS

E. O. Hulburt Center for Space Research, Naval Research Laboratory, Washington, D.C. 20375, U.S.A.

K. L. HARVEY*

Solar Physics Research Corporation, Tucson, Ariz. 85718, U.S.A.

and

J. W. HARVEY

Kitt Peak National Observatory**, Tucson, Ariz. 85726, U.S.A.

Abstract. This paper presents some of the results that have been obtained from the Kitt Peak observations of coronal holes and the NRL observations of coronal transients during the recent years near sunspot maximum (1979–1981). On the average, low-latitude coronal holes of comparable size contained 3 times more flux near sunspot maximum than near the previous minimum. In the outer corona, transients occurred at the observed rate of at least 2 per day, and quiet conditions persisted during less than 15% of the observed days. We describe a sample of the more than 800 events that we have observed so far, including the observation of a comet apparently colliding with the Sun.

1. Introduction

The objective of this review is to list a few recent observations that add to our knowledge of coronal structure near sunspot maximum. The first subject concerns the discovery by Harvey *et al.* (1981) that coronal holes had greater average field strengths at their bases in 1979 and 1980 than they did during 1973–1978. This result may simply reflect the fact that there was more magnetic flux on the Sun at sunspot maximum than at minimum. Nevertheless, such a sunspot-cycle variation of coronal hole field strength was not anticipated in 1975–1976 when coronal holes were being studied intensively at the first Skylab Workshop Series (Zirker, 1977).

The second subject concerns the great number and variety of coronal transient phenomena that have been observed by the NRL Earth-orbiting coronagraph during 1979–1981. With only 40% of the data yet reduced, routine observations of the white-light corona between 2.5 and $10.0R$ have already shown 495 mass ejections on a time scale of a few hours or less and 332 additional, slower events. With processed images for parts of 390 days, this corresponds to an average of at least 2 events per day. Truly quiet days were rare with only 63 days (15%) showing slow evolution without transients.

---

† Paper presented at the IX-th Lindau Workshop 'The Source Region of the Solar Wind'.
* Visiting Astronomer, KPNO.
** Operated by the Association of Universities for Research in Astronomy, Inc., under contract with the National Science Foundation.

Space Science Reviews **33** (1982) 219–231. 0038–6308/82/0332–0219$01.95.

2. Magnetic Measurements of Coronal Holes

During the Skylab Workshop Series on Coronal Holes, it was suggested that the average magnetic field strength at the base of a coronal hole might be as large as 10 G (see for example, Hundhausen (1977), and references contained therein). This conclusion was based on the conservation of magnetic flux within a hypothetical tube that extended from the base of a coronal hole at the Sun's surface to a corresponding feature in interplanetary space. Thus, by measuring the interplanetary field strength, B_{ip}, and estimating the ratio of coronal hole area, A_{ph}, to the assumed area of the corresponding interplanetary feature, A_{ip}, one could deduce the average field strength at the photosphere, B_{ph}, from

$$B_{ph} = B_{ip} \left(\frac{A_{ip}}{A_{ph}} \right). \tag{1}$$

In this way, the interplanetary measurements led to values of $B_{ph} \approx 10$ G for the coronal holes that were observed in 1973–1974 during the Skylab mission. However, direct measurements of the photospheric magnetic flux within these same coronal holes gave average field strengths in the range 0.3 – 7.2 G, with 2–3 G being the most commonly occurring values (Howard and Harvey, 1977; Bohlin and Sheeley, 1978).

Since the end of the final Skylab mission in February 1974, direct observations of coronal holes against the solar disk have been limited to relatively few X-ray images obtained from one or two rocket flights per year. However, helium images show the location of these X-ray holes as regions where the chromospheric network is weak or absent (Tousey *et al.* 1973; Harvey *et al.* 1975a, 1975b; Harvey and Sheeley, 1979), and He I 10830 Å images have been obtained almost daily at Kitt Peak since February 1974. Combined with the Skylab images during 1973–1974, this sequence of helium images provides high-quality synoptic observations of coronal holes over a 9-year interval that spans most of a sunspot cycle.

In the process of analyzing these long-term synoptic observations, Harvey *et al.* (1982) noticed that some holes in 1979 and 1980 seemed to contain considerably more net flux then the holes that occurred during the declining phase of the cycle and near sunspot minimum (cf. Harvey and Sheeley, 1980).

They also noticed that the central meridian passage dates of the new low-latitude holes often correlated with the days of enhanced mean solar magnetic field as observed from Earth, just as Scherrer and Svalgaard (1977) had found earlier. This latter result suggested that coronal holes may sometimes contribute a significant fraction of the Sun's net flux as seen from Earth. By 1979 the Sun's mean field strength often reached 0.5 G and even exceeded 1.0 G on some days. This latter value corresponds to a net flux of 15×10^{21} Mx averaged over the solar disk. Harvey *et al.* noted that to account for even half this amount of flux, a typical coronal hole of area 4×10^{20} cm^2 would have an average field strength of 19 G.

In pursuit of this idea, Harvey *et al.* (1982) measured the areas and magnetic fluxes at the base of 33 coronal holes on 63 separate days during 1975–1980. Also, to test the consistency of their measurement technique, they remeasured the fluxes and areas of some of the earlier holes in 1973–1974.

We shall describe their general technique and typical results with the aid of Figure 1. First, coronal holes were identified on the helium images (left) as regions where the chromospheric network was weak or absent. In this figure, the holes are outlined by white contours. Second, these contours were transposed to the corresponding photospheric magnetograms (right). In the magnetograms, lighter-than-average features indicate positive line-of-sight components of field (toward the observer) and darker-than-average features indicate negative fields. Third, the net fluxes and areas within each contour were measured, and the mean field strength was computed from their ratio. As the caption indicates, the resulting mean field strengths ranged from 5.2 to 25.7 G for these four coronal holes.

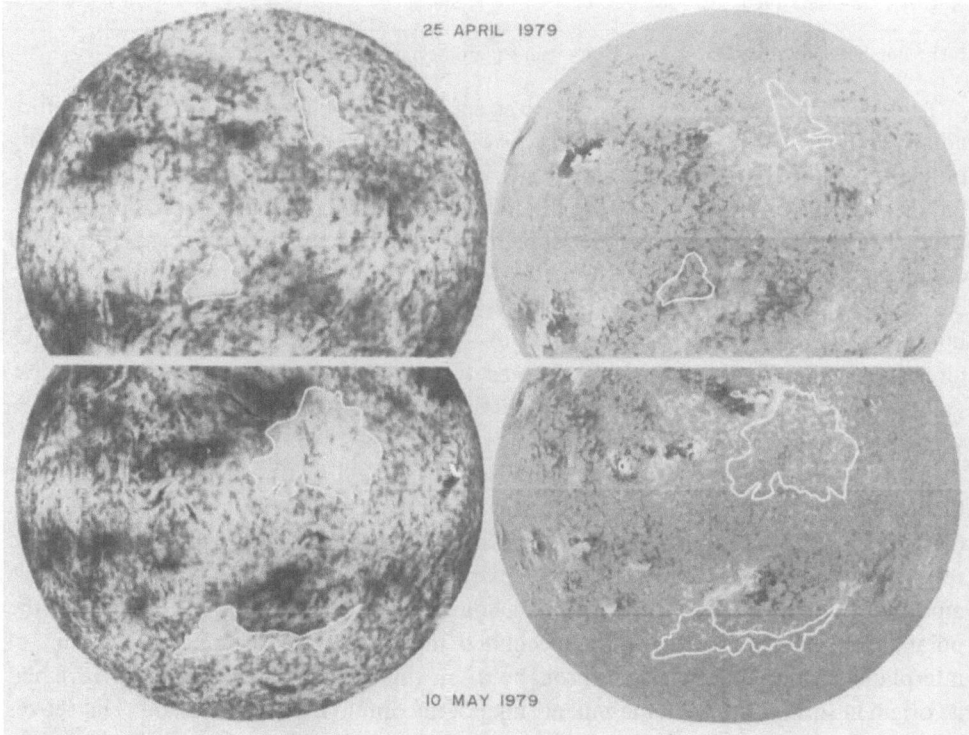

Fig. 1. He I 10830 Å images (left) and photospheric magnetograms (right) illustrating the variety of coronal holes that contributed to the broadening as well as the shifting of the coronal hole field strength distribution in 1979 and 1980. The fluxes and average field strengths within these holes are:

April 25			May 10		
N	16.9 G	3.6×10^{21} Mx	N	12.7 G	12.9×10^{21} Mx
S	-25.7 G	-3.7×10^{21} Mx	S	5.2 G	2.9×10^{21} Mx

(from Harvey et al., 1982).

Harvey *et al.* found that prior to July 31, 1978 all of the measured average field strengths lay within the 0.3–7.2 G range that Bohlin and Sheeley (1978) had reported for holes during 1973–1974. However, after July 31, 1978 most of the measured field strengths exceeded 7.2 G. During 1979, five independent holes had average field strengths exceeding 20 G, and one 'transient' hole had a strength of 36.3 G.

TABLE I

Average values of field strength, flux, and area of a sample of coronal holes during intervals before and after July 31, 1978

Interval	Number of measurements	Field strengths (G)	Fluxes (10^{21} Mx)	Areas (10^{20} cm^2)
February 1975–July 1978	30	4.1(1.7)	1.6(1.1)	4.0(2.4)
August 1978–June 1980	38	11.9(7.5)	5.0(3.9)	4.5(2.7)

[a] The numbers in parenthesis are the root-mean-square deviations from the averages.

Table I contains average values of field strength, flux, and area of coronal holes before and after July 31, 1978. The average areas of the holes that they measured changed by only a factor of 1.1 from 4.0 to 4.5 × 10^{20} cm^2. In contrast, the fluxes increased by a factor of 3.1 from 1.6 to 5.0 × 10^{21} Mx. The average field strengths also increased by a factor of 3.1 from 4.1 to 11.9 G. As shown by the numbers in parenthesis, the standard deviations of flux and field strength also increased substantially from 1.1 to 3.9 × 10^{21} Mx and 1.7 to 7.5 G, respectively. Harvey *et al.* concluded that the distributions of flux and field strength shifted to higher values and broadened. As we have already seen in Figure 1, these broadened distributions include some relatively low values that are comparable to the ones obtained earlier as well as some values that are very much higher than the earlier ones.

Harvey *et al.* supposed that this increase in coronal hole field strength simply reflected the fact that there was more flux on the Sun near sunspot maximum then near minimum (3 times more flux according to Howard and LaBonte, 1981), and that the coronal holes received their proportionate share of the extra flux. They also noted that a comparable amount of open flux from the large polar holes had disappeared by 1979–1980 as the polar fields vanished. Thus, they speculated that the relatively constant amount of interplanetary flux that persisted throughout this time (King 1979, 1981) may have had its origin in smaller areas on the Sun at sunspot maximum than at minimum. This leaves us with the question of whether or not such an increased flux tube divergence might account for the relatively low speeds of solar wind streams at Earth during sunspot maximum. Evidently the theoretical treatments of this problem have given contradictory results (Kopp, 1977).

3. Coronal Transients During 1979–1981

The NRL white light coronagraph has been obtaining full field images of the Sun's outer corona (2.5–10.0R) routinely since it began operating in Earth-orbit on March 28, 1979.

With few interruptions, this instrument has obtained images at 10min intervals during the 1 h sunlit portion of each 97 min satellite polar orbit. This instrument and its initial observations have been described in detail elsewhere (Sheeley *et al.*, 1980a).

Observations have been processed for 390 days since March 28, 1979. The processing consists of constructing one ordinary coronal image for each satellite orbit and one difference image which shows the change that has occurred between that orbit and the start of that day (or mid-day for orbits in the second half of each day). When obvious changes are detected, the remaining 4 difference images in each orbit are processed to show the temporal evolution of each change.

With data reduced for parts of 1979, 1980, and 1981, we have observed 495 mass ejections on a time scale of a few hours or less and 332 additional events on a time scale of 6–12 h. As we have described earlier (Sheeley *et al.*, 1980b), during sunspot maximum coronal transients, like coronal streamers, occurred at all position angles. Hildner *et al.* (1981) have also noted this result for observations during March–September 1980 with the coronagraph on the Solar Maximum Mission Satellite. Also in accord with the SMM observations, we found that the coronal transient angular distribution had maxima at the sunspot belts during 1979–1981. Some preliminary statistics such as speeds (150–900 km s^{-1}) and masses (7 × 10^{14} – 2 × 10^{16} g) were described by Poland *et al.* (1981) using relatively few events. We have not yet extended the measurements of ejected coronal masses, but we have seen both lower (50 km s^{-1}) and higher (1200 km s^{-1}) speeds in our larger sample of events.

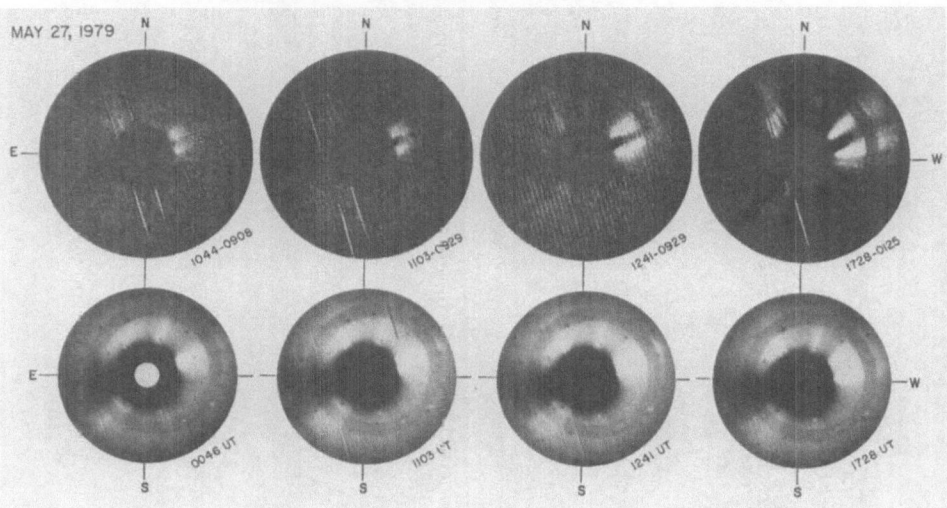

Fig. 2. Coronal mass ejection associated with the disruption of a pre-existing coronal streamer. In this figure and all subsequent figures, a small white disk in the center of the field indicates the size and location of the Sun's photosphere.

At least 151 (30%) of the 495 primary transients have a characteristic spike structure in our 2.5–10.0R field of view. Figure 2 shows a 2-pronged spike transient which was the origin of an interplanetary shock at Helios 1 on May 28 (Schwenn, 1981). Sometimes such 2-pronged events show a complete loop structure as they begin to rise through our field of view, but this one did not. Comparisons of similar events obtained simultaneously with other observations at smaller radial distances show that some of our spike transients began as bright loops in the lower corona (Wagner *et al.* 1980), and some did not (Fisher, 1981). In the latter case, the events began as expanding depletions that developed bright spikes along their sides (cf. Fisher, 1982). In Figure 2 a comparison of the ordinary images (below) with the subtracted images (above) suggests that this May 27, 1979 event involved the splitting of a bright, pre-existing coronal streamer. Such 'streamer blowouts' are relatively common.

Figure 3 shows a much smaller 2-pronged event on the morning of June 9, 1979. One can see that this event, unlike the May 27 event, began in our field of view as a complete loop, but eventually lost its top as it progressed outward. However, like the May 27 event, this June 9 one has a darker-than-average region at its base, indicating a coronal depletion. It too may have been a 'streamer blowout'. This was one of the 13 events (3%) that we called 'small loops' rather than 'spikes'.

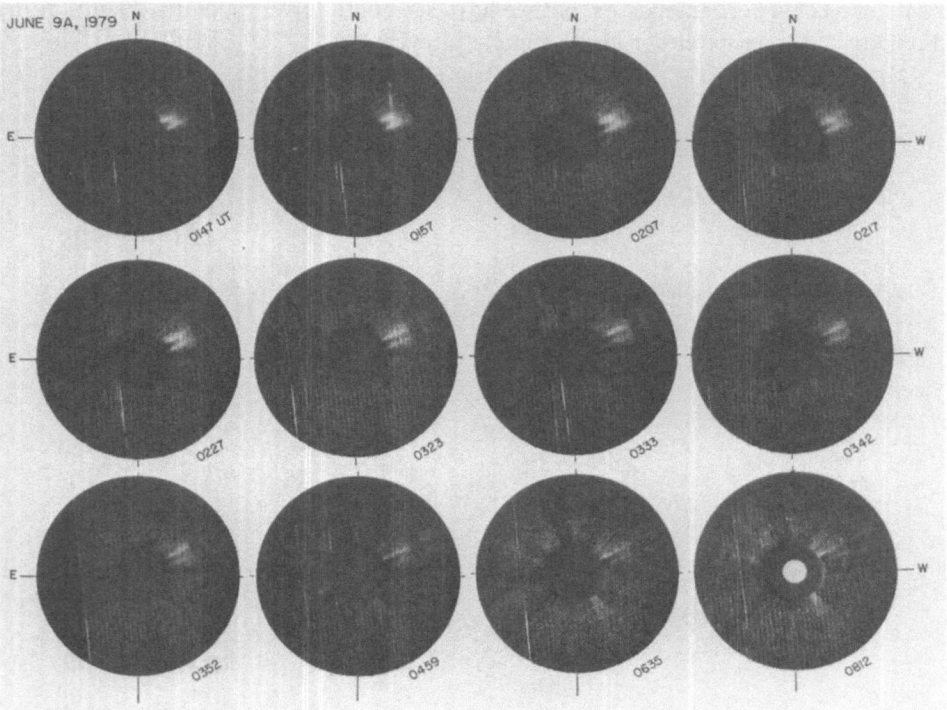

Fig. 3. Evolution of a coronal transient from a 'small loop' to a '2-pronged spike' configuration.

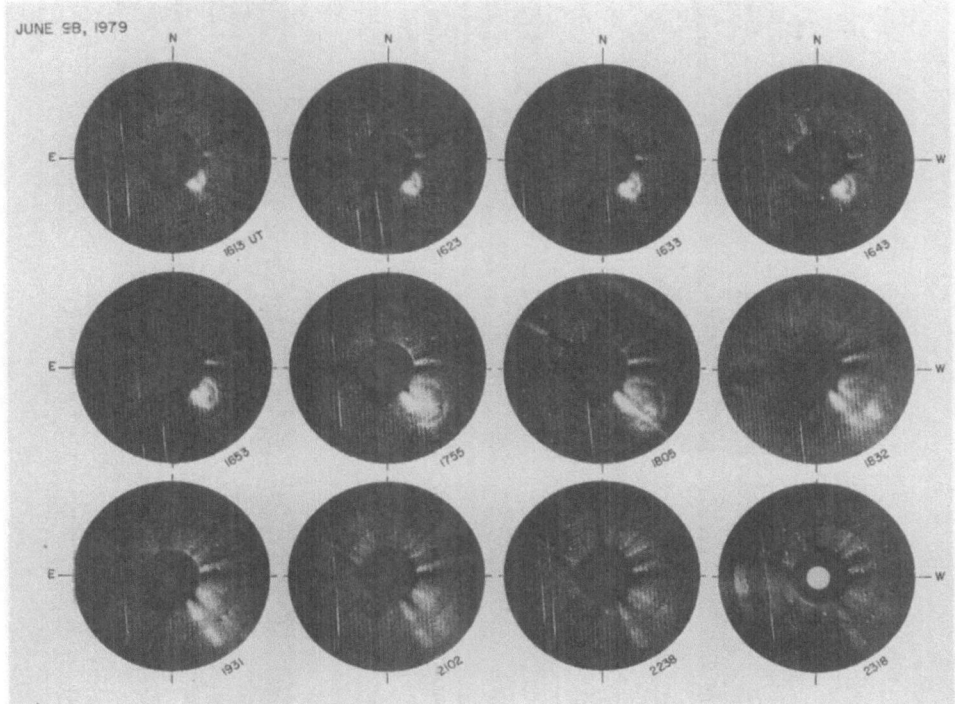

Fig. 4. Evolution of a 'big loop' coronal mass ejection (in the 18 : 05 UT image, the narrow streak extending SW to the edge of the field of view is an artifact).

Figure 4 shows a mass ejection during the afternoon of June 9, 1979. We call this event a 'big loop', although there are certainly some spike structures that follow the loop out into the field of view. We have classified 55 (11%) of our 495 major events as big loops, but this event is one of the few that was sufficiently well defined to show definite curvature in its height-time plot. Between 16 : 13 UT and 18 : 05 UT the front of the loop accelerated from approximately 275 to 600 km s^{-1} as it progressed from 3.8 to 7.3R. This event also seems to have been the source of an interplanetary shock at Helios 1 (Schwenn, 1981).

Figure 5 shows another loop-shaped mass ejection that was sufficiently well-defined to show definite curvature in its height-time plot. In this May 24, 1979 event, the loop accelerated rapidly from 375 to 1000 km s^{-1} in only 41 min. This trailing prominence material accelerated more slowly from 400 to 950 km s^{-1} in 127 min as it traversed the field of view. We have seen obvious prominence material in only 8 (2%) of our 495 major events.

Figure 6 shows a very large event that was already in progress at 07 : 35 UT on September 1, 1980 when we first observed it. We have called 9 (2%) such events 'quadrant fillers', but many of the events that we have called 'big loops' eventually

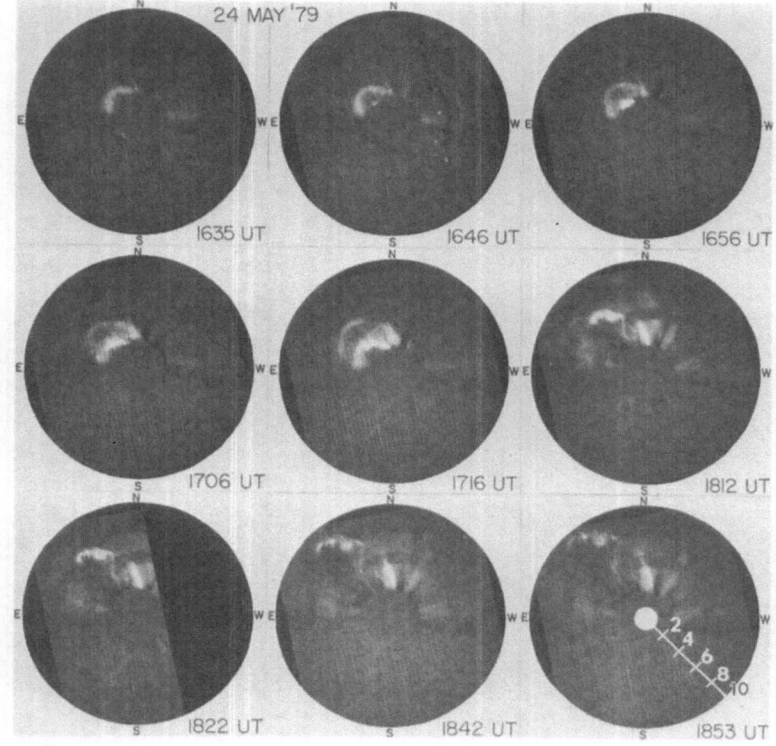

Fig. 5. Evolution of a 'big loop' followed by an eruptive prominence.

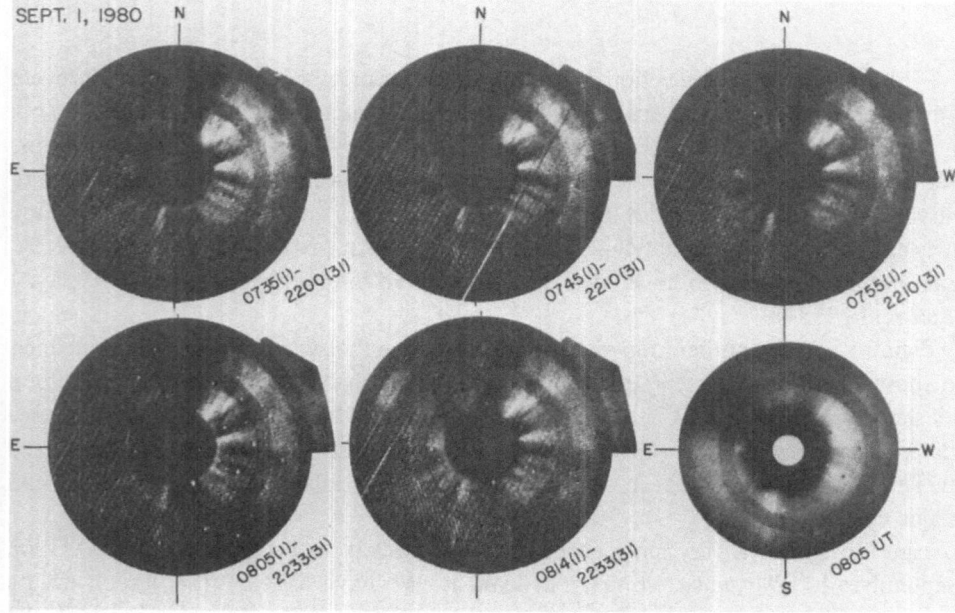

Fig. 6. A large 'quadrant filling' mass ejection in progress.

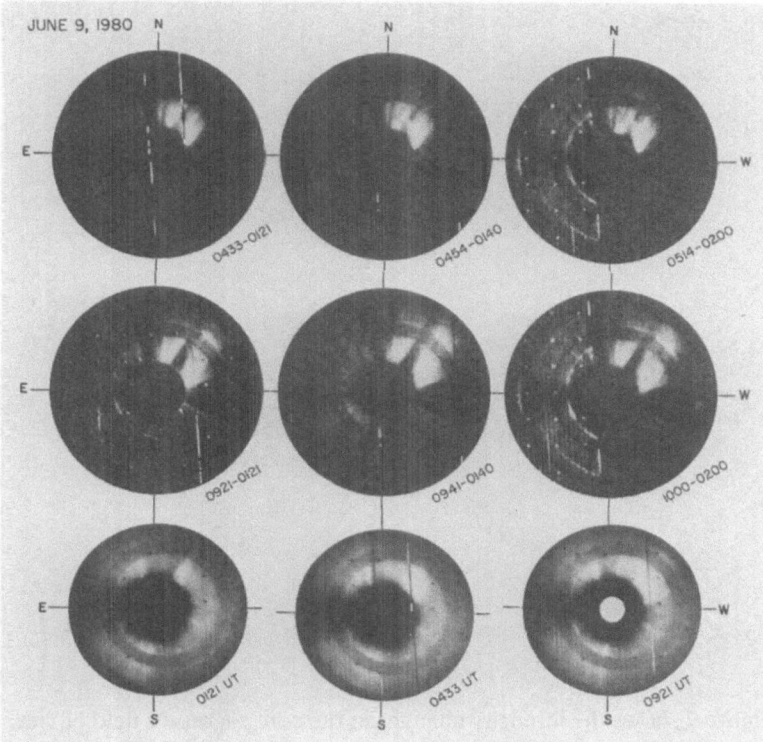

Fig. 7. A high-latitude mass ejection associated with the disruption of a pre-existing coronal streamer.

developed into such quadrant fillers. An interesting feature of this September 1, 1980 event is the dark hole that can be seen progressing radially out through the field of view during 07:35–08:14 UT. Its speed in the plane of the sky was approximately 1200 km s^{-1}. Like nearly all 'quadrant fillers' and 'big loops', this event was the origin of an interplanetary shock (Schwenn, 1981).

As we have already mentioned, near sunspot maximum, coronal transients occur at all position angles. Figure 7 shows a 2-pronged event centered at approximately N 60° on the west limb on June 9, 1980. The depletion that is visible in the difference images during 09:21–10:00 UT corresponds to the pre-event streamer that is visible in the ordinary image at 01:21 UT. The depletion indicates that this streamer was destroyed during the mass ejection. Figure 8 shows another transient projecting nearly over the north pole a year earlier on June 10, 1979. This spike event has no depletion, and consists of diffuse emission bordered by a pair of bright spikes. At present we do not know whether such events usually originate at high latitudes in association with (for example) eruptive polar prominences, or whether they are associated with low-latitude active regions (for example) on the front side or back side of the Sun and simply appear to be at high latitudes when they are seen in projection against the plane of the sky.

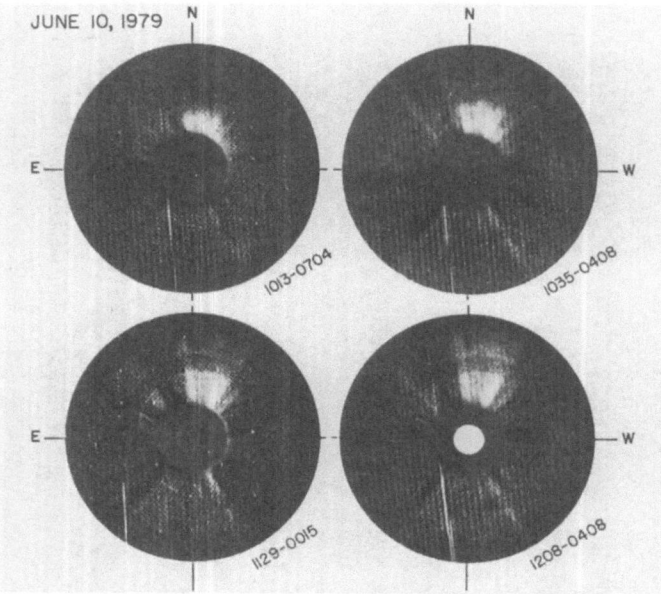

Fig. 8. A high-latitude mass ejection with a characteristic spike/fan structure.

Finally Figure 9 shows the intensity changes in the coronagraph's field of view during August 30–31, 1979 when a comet (Howard-Koomen-Michels 1979 XI) approached and apparently collided with the Sun (Michels *et al.*, 1981, 1982). Preliminary calculations based on the pre-perihelion images (left) suggest that this comet was a member of the Kreutz group of sungrazers (Kreutz, 1888, 1891, 1901; Marsden, 1967, 1981), and that it collided with the Sun during the interval 22 : 00–23 : 00 UT on August 30, 1979. The difference images in Figure 10 show the intensity changes that occurred since 23 : 44 UT on August 30, and suggest that the evolving pattern had 2 components.

One component consists of diffuse emission spread over a broad range of position angles from the northeast to the southwest. It first appeared beyond the occulting disk at 03 : 06 UT and spread in a northwest direction to eventually fill the entire northwest quadrant before it faded substantially by 23 : 48 UT. This pattern is unlike any coronal structure we have ever seen. However, it does match reasonably well the projected patterns that we have calculated for cometary dust assumed to be released from the inbound nucleus and driven into hyperbolic trajectories by the combined effect of solar radiation pressure and gravity. This fit is obtained for plausible values of the orbital elements (Michels *et al.*, 1982) and effective dust sizes ($G_{dust}/G_{gravity} \approx 2-3$).*

The other component consists of a relatively narrow fan of emission extending almost due northward and resembling the coronal mass ejection in Figure 8. The intensity of

* Since the presentation of this paper, Keller and Richter (1981) and Sekanina (1981, 1982) have reached essentially the same conclusion from their own calculations of the projected pattern of cometary dust.

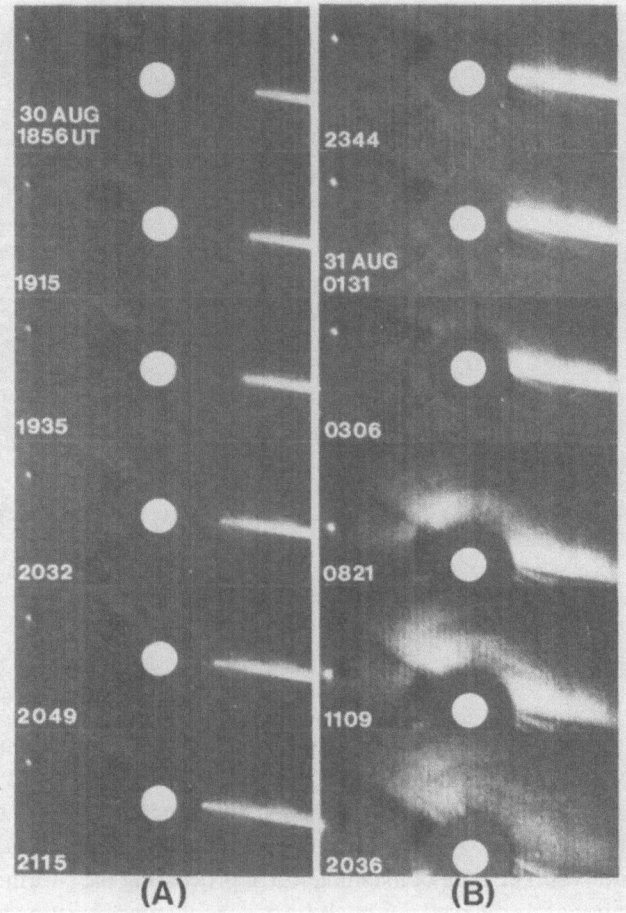

Fig. 9. Intensity changes in the coronagraph's field of view as comet Howard-Koomen-Michels 1979 XI
approached the Sun (left) and after perihelion/collision (right). Venus is at the extreme left.

this component was greatest during 10 : 59–14 : 11 UT and decreased relatively slowly
so that by 23 : 48 UT at least part of it dominated the fainter diffuse component. Like
the June 10, 1979 coronal transient in Figure 8, this narrow component was broken by
the polarizing ring at $5R$ which suggests that this radiation was produced by Thomson-
scattered emission from electrons in the plane of the sky. In contrast, in the northward
direction the diffuse component was not cut off by the polarizing ring. These facts (and
the knowledge that this northward location was the site of unusually high transient
activity during the few days prior to the arrival of the comet) lead this writer to suspect
that the narrow component was a coronal transient unrelated to the comet and that the
broad diffuse component was cometary dust. Work is in progress to resolve this
question.

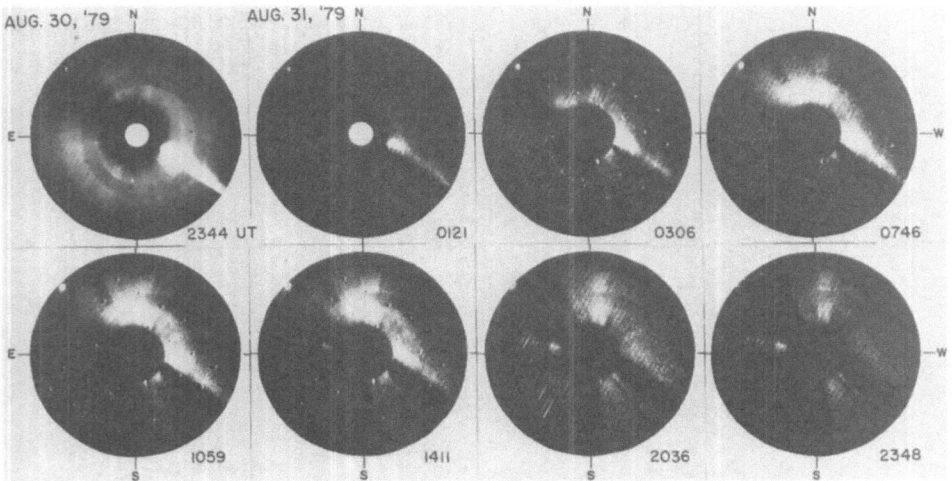

Fig. 10. Intensity changes in the coronagraph's field of view since 23:44 UT August 30 (shortly after perihelion/collision). Venus is at the upper left.

4. Conclusion

All of the results presented here were derived from routine observations over a long period of time ranging from at least 3 yr for the NRLcoronagraph to at least 8 yr for the KPNO magnetograph and spectroheliograph. A characterization of coronal holes and their interplanetary effects over the sunspot cycle is just beginning as the data begin to span an 11-yr interval. The time-consuming job of processing the enormous data base of coronal observations has delayed our analysis of coronal mass ejections near sunspot maximum. In that respect, the current review could contain only those qualitative results that are immediately obvious. Of course, the exciting discovery of the comet was an unexpected result that adds even more emphasis to the importance of synoptic observations.

Although we have only begun to analyze these coronal hole and coronal transient observations both separately, together, and with other data such as the Helios and ISEE interplanetary measurements, one general result is clear. During the recent years near sunspot maximum, the corona was characterized by strong magnetic fields and incessant change. We do not yet know whether this change simply reflected the high rate of magnetic flux emergence at this phase of the cycle, or whether it was related to the solar flares and eruptive prominences that intermittently accompanied the flux emergence. We are beginning to learn that these coronal changes lead to relatively slow solar wind streams (at least in the ecliptic plane) and an increased frequency of interplanetary shocks. The next step is to examine these new observations in detail in order to understand the physical processes in the corona near sunspot maximum.

References

Bohlin, J. D. and Sheeley, Jr., N. R.: 1978, *Solar Phys.* **56**, 125.

Fisher, R. R.: 1981, private communication of *MK-III K-coronameter Observations from Mauna Loa*.

Fisher, R. R.: 1982, *Space Sci. Rev.* **33**, 9 (this issue).

Harvey, J. W. and Sheeley, Jr., N. R.: 1979, *Space Sci. Rev.* **23**, 139.

Harvey, K. L. and Sheeley, Jr., N. R.: 1980, *Bull. Am. Astron. Soc.* **12**, 918.

Harvey, J., Krieger, A. S., Timothy, A. F., and Vaiana, G. S.: 1975a, in G. Righini (ed.), *Osserv. Mem. Oss. Arcetri* **104**, 50.

Harvey, J. W., Krieger, A. S., Davis, J. M., Timothy, A. F., and Vaiana, G. S.: 1975b, *Bull. Am. Astron. Soc.* **7**, 358.

Harvey, K. L., Sheeley, Jr., N. R., and Harvey J. W.: 1982, *Solar Phys.* **79**, 149.

Hildner, E., Illing, R. M. E., Wagner, W. J., House, L. L., Sawyer, C. B., and Hyder, C. L.: 1981, *Bull. Am. Astron. Soc.* **13**, 755.

Howard, R. F. and Harvey, J. W.: 1977, personal communications to the *Skylab Workshop Series on Coronal Holes*.

Howard, R. and Labonte, B. J.: 1981. *Solar Phys.* **74**, 131.

Hundhausen, A. J.: 1977, in J. B. Zirker(ed.), *Coronal Holes and High-Speed Wind Streams*, Col. Assoc. Univ. Press, Boulder, p. 225.

Keller, H. U. and Richter, K.: 1981, Private communication.

King, J. H.: 1979, *J. Geophys. Res.* **84**, 5938.

King, J. H.: 1981, *J. Geophys. Res.* **86**, 4828.

Kopp, R. A.: 1977, in J. B. Zirker (ed.), *Coronal Holes and High-Speed Wind Streams*, Col. Assoc. Univ. Press, Boulder, p. 179.

Kreutz, H.: 1888, *Publ. Sternw. Kiel* **3**.

Kreutz, H.: 1891, *Publ. Sternw. Kiel* **6**.

Kreutz, H.: 1901, *Astron. Abhandl.* **1**.

Marsden, B. G.: 1967, *Astron. J.* **72**, 1170.

Marsden, B. G.: 1981, *IAU Circ.*, No. 3640.

Michels, D. J., Koomen, M. J., Howard, R. A., and Sheeley, Jr. N. R.: 1981, *Trans. Am. Geophys. Union (EOS)* **62**, 376.

Michels, D. J., Sheeley, Jr., N. R., Howard, R. A., and Koomen, M. J.: 1982, *Science* **215**, 1097.

Poland, A. I., Howard, R. A., Koomen, M. J., Michels, D. J., and Sheeley, Jr., N. R.: 1981 *Solar Phys.* **69**, 169.

Scherrer, P. H. and Svalgaard, L.: 1977, *Trans. Am. Geophys. Union (EOS)* **58**, 769.

Schwenn, R.: 1981, private communication.

Sekanina, Z.: 1981, in B. G. Marsden (ed.), *IAU Circ.*, No. 3647.

Sekanina, Z.: 1982, *Astron. J.* (submitted).

Sheeley, Jr., N. R., Michels, D. J., Howard, R. A., and Koomen, M. J.: 1980a, *Astrophys. J. Letters* **237**, L99.

Sheeley, Jr., N. R., Howard, R. A., Koomen, M. J., Michels, D. J., and Poland A. I.: 1980b, *Astrophys. J. Letters* **238**, L161.

Tousey, R., Bartoe, J.-D. F., Bohlin, J. D., Brueckner, G. E., Purcell, J. D., Scherrer, V. E., Sheeley, Jr., N. R., Schumacher, R. J., and Van Hoosier, M. E.: 1973, *Solar Phys.* **33**, 265.

Wagner, W. J., Sawyer, C., Illing, R. M. D., House, L. L., Querfield, C. W., Sheeley, Jr., N. R., Howard, R. A., Koomen, M. J., Michels, D. J., and Smartt, R. N.: 1980, *Bull. Am. Astron. Soc.* **12**, 902.

Zirker, J. B.: 1977, *Coronal Holes and High-Speed Wind Streams*, Colorado Associated Univ. Press, Boulder.

CORONAL TRANSIENT PHENOMENA*

M. DRYER

Space Environment Laboratory, NOAA/ERL, Boulder, Colo. 80303, U.S.A.

Abstract. Solar coronal transients, particularly those caused by flares and eruptive prominences, play a major role in the fields of solar-terrestrial physics and astrophysics. In the former field, coronal transients and their associated interplanetary disturbances are responsible for solar and galactic cosmic ray modulations, as well as planetary magnetospheric and ionospheric disturbances. In the latter field, supernovae remnants are scaled-up manifestations of such disturbances; that is they are *stellar*, rather than solar, coronal transients. Study of the more accessible *solar* transients is proving invaluable in both fields and is, therefore, selected for attention in this paper.

A series of coronal transient observations is discussed in the spirit of a representative overview following some introductory remarks on the background solar wind. One of these observations is chosen because its interplanetary signature – the shock wave – was detected by two spacecraft at different heliocentric radii. Other cases are chosen because of the extended observations of embedded eruptive prominences. Progress is also being made in the interdisciplinary areas of optical imagery complemented with radio astronomical techniques.

Finally, several recent theoretical models and MHD computer simulation studies are summarized. It is suggested that further comparison of specific events with such models promises a rich harvest of physical understanding of the origin, structure and interplanetary progeny of coronal transients.

1. Introduction

The coronal transient, manifested by unambiguous optical imagery, is one of the outstanding physical phenomena under scrutiny by a wide-ranging group of interdisciplinary physicists. This topic is important and crucial in its own right because of: (i) its obvious link between solar, interplanetary and magnetospheric physics via an explicit physical mechanism; as well as, (ii) its role as a testing ground for theories of energetic astrophysical phenomena such as novae. It is suggested that a spectrum of energetic phenomena ranging from impulsive tests in simple (shock tubes) and complex (beam-pellet) laboratory devices to solar coronal transients and even supernovae has a commonality in terms of dynamic characteristics. The coronal transients caused by solar flares are manifestations of energy releases within the range of $10^{29}-10^{33}$ erg. Thus, it is imperative from the view of the solar-terrestrial system, such as the geomagnetic storm and sun-weather physical relationships, that we learn more about these transients. Also, despite the disparity in energy magnitudes when compared with supernovae ($\sim 10^{50}$ erg), we will undoubtedly learn much about the latter from a study of the solar case where tests are accessible.

During the Solar Maximum Year (August 1979 to May 1982) as well as during the post-SMY analysis phase, the white light imagery from the spacecraft, Solar Maximum Mission and P78-1, is being examined in the context of observations from complementary techniques. Among these are: (i) the ground-based radio observations at Culgoora, Fort Davis, Clark Lake and Meudon of type II and IV radio bursts; (ii) the ground-

---

* Paper presented at the IX-th Lindau Workshop 'The Source Region of the Solar Wind'.

Space Science Reviews **33** (1982) 233–275.   0038–6308/82/0332–0233$06.45.

based K-coronagraph observations at Mauna Loa of phenomena below the orbiting spacecraft's occulting disks; (iii) the conventional Hα observations from the ground, augmented now by additional lines such as Fe X at Sacramento Peak; (iv) the unconventional spectral broadening, Faraday rotation, and Doppler shifts of Deep Space Network radio telemetry signals at discrete frequencies to and from distant spacecraft at and near superior conjunction; (v) the exciting 'tracking' of type II shocks from the Sun to ISEE-3 and Prognoz 8; and ultimately, (vi) the unambiguous association of coronal transient-produced shock waves with *in situ* shock observations.

Finally, a more complete understanding of the nonlinear dynamic phenomena produced by solar flares and their coronal transient manifestations is provided by temporal magnetohydrodynamic computer simulations. These codes, while still lacking physical dissipative effects (radiation, resistivity, and thermal conduction), provide great insight into global mass, energy, and momentum transfer via coupled large amplitude MHD wave propagation by all fast, slow and Alfvén modes. Future work with the dissipative effects will now rest upon a firm basis. Additional insight is provided by the study of the buoyancy effect upon loop structures such as eruptive prominences and, at the lower energy part of the spectrum, the loss of static equilibrium possessed by these structures within non-potential magnetic field configurations. At the risk of over-simplification, coronal transient phenomena can be characterized as the more apparent changes that are observed (usually by white-light orbiting coronagraphs) between one quasi-steady coronal state and another. By this definition, it is clear that such changes are temporal, dynamic phenomena. Diagnostically, these changes were first observed by the imagery obtained by the Naval Research Laboratory's white light coronagraph on OSO-7 (cf., Howard *et al.*, 1975). The 'changes' were observed as localized travelling electron enhancements produced by Thomson scattering of photospheric emission by free electrons. More extensive observations were obtained by the High Altitude Observatory's white light coronagraph on Skylab and summarized by Hildner (1977). The latter observations (77 mass ejections) were associated with solar surface phenomena by Munro *et al.* (1979) who found that 40% of the coronal transients were associated (during 227 days of observations during solar minimum, 1973–74) with flares; 50%, with eruptive prominences having no apparent flare association at or near the limbs; and 70%, with eruptive prominences or disappearing Hα filaments (with or without flares). In a succinct summary, MacQueen (1980) noted that a subjective classification of the Skylab observations has been made by Munro (1977). In this classification, outwardly-expanding loops comprise nearly one-third of all observed transients; clouds or amorphous blobs comprise about one-quarter of the sample; and, quoting MacQueen (1980): 'The remainder defy specific classification'. A recent observation of a transient directed toward Earth by Michels *et al.* (1982) suggests that (in general) many of the 'loop' category should more properly be included within the 'cloud or amorphous blob' category.

In terms of gross statistics, Hildner *et al.* (1981) reported recently that SMM's C/P (Coronagraph/Polarimeter) observed 50 mass ejections during 138 observing days as compared with 77 observed by a similar HAO instrument on 227 observing days on

Skylab, as noted above. During the period leading up to and including the Solar Maximum Year (August 1979 – May 1982), the SCOSTEP project STIP (Study of Travelling Interplanetary Phenomena) encouraged the interdisciplinary observation of such events with the broad spectrum of diagnostic instruments. Fortunately, several newer orbiting coronagraphs were available: the HAO coronagraph/polarimeter (C/P) on the Solar Maximum Mission and the NRL coronagraph (Solwind) on the USAF spacecraft, P78-1*.

In the following Discussion, I will attempt an overview of the following topics: (i) the background solar wind; (ii) several 'typical' white light coronal transients; (iii) some complementary observations by optical, radio, and *in situ* techniques from the ground and from spacecraft; and (iv) a status report of the theoretical progress in our understanding of major aspects of the coronal transient phenomenon.

2. The Background Solar Wind

It was suggested in the Introduction that, broadly speaking, the coronal transient may be characterised as the transitory process between one quasisteady state (long, compared with the 1–2 h time scale of the primary phenomenon of interest) and another such state. Thus, tenuous changes such as those observed prior to a solar flare in Hα (Martin and Ramsey, 1972; and Hansen *et al.*, 1974), Fe XIV green line (Bruzek and DeMastus, 1970), and the white light forerunners (Jackson and Hildner, 1978; and Jackson, 1981) will not be considered. In a sense, then, I will confine attention only to the more obvious, clear-cut cases of coronal transients that are representative of the higher solar power releases, as in solar flares, that may in many cases, be associated with eruptive prominences. Thus, the rate of energy release is an important consideration that must always be kept in mind.

It was pointed out in an earlier review (Anzer, 1980), as well as in one of the MHD studies (Wu *et al.*, 1981), that the background solar wind must also be kept in mind. This point is important when we consider the coronal transient's initiation at the sun and its propagation into the interplanetary medium where its consequences (such as shock waves, flare ejecta) may be identified by spacecraft remote and *in situ* observations as well as by scintillation (IPS and telemetry) techniques. We must ask, then: what is the solar wind velocity (in particular) *near* the Sun?

An observational answer to this fundamental question is now available. Radial dependence of the solar wind velocity has been deduced from Helios 1/2 and Pioneer 10/11 observations by Woo (1978); from Mars-2, Mars-7, and Venera 10 by Yakovlev *et al.* (1980); and from Viking and Mariner 10 (Tyler *et al.*, 1981) from radio telemetric scattering observations. Generally, these observations show a velocity change from ≤ 100 km s^{-1} within or at $15 R_{\odot}$ (where $R_{\odot}$ is the solar radius, 6.96×10^5 km) to ≥ 200 km s^{-1} outside this distance. Since the models (such as those noted below)

---

* The Solwind instrument has continued to operate, starting on 28 March, 1979, to the date of this writing. The C/P instrument has operated from shortly after launch on 14 February 1980 to about September 1980 and is scheduled for refurbishment as part of a Shuttle mission for SMM repair in 1984.

suggest that solar wind acceleration may take place very low in the inner corona, it is also important to note that both early radar studies (James, 1966), as well as ultraviolet studies within and beyond several coronal hole transition regions (Rottman *et al.*, 1981; Withbroe *et al.*, 1982), provide velocities from 3–5 km s^{-1} (transition region) to about 85 km s^{-1} ($\sim 4R_\odot$). In fact, the more recent radio scattering observation by Woo (1978) provides a value of 24 km s^{-1} at $1.7R_\odot$ that is accurate to within a factor of 2–2.5. We note also that the idea of a 'turbulent envelope' around the sun, bounded at $\sim 15R_\odot$, suggested by Ekers and Little (1971), is consistent with their radio observations. They also noted that this outer boundary is rather sharp, at least under solar minimum conditions. Despite this possible anomaly (that suggests a selective form of wave damping analogous to that possibly associated with the steep temperature gradient within the transition region), it is still remarkable that even the isothermal model of Parker (1963) provides velocities similar to those mentioned above.

The trend indicated by the observations summarized above strongly suggests that the basic result first proposed by Parker (1958) is still valid – even fairly close to the sun. That is, the velocity near the sun is small relative to higher values that have been proposed in a plethora of relatively-recent studies. Moreover, even the early one- and two-fluid solar wind models with phenomenologically-chosen radial dependence of conductive properties (Cuperman and Harten, 1970a, b, 1971; Wolff *et al.*, 1971) provide velocities that are appropriate to the observations given above.

An empirically-derived solar wind model, appropriate for a representative coronal density distribution (optically derived by Newkirk, 1967) and solar wind, is given in Table I (from Dryer and Cuperman, 1972). This model is simply based on mass conservation (of protons, n_i, and electrons, n_e) of a spherically-expanding neutral plasma. The convective acceleration, relative to the gravitational acceleration, is shown in the last three columns of Table I. This acceleration is due to the force caused by the pressure difference between that at the base of the inner corona, p_c, and that in interstellar space, p_i. Here, representative values of density and temperature in the inner corona (interstellar space*) are 4×10^8 cm^{-3} and 1.5×10^6 K (10^{-1} cm^{-3} and 10^4 K) giving, thereby, a many-fold pressure difference:

$$p_c - p_i \simeq (10^{-1} - 10^{-13}) \text{ dyn cm}^{-2}. \tag{1}$$

It is interesting to note in Table I that a non-monotonic velocity change takes place between 8–15$R_\odot$. This result appears to anticipate the 'turbulent envelope' observation of Ekers and Little (1971), Armstrong and Woo (1981), and Tyler *et al.* (1981). Although the original 1972 footnote provides a caveat for this simple result, it is likely that the Newkirk (1967) density distribution for a representative 'quiet-Sun' is correct and not, after all, an 'apparent anomaly' as originally suggested in Table I. More recently, Coutourier *et al.* (1980) have derived similar results with a more sophisticated model that considers mechanical flux at the photospheric level and the achievement of the transition zone temperature increase via assumed shock wave damping. Also listed

---

* The presence of interstellar neutral gas penetration and of a standing heliospheric shock wave is of secondary importance for the purposes discussed here.

TABLE I

Representative 'Inner' and 'Outer' solar corona properties (based on Dryer and Cuperman, 1972, observed velocities have been added)

| $V_{observed}$ (km s⁻¹) | | | | $r/R_⊙$ | n_e^a (cm⁻³) | $(n_e + n_i) r^2$ | V (km s⁻¹) | $V\,dV/dr$ (km s⁻²) | $g_⊙$ (km s⁻²) | Ratio of convective to gravitational acceleration |
Rottman et al. (1982)[f]	Yakovlev et al. (1980)[e]	Woo (1977)[e]	Tyler et al. (1981)[d]							
7				transition	—	—	—	—	—	—
12				1.02	4.0×10^8	4.0×10^{30}	0.25	1.5×10^{-6}	2.6×10^{-1}	6×10^{-5}
				1.2	7.0×10^7	1.0×10^{30}	1.0	1.0×10^{-5}	1.9×10^{-1}	5×10^{-4}
	~16			1.6	1.0×10^7	2.5×10^{29}	4.0	6.0×10^{-5}	1.0×10^{-1}	6×10^{-4}
	~20			2.0	2.8×10^6	1.2×10^{29}	8.3	2.2×10^{-5}	7.0×10^{-2}	3×10^{-4}
	20			2.5	9.0×10^5	5.6×10^{28}	17.8	5.0×10^{-4}	4.0×10^{-2}	1×10^{-2}
	30			3.0	4.0×10^5	3.5×10^{28}	27.5	1.0×10^{-3}	3.0×10^{-2}	3×10^{-2}
				4.0	1.2×10^5	1.9×10^{28}	52.5	1.4×10^{-3}	1.7×10^{-2}	8×10^{-2}
		~100		6.0	3.1×10^4	1.1×10^{28}	91	2.2×10^{-3}	7.5×10^{-3}	3×10^{-1}
				8.0	1.3×10^4	8.0×10^{27}	125	~ -2.2×10^{-3}	~ -5.0×10^{-3}	4×10^{-1}
		190	≲100	10.0	9.8×10^3	9.6×10^{27}	102^b	2.3×10^{-3}	2.7×10^{-3}	9×10^{-1}
				15.0	2.5×10^3	5.5×10^{27}	182	5.1×10^{-4}	1.2×10^{-3}	4×10^{-1}
		250	≳200	20.0	1.3×10^3	5.2×10^{27}	192	1.6×10^{-3}	7.0×10^{-4}	2.3
				30.0	4.6×10^2	4.0×10^{27}	250	3.0×10^{-4}	3.0×10^{-4}	1.0
				215.0	~5^c	~2.5×10^{27}	~320^c	—	—	—

[a] From optical (except as noted) measurements summarized by Newkirk (1967) for equatorial regions. Billings (1966) gives $n_e = 2.3 \times 10^8$ cm⁻³ at $1 R_⊙$.

[b] It is generally believed that the solar wind velocity increases monotonically with distance. Therefore, this figure is too low because of an apparent anomaly in the optically derived electron density. (See text for a discussion of this point on the basis of recent radio scattering and ultraviolet observations).

[c] Direct measurement equal to proton number density. Heavier ion species are neglected. Electrons move with the same velocity, also noted in the last row, as the protons and heavier ions when the velocity is of the order shown. Alpha particles are sometimes lower and, in the case of high speed streams, higher by about 1% than the proton and electron bulk velocity (see Feldman et al., 1977, for a review of statistical properties of 'fast' and 'slow' solar wind flows).

[d] Radio-wave scattering observations from ~6 to 44 $R_⊙$ with a heliolatitude range from $-17°$ to $+7°$.

[e] Spectral broadening observations. In more resent and precise observations (multiple station intensity scintillations of spacecraft signals from Helios 1, Helios 2, and the Viking orbiters in 1976–78), Armstrong and Woo (1981) obtained a more complete set of rms solar wind velocities as far as 30 $R_⊙$.

[f] From coronal hole blue-shifted EUV emission lines of O v and Mg x.

in Table I are some recent solar wind velocity observations by various techniques (radio, EUV) in the lower corona. The observations, with the exception of those of Rottman *et al.* (1982), are reasonably similar to the analysis.

On the whole, then, it appears that some recent studies (Munro and Jackson, 1977; Suess *et al.*, 1977; Kopp and Holzer, 1976) that suggest a general importance of non-radial expansion and/or momentum transfer of the solar wind are inappropriate. Other studies that suggest a zeroth-order effect of momentum transfer by Alfvén wave damping (Barnes and Hartle, 1972; Hollweg, 1978) are also primarily of academic interest only for the solar wind. Of course, there are very specialized models that recognize the local importance of solar magnetic field topology (cf. Suess, 1979; Steinolfson and Wu, 1980; Steinolfson *et al.*, 1982). Excellent summaries of the single- and multifluid models of the solar wind, as well as the present basis of further plasma kinetic studies are given by Cuperman (1980) and Wu (1980).

Finally, these introductory remarks and, particularly, the background solar wind velocity, V, given in Table I should be kept in mind when one examines the experimental coronal transient observations (cf. Hildner, 1977; MacQueen, 1980; Stewart, 1980; and Maxwell and Dryer, 1981), on the one hand, and the theoretical results (cf. the review by Anzer, 1980) on the other hand. The latter, as will be discussed later, fall into several groups: (i) MHD models wherein the transient is produced initially by a physically-unidentified pulse at or near the base of the corona; (ii) force imbalance introduced by self-induced or externally-induced magnetic pressure; (iii) magnetohydrodynamic buoyancy forces; and (iv) transient motion inferred from the loss of equilibrium of non-potential magnetic structures. Thus, the *steady-state* plasma pressure gradient (plus the inferred magnetic) imbalance and the convective acceleration given in Table I are also important because they represent background 'buoyancy' considerations for any additional ejecta and waves.

3. Coronal Transient Observations

As noted earlier, summaries of some of the Skylab and OSO-7 white-light observations during the minimum (October 1971 to February 1974) of the last solar cycle (i.e., cycle 20) have been presented elsewhere. Since that time a wealth of new data on coronal transients has been accumulating from the coronagraphs on P78-1 and SMM at a rate of more than one per day on the average. Thus, the present status during pre-maximum years is represented with a variety of data compilations and interpretative views (optical, radio, and theoretical) in surveys by Hildner (1977), MacQueen (1980), Anzer (1980), Dulk (1980), Stewart (1980), Wu (1980), and Maxwell and Dryer (1981).

Parenthetically, it should be noted that *magnetic transients* have recently been observed (Zirin and Tanaka, 1981) by the videomagnetograph at the Big Bear Solar Observatory in spatial and temporal coincidence with two substantial Hα flares. One might first speculate that these magnetic transients (that are observed close to the photosphere) are observational parts of the evidence that fit into the scenario (Tanaka and Nakagawa, 1973; Nakagawa, 1976) of preflare magnetic energy build-up, storage of this energy in departures from potential fields, the flare process (and all that this entails, viz., particle acceleration, X-rays, etc.), and the ensuing coronal and solar wind responses. As yet, however, there is no evidence that links the so-called magnetic transient with the white light transient. A promising MHD theoretical, quasi-3D, model discussed by Nakagawa *et al.*, (1980, 1981) and Wu (1982a) may provide additional insight for this scenario.

Returning to the newer coronagraph imagery, photograph subtraction techniques (wherein a selected pre-event image is subtracted, via pixel counts at each pair of points, from a sequence of temporal event images) are now used more extensively than in the past. An example of P78-1 SOLWIND coronagraph images is shown in Figure 1 (from

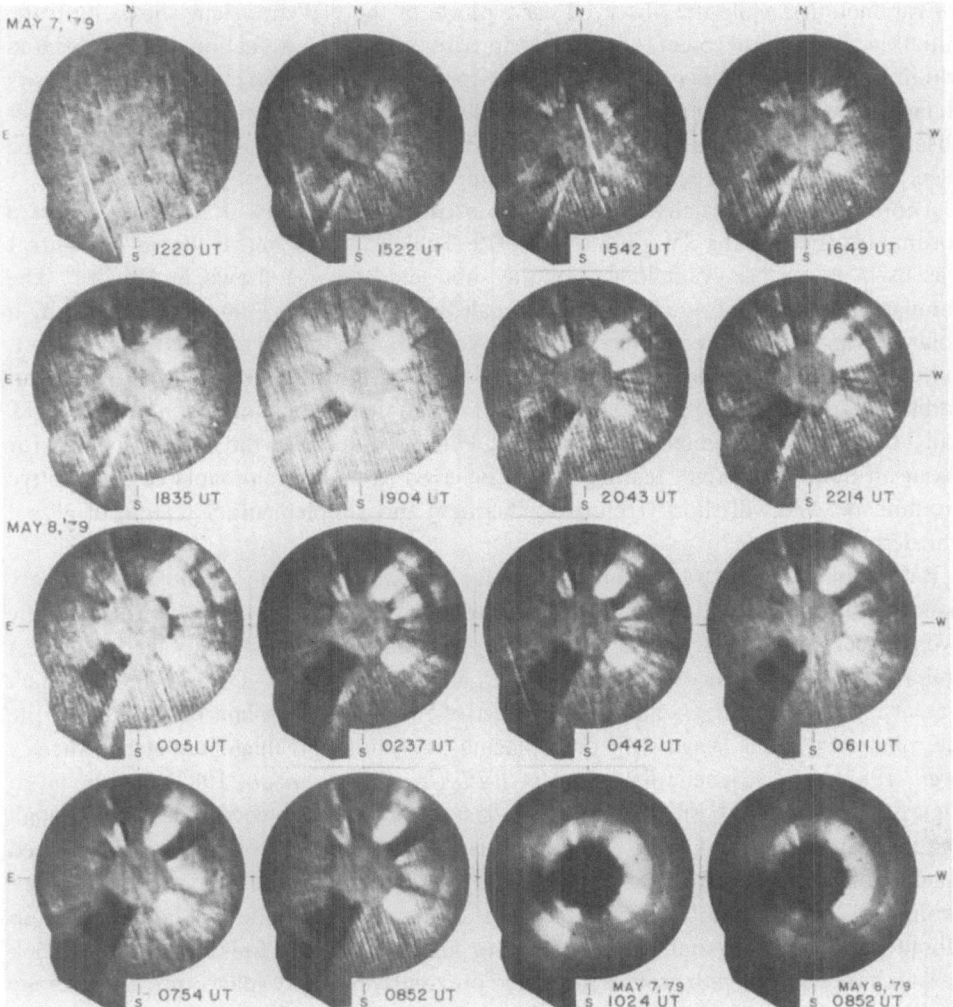

Fig. 1. White light coronal transients (SE and NW) observed on 7 May 1979 by the Naval Research Laboratory coronagraph SOLWIND on P78-1 (Sheeley *et al.*, 1980).

Sheeley *et al.*, 1980). In this sequence, the 'pre-event' image is shown in the 10 : 24 UT frame (lower right) for 7 May, 1979. After subtraction from subsequent images at 12 : 20 UT, 15 : 22 UT, 15 : 42 UT, etc., the final subtracted images are shown, starting at the upper left corner until 08 : 52 UT on 8 May, 1979. Comparison of this latter subtracted image with its original image shows that a new, more-or-less quiescent coronal structure has been achieved after the major coronal transients in both the NW as well as the SE quadrants. The latter was in progress during the first half of the day 7 May; the former took place during the second half of that day. The central occulting disk is located at the equivalent of $2.6R_\odot$; the outer limit of the field-of-view is $10R_\odot$. It is seen that the observed later phase of the SE transient displays an electron depletion (or rarefaction), while the observed early phase of the NW transient shows both the initial electron enhancement (or compression) followed by a rarefaction that was located within the transient closer to the sun. This spatial and temporal characteristic density change is one of the most important features that was immediately apparent in the earlier observations. As discussed later, this characteristic is an essential feature that must be used as a key test for theoretical interpretations.

Another example (Michels *et al.*, 1980) is shown in Figure 2 that clearly shows a coronal transient in the SW on 8 May, 1979. The 08 : 52 UT image shown in Figure 1 was used as the pre-event image for the subtracted images shown in Figure 2. The annular dark region at $\sim 5R_\odot$ (note the scale on the 12 : 46 UT image) is caused by a polarizer ring as discussed in the figure title. Noting (and allowing for) this technical fact, the compression and rarefaction are again seen very clearly. This time, however, an additional brighter feature is seem embedded within the rarefaction region at 10 : 58 UT and 11 : 07 UT, after which the spacecraft, P78-1, passed into the earth's shadow for about an hour. The bright feature is now believed to be the remnants of an eruptive prominence. We will shortly return to this new and supplementary feature of global transient behavior.

Before following the trail of the prominence, it is worth noting that the interplanetary signature of the 8 May, 1979 coronal transient, namely its shock wave, was detected by two spacecraft. Helios-2 (at 0.3 AU – $64R_\odot$ – during its perihelion passage at the west limb relative to an observer at earth), as well as Pioneer–Venus–Orbiter (PVO), which was at 0.7 AU – $156R_\odot$ – about $110°$ west of the Sun–Earth line both detected the interplanetary shock wave with their plasma analyzers and magnetometers (Sheeley *et al.*, 1982). The magnetic field data at PVO (from Dryer *et al.*, 1982) are shown in Figure 3 at 04 : 39 UT on 10 May, 1979 just before that spacecraft passed through periapsis above the sunlit side of Venus. It is of some interest to note that the induced magnetic field produced by the solar wind's interaction with the Venusian ionosphere (from $\sim 21 : 00-22 : 00$ UT) is of the same order of magnitude as the shock jump (about $15-20\gamma$). The continued increase to about 80γ some four hours after shock passage is consistent with profiles measured previously by many spacecraft. The average rectilinear velocity of the portion of the asymmetrical shock that was produced by the SOLWIND-observed coronal transient is about 700 km s^{-1}. Its 'birth' was assumed to have taken place at $\sim 08 : 10$ UT on 8 May, 1979 when a large Hα solar eruption took

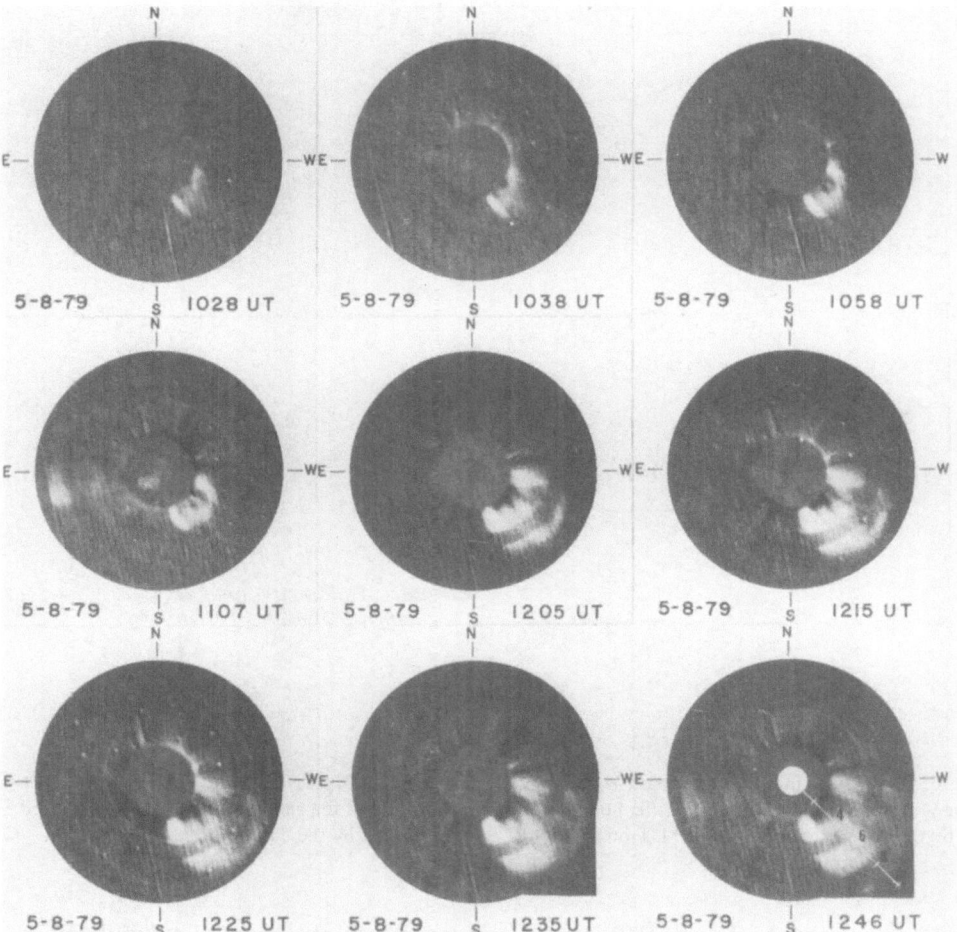

Fig. 2. White light coronal transient observed on 8 May 1979 by the Naval Research Laboratory corona-graph SOLWIND on P78-1. 'The slightly off-center ring at $\sim 5 R_\odot$, darker than its surroundings, is caused by a focal plane polarizer with its axis everywhere approximately radial; the remainder of the focal plane contains a concentric polarizer with its axis everywhere at right angles to the solar radius, thus discriminating in favor of K-coronal radiation' (Michels *et al.*, 1980).

place (Sheeley *et al.*, 1980). For a discussion of other interplanetary signatures of solar flares (and other coronal transients) as detected at large and small heliocentric radii, respectively, the reader is referred to Intriligator (1980) and to Howard *et al.* (1981)*.

---

* In the recorded discussion of Intriligator (1980), the author referred to a significant solar flare on the east limb on 18 August, 1979, just a few days before her presentation at the IAU Symposium No. 91. This event will be discussed later in the present paper because of its relevance to the question of the globally-restricted extent of a transient's shock wave. In the preliminary report of Howard *et al.* (1981), it was reported that coronal transients, on an average, occurred at a rate of 1.2 per 24-hr of observing time during the period 24 February, 1979 (launch) – 30 September, and from 24 February, 1980 – 30 June, 1980. When either Helios 1 or Helios 2 was within 30° of the solar limb, approximately 14 interplanetary shocks were observed and compared with coronal transient events.

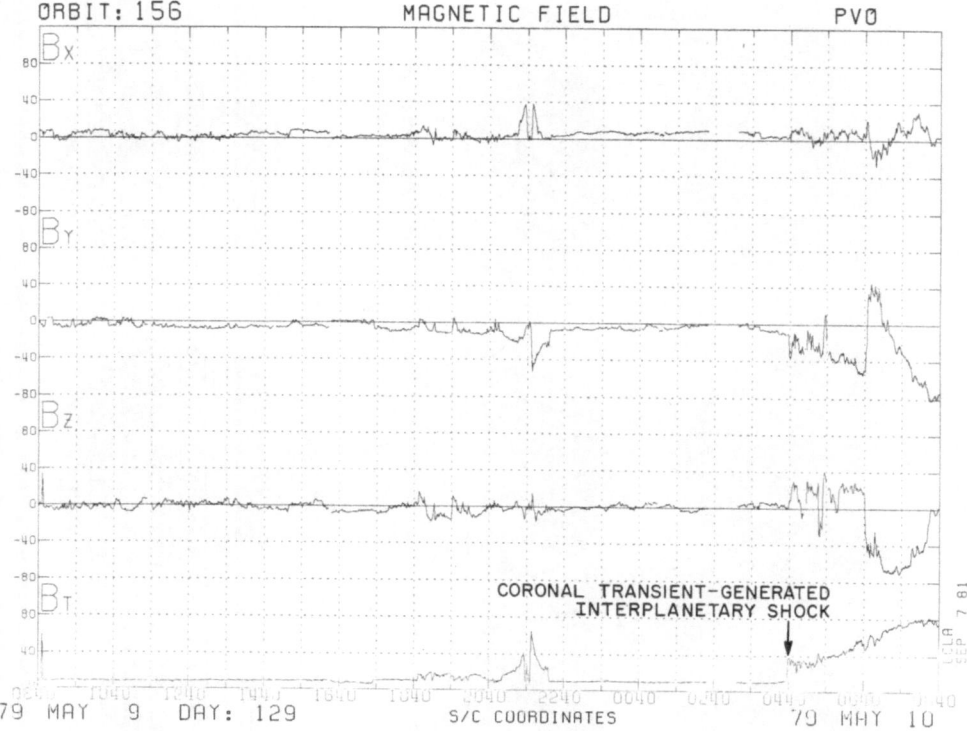

Fig. 3. Solar wind magnetometer data from Pioneer-Venus-Orbiter (Dryer *et al.*, 1982) on 9–10 May, 1979. The jump at 04 : 39 UT (and thereafter) on 10 May is the signature of the shock wave that was generated by the solar eruption at 08 : 10 UT on 8 May, 1979 and followed by the coronal transient shown in Figure 2.

We return now to the subject of the prominence that is often seen, first lifting off (in Hα) from the Sun and then, later is seen in white light as well as Hα over a larger altitude range. Figure 4 shows an SMM-observed coronagraph/polarimeter observation at 10 : 45 UT on 5 May, 1980 of an eruptive prominence that is, itself, embedded within an expanding loop-like coronal transient. The eruptive prominence, or flux rope, was observed to unwind – a phenomenon that has a theoretical explanation discussed by Yeh (1982a, b). This event is seen at a later time (12 : 05 UT) in Figure 5. This case, as well as another from the same active region on the west limb at 09 : 47 UT on 6 May, 1980, is discussed from an observational viewpoint by House *et al.* (1981a, b) in terms of the unwinding of the flux rope as well as the question of survivability of the cool neutral hydrogen indicated in Figure 5. The latter question may have a physical explanation, supported to some extent by numerical 1-D simulation of cold, dense plasma 'bullets' ($T_e \sim 10^5$ K, $n_e \sim 10^{10}$ cm^{-3}). It has been suggested by Karpen *et al.* (1981), who considered radiation and thermal conduction in their computations, that these shock-producing pistons (blobs or bullets) can survive within the hot corona for about one minute. While such effects may succeed, on the small scale, in heating the

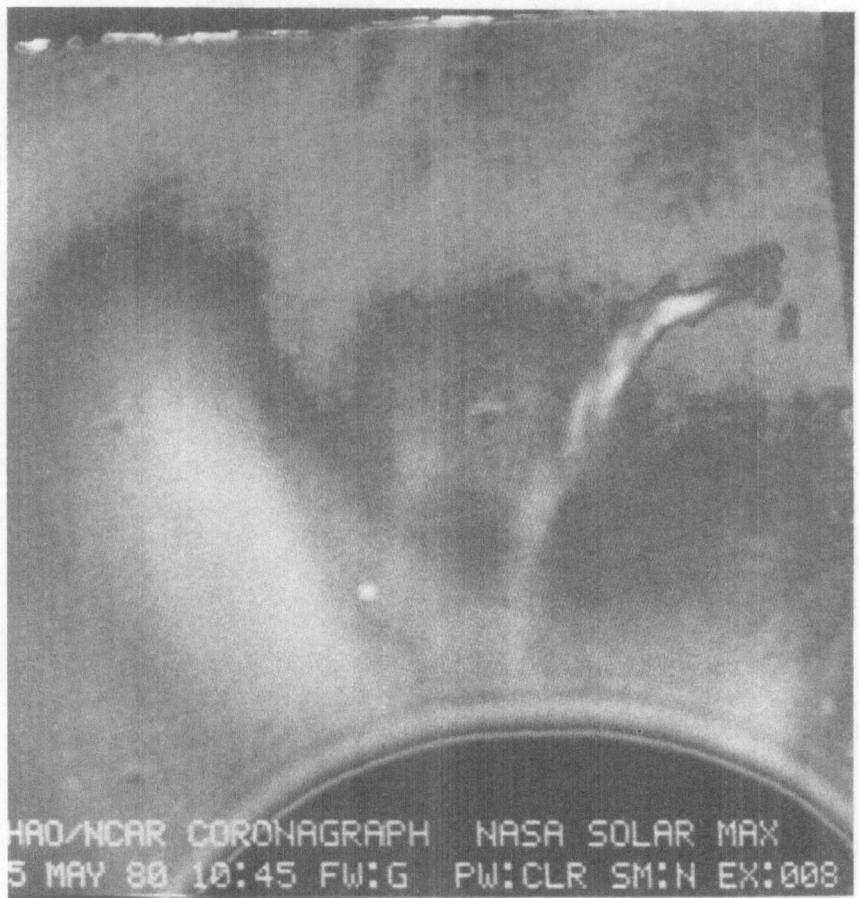

Fig. 4.   Coronal transient observed on 5 May, 1980 (10 : 45 UT) by the High Altitude Observatory corona-graph/polarimeter on the Solar Maximum Mission spacecraft. The Hα prominence was observed at Wrocław Observatory to lift off at 08 : 48 UT. The occulting disk's rim is at $1.5R_\odot$. The 'whip-lash' eruptive prominence is clearly seen within an apparent-helically-shaped magnetic loop which, itself, is embedded within a loop-like coronal transient that subtended an angle of ~ 48° above the west limb on 5 May, 1980. The transient achieved a velocity of 520 km/sec within the field of view to $4.7R_\odot$. The image shown here was observed with a green spectral filter that had a 5014 to 5328 Å bandwidth, a clear polaroid, and an 8 s exposure (NCAR/HAO C/P team, private communication, 1981).

base of the corona via dissipation of their shocks, it is unlikely that the neutral hydrogen in the eruptive prominences of the type shown, on a larger scale, in Figures 4 and 5 can survive for a length of time several orders of magnitude longer than one minute. It would be expected, for example, that the inefficient thermal heat conduction transverse to the helical magnetic field (not considered by Karpen *et al.*, 1981) would play a significant role in the survivability of the neutral hydrogen.

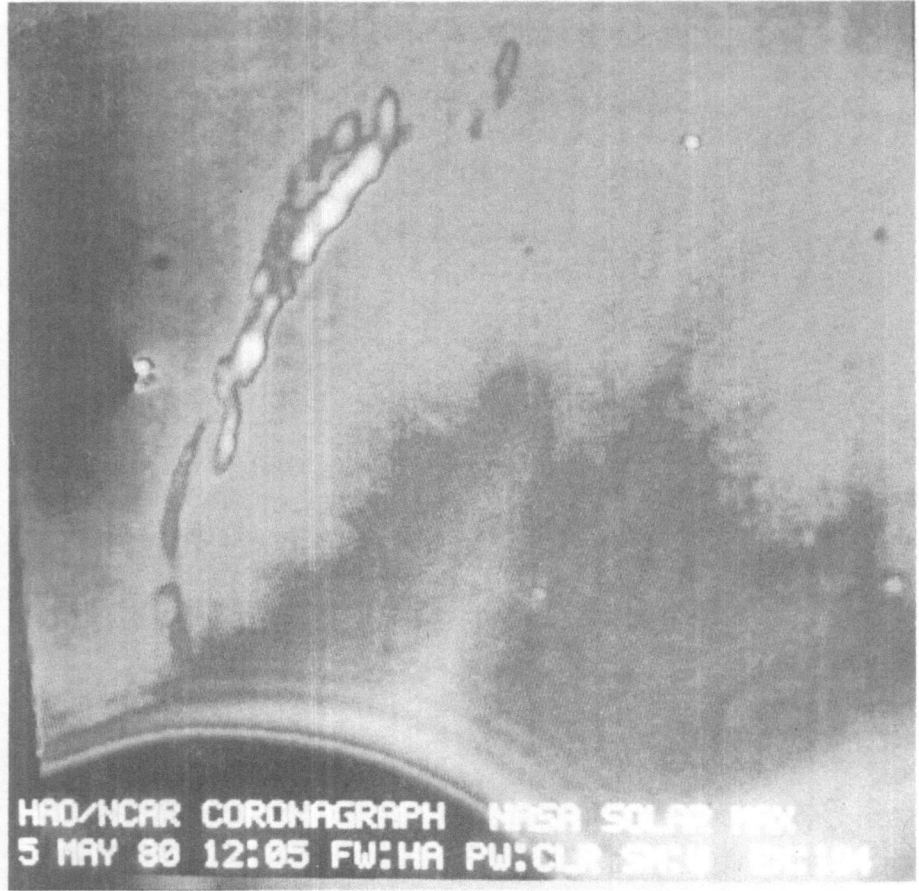

Fig. 5.   Eruptive prominence shown in Figure 4 is shown here at a later time (12 : 05 UT, 5 May, 1980). This
image was observed with an Hα spectral filter that had a 6385 to 6543 Å bandwidth, a clear polaroid, and
a 184 s exposure. This Hα prominence achieved a velocity of 322 km s^{-1}. (House *et al.*, 1981a, b).

Another example of a 'prominence within a loop-like' transient is shown in Figure 6
(NCAR/HAO C/P team, private correspondence, 1981). This event took place directly
above the solar north pole during the period 04 : 49–07 : 21 UT on 14 April, 1980. This
case is of interest because of the larger field of view of the complete coronal transient
and its embedded, outward-moving erupting prominence. It is very likely that these
erupting prominences continue to move behind the main transient beyond the C/P
field-of-view and into that of SOLWIND on P78-1 as suggested earlier during the
discussion above concerning the 8 May, 1979 event. This suggestion was indeed
confirmed by a remarkable series of three coronagraph observations (Wagner *et al.*,
1980) on 15 April, 1980 of a single coronal transient (on the east limb) that followed
an eruptive prominence that was first seen in Hα emission, followed by a bubble-shaped
Fe x brightening above the Sacramento Peak's occulting disk at $1.2 R_{\odot}$. This was then

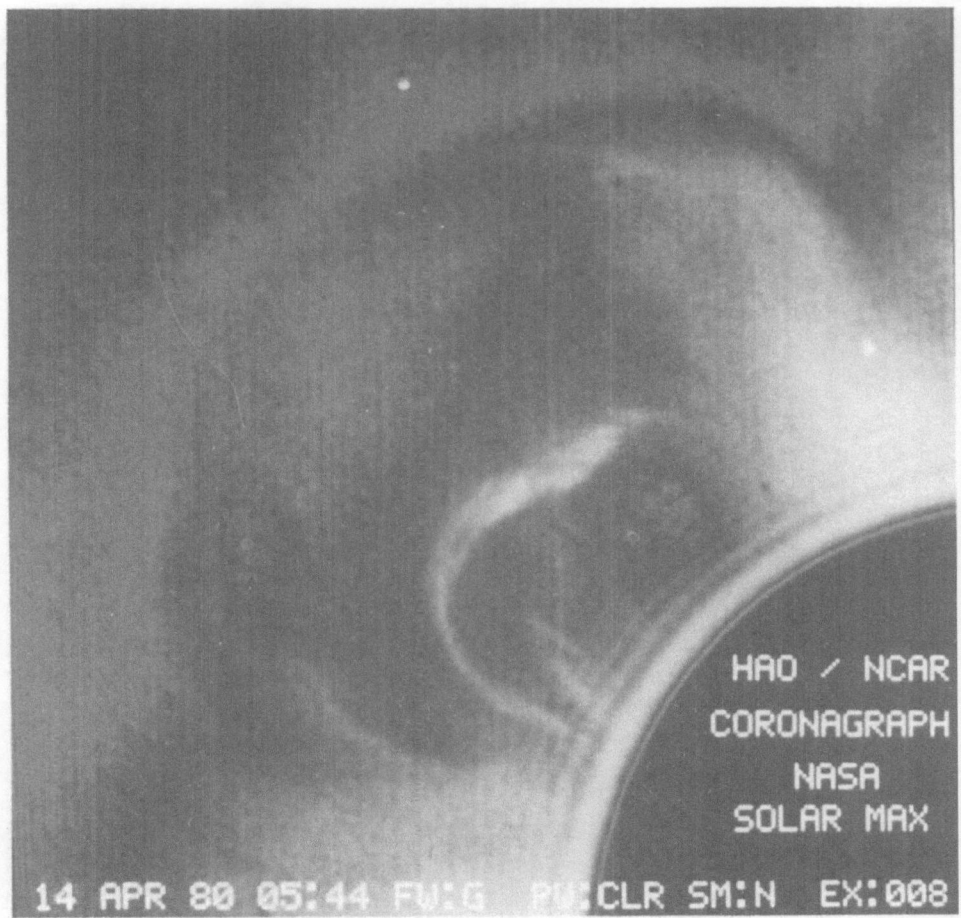

HAO / NCAR
CORONAGRAPH
NASA
SOLAR MAX

14 APR 80 05:44 FW:G PW:CLR SM:N EX:008

Fig. 6. Another example of a 'prominence within loop-like' coronal transient at 05 : 44 UT on 14 April, 1980. The green spectral filter is used in this case (as in the case of Figure 4). The direction of motion of the axis of symmetry is within 5° of the solar north pole. The radius of the occulting disk, for scaling purposes, is $1.5R_\odot$. (NCAR/HAO C/P team, private communication, 1981).

followed, in turn, by a white light loop-like transient that was observed by SMM's C/P, then by P78-1's SOLWIND to $10R_\odot$. Again, the prominence material was surrounded by the transient.

An example of a C/P subtracted image is shown in Figure 7 for a west limb coronal transient. This image was obtained at 23 : 20 UT during a 38 min interval of observation (23 : 16–23 : 54 UT) on 9 April, 1980. This was an unusual transient that consisted of a series of outward-moving multiple loops. It is the present writer's opinion that these loops represented a moving wave train that consisted of alternate compressions (i.e., 'loops' of enhanced electron density), each followed by a rarefaction. The subtracted image in Figure 7 was obtained by subtracting a frame at 00 : 52 UT on 10 April, 1980

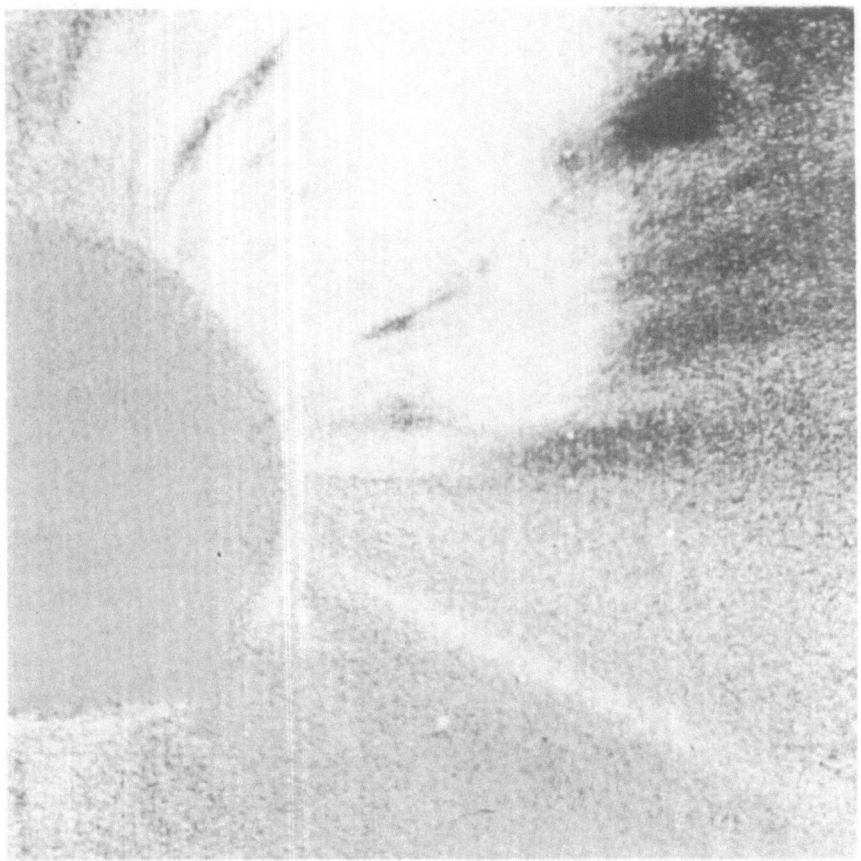

Fig. 7. An example of a subtracted image, bubble-like, coronal transient at 23 : 20 UT on 9 April, 1980 at the west limb. The shadow of the occulting disk's support blocks the corona at the Sun's south pole. This image was formed by subtracting another one 92 min later from the one taken at 23 : 20 UT. (NCAR/HAO C/P team, private communication, 1982).

from the earlier frame at 23 : 20 UT on 9 April, 1980 during the transient process. Thus, the fact that what is thought to be a post-event image was used in the subtraction process (to bring out changes in coronal intensity and structure), rather than a pre-event one, suggests that a substantial rarefaction region (as in the cases of Figures 1 and 2) existed in the latter image. After subtraction from the 23 : 20 UT image, then, one can observe several alternate bands of enhancement (compression) and depletion (rarefaction) of electrons in a concentric bubble-like geometry. Wave trains of this type have been demonstrated in 1-D polytropic plasma flows in the solar wind (Wu *et al.*, 1976) and in the lower corona (Suess, 1981). In the latter case, it was also shown that the amplitudes of the periodic wave trains are significantly damped if a classical Spitzer thermal conductivity were introduced to the flow. A bimodel wave train was also obtained by Dryer *et al.* (1979) who used Skylab emission measures (soft X-ray) for the flare pulse and the 2-D MHD model of Steinolfson *et al.* (1978). In another study that included 1-D MHD flow from within the lower corona, through the critical points, and thence outward into the supersonic solar wind, Steinolfson and Dryer (1982) show that these MHD wave trains persist into the interplanetary medium only if they are generated from within the lower corona where the flow is both subsonic and subalfenic. Conversely, the waves described by Suess (1981) are not present behind the leading shock wave if the generating pulse takes place (probably unrealistically) outside the critical points of an (ideal) solar wind.

Thus far, I have discussed *coronal transients* only within the context of the broad band optical observations that gave rise to the use of that term. In the next section, an overview is given in terms of an interdisciplinary coupling (observed as well as inferred) of the optical observations with radio observations.

4. Coupling of Optical/Radio Observations

Data on optical ejecta and radio bursts of types II and IV have been summarized by MacQueen (1980), Stewart (1980), Dulk (1980), and Maxwell and Dryer (1981, 1982). Several of these authors summarized their interpretations concerning mass ejection events in terms of one-dimensional radius-time and two-dimensional radius-latitude (or longitude) diagrams. In the following summary of several more recent observations, it is useful to repeat these diagrams in order to give the reader some perspective of spatial and temporal interdisciplinary observations.

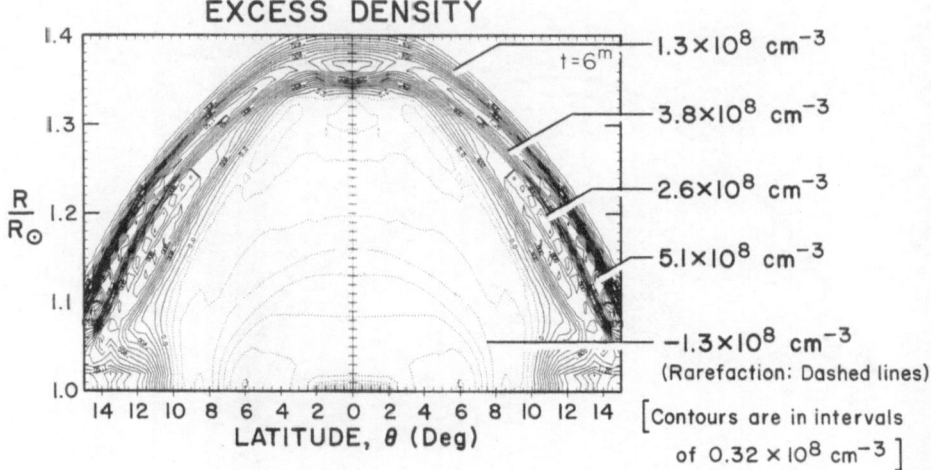

Fig. 8. Representative coronal density enhancement (compression) and depletion (rarefaction) computed by a 2-D MHD model. The solid lines are the excess $(n-n_0)$ density contours as labeled; the dashed lines are the depletions. The absolute values are based on an exponentially-decreasing density with $n_0 = 3 \times 10^8$ cm^{-3} at the base of the equilibrium atmosphere. (After Dryer and Maxwell, 1979). The rarefaction near the axis of symmetry is the result of a simulated evaporation produced by the intense heat pulse which is used as input to simulate the flare energy release and subsequent loop (or dome)-like coronal transient represented by the solid contours. This interpretation is extended in Figure 9.

We start with the theoretically-proposed calculated MHD properties (Nakagawa *et al.*, 1978; Wu *et al.*, 1978; Steinolfson *et al.*, 1978) of the outwardly moving material that, according to MacQueen (1980), are qualitatively in agreement with observed transient properties. Figure 8 (after Dryer and Maxwell, 1979) shows the calculated excess density, $n-n_0$, in two-dimensional space at a specific time after the introduction of a thermal pulse which is assumed to represent release of energy in a flare. These calculations, made with the MHD code discussed by Steinolfson *et al.* (1978), show both the compression above the local ambient values of density, n_0, (solid lines) as well as the rarefaction (dashed lines). The originally undisturbed density at the coronal base was taken as 3×10^8 cm^{-3} with an exponentially-decreasing stratification above that level. At a representative time of six minutes after the introduction of a ten minute thermal pulse (equal to a forty-fold increase of the original coronal temperature of 1.6×10^6 K), one can easily discern in Figure 8 the coronal transient's loop-like (more realistically: dome-like) *enhancement*, $n-n_0$, ranging from $\sim 5 \times 10^8$ within the legs to $\sim 1 \times 10^8$ cm^{-3} close to the leading edge's outermost extent. The innermost rarefaction to values about one-tenth the original coronal density is also shown.

The montage shown in Figure 9 presents an extension of the interpretation noted above. Superimposed upon the previous figure, now, are the following schematics:

(i) The shock front, diagnosed by the type II plasma emission, is presumed to be located at the leading edge of the dome-like expanding compression. This front may (possibly) also be diagnosed in some cases as a faint white light enhancement called 'forerunners' as discussed by Jackson and Hildner (1978) and Jackson (1981). Its 'skirt' in the chromosphere is possibly the well-known Moreton-wave detected in Hα.

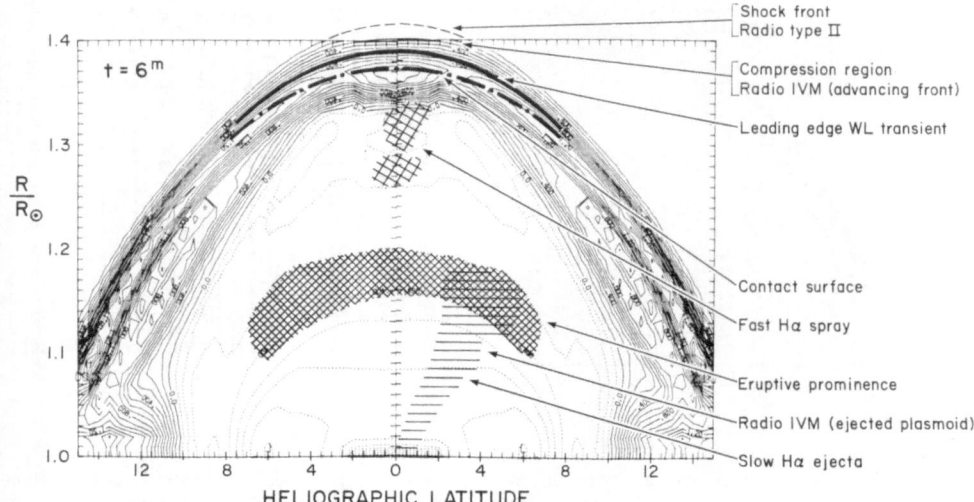

Fig. 9.   Montage of physically-significant boundaries (shock, etc.) superimposed on representative MHD computations of excess density contours shown in Figure 8. (Maxwell and Dryer, 1981). The magnetic field was 'open' prior to the pulse onset and did not consider any complex topologies in the vicinity of the 'flare'. As the transient moves outward, the field lines are pushed aside, returning only after the transient leaves the field-of-view.

(ii) The post-shock compression is seen within the entire enhanced (i.e., positive values of $n-n_0$) dome-like region. Most of this region is representative of what is commonly called 'coronal transient'.

(iii) The leading edge of the white-light coronal transient is determined by observed brightness enhancements following image subtraction techniques as discussed earlier. This is the heavy solid line drawn arbitrarily on the otherwise-computed contours. Determination of the leading edge is a strong function of instrument calibration and experimental experience with interpretation of the subtracted brightness profiles. Thus, most likely it would be behind the actual shock, although its juxtaposition in some special cases should not be precluded.

(iv) The dash-dot line (drawn behind the solid line) is the contact or piston surface that marks the plasma that was originally located at the coronal base. The mathematical and physical basis for this surface has been discussed in a very wide context of laboratory to cosmological applications by Glass (1977) and in the solar flare and interplanetary situations by Dryer (1981). Hopefully, the characteristics of the plasma along this surface will retain some of those in the flare ejecta (cf. Hirshberg, 1968). An unsolved problem in interplanetary physics is the unambiguous identification of *in situ* plasma data with flare-ejected plasma. The velocity of the contact surface can be as high as 80% of the shock's velocity as discussed by Maxwell and Dryer (1981).

(v) The remaining cross-hatched regions in Figure 9 are schematics (or 'cartoons') intended to represent, more-or-less, the spatial juxtaposition of Hα observations (fast sprays, loop-like helical eruptive prominences, and slow ejecta that either are heated and vanish or fall back to the surface) within the larger spatial extent of the coronal transient.

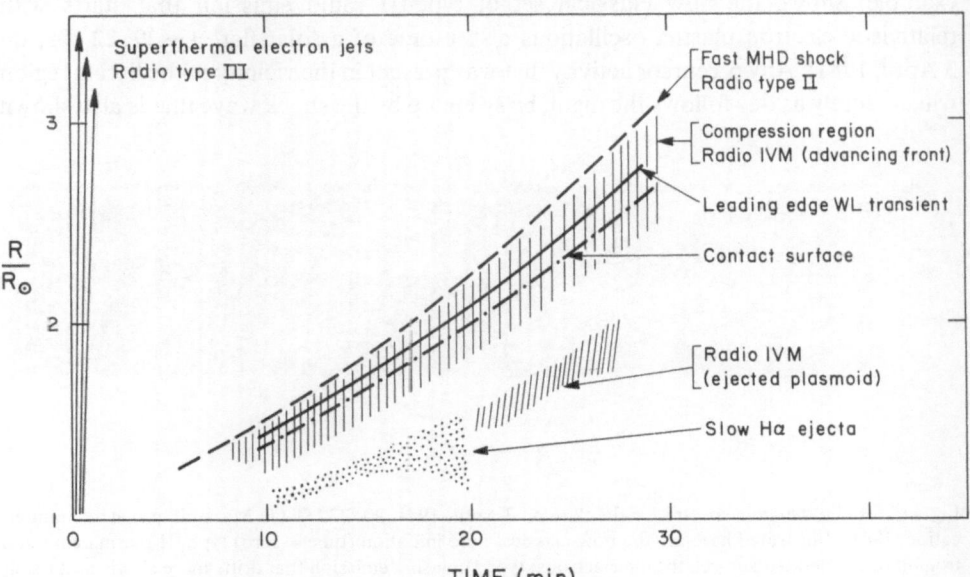

Fig. 10.   Schematic radius-time diagram illustrating the relationship among the various diagnostic techniques and their physical interpretations (Maxwell and Dryer, 1981).

Also suggested within this sketch is a separate class of type IVm radio bursts that have been colocated with white light plasmoids for one event by Stewart *et al.* (1982) who used simultaneous SMM C/P and Culgoora radioheliograph data.

Another useful tool for the organization of both observational and theoretical data (and, of course, their comparison) is the radius-time plot. The conventional format is that shown in Figure 10 which is simply a temporal history of all interdisciplinary data. Their interpretations by Maxwell and Dryer (1981), as shown therein, are generally in agreement with those of MacQueen (1980). It should be noted that all events are assumed to originate at a 'time-zero' when microwave radio bursts, superthermal electron jets, and type III radio emission occur more-or-less simultaneously with an increase of Hα emission in a flare. This assumption can be taken seriously only for the most pathological phase of solar activity – the solar flare to which most of our attention is directed. But we must not delude ourselves that '... a gas like the Sun would be entirely placid, the epitome of celestial tranquility' (Parker, 1977). Thus, it should not come as any surprise that pre-flare 'activity' should be detected by careful examination of optical phenomena in various lines, including white light, in the visible (e.g., Bruzek and Demastus, 1977; Martin and Ramsey, 1972; and Jackson, 1981). It should be understood, then, that our main attention is directed to the physics that follows the main act, as it were, and not the precursor activity.

We have, therefore directed attention to the major coronal transients that follow solar flares and that may (or may not) be associated with eruptive prominences. Thus, to return to our notion of 'time-zero', I remind the reader of the well-known dynamic spectra observed by ground-based radio astronomical observatories. Figure 11, for example, shows the now classical set of type III radio emission that starts with relativistic electron plasma oscillations at the time of a solar flare (~ 20:22 UT, on 3 April, 1981). Any precursor activity that was present in the vicinity of the flaring region will, as surely as day follows the night, be swept up by the shock wave that is also shown

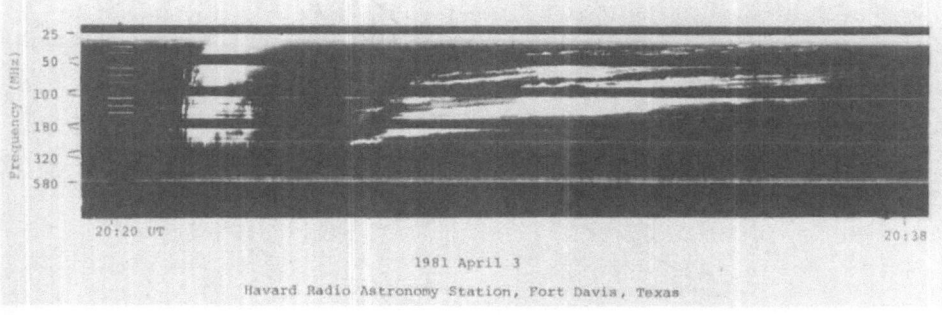

Fig. 11.   Radio dynamic spectra for the flare on 3 April, 1981, 20:22 UT (A. Maxwell, private communication, 1981). Illustrated here are the now-classical flare initiation (time = zero) type III plasma emission from flare-accelerated sub-relativistic electrons; type II plasma emission that drifts more slowly as a result of shock motion upward through the corona; and type IV broad band emission that can be due to either plasma emission from density enhancement regions behind the shock or to gyro-synchroton emission from < 500 keV electrons.

in Figure 11 as the type II split-band fundamental and second harmonic. This split-band structure has been interpreted by Smerd *et al.* (1975) to be caused by simultaneous plasma emission from the plasma ahead of and behind the shock front. These authors suggested that the shock's Alfvén Mach number, M_A, could be given, approximately, as:

$$M_A \simeq 1 + \frac{4}{3} \frac{\Delta f}{f_l} + \frac{2}{3} \left(\frac{\Delta f}{f_l} \right)^2, \tag{2}$$

where f_l is the (lower) plasma frequency ahead of the shock wave and $\Delta f = f_u - f_l$, where f_u is the (upper) plasma frequency behind the shock. Implicit in this derivation is the assumption of a perpendicular shock wave wherein the density change across the shock is given as:

$$\frac{n_1}{n_0} = \left(\frac{f_u}{f_l} \right)^2 = \frac{4 M_A^2}{3 + M_A^2}, \tag{3}$$

where

$$f = 9 \times 10^{-3} n^{1/2} \text{ (MHz)}. \tag{4}$$

From Equation (2), then, it is seen that the Alfvén Mach number is a function of the band-splitting, Δf. Smerd, Sheridan and Stewart estimated (using nine cases of split-band type II bursts) that $1.2 \lesssim M_A \lesssim 1.7$. Also, by using the drift rate, df_l/dt, and a coronal density model equal to either an early 1961 Newkirk model (Figure 6 in Newkirk, 1967), or twice those values, they estimated the range of magnetic field magnitudes to be within the range $0.3 \lesssim |\mathbf{B}| \lesssim 4$ G. In a more recent study that presents the most complete set of simultaneous radioheliograph and green continuum (5014–5328 Å) coronagraph data available to date (for 27 April, 1980), Stewart *et al.* (1982) estimated $B > 0.7$ G at a time when a type IVm plasmoid was at $2.5 R_\odot$. In order to arrive at this estimate, they assumed that the plasmoid was confined by a magnetic field such that the external magnetic field pressure could be equated to the internal plasma pressure, $2 nkT$ (where T was assumed to be 10^6 K). The density, n, was taken as 10^8 cm^{-3}, some 20–46 times larger than the background. Their additional assumption was that the plasma density, apparently this large, was high enough to produce 80 MHz radiation by emission at the second harmonic, rather than by the usually-assumed gyro-synchrotron mechanism. It is of interest to note that the Stewart *et al.* (1982) estimate falls within the range suggested by Smerd *et al.* (1975) at similar distances from the Sun's center.

The reader ought not to be left with the impression that a permanent shock wave will always be formed. Another dynamic spectrum from the Harvard Radio Astronomy Station at Fort Davis is shown in Figure 12 for the last (18 : 21 UT) of three west limb flares on 29 June, 1980. Although the microwave type III emission clearly marks the flare's onset, the shock was apparently delayed in its formation until ~ 18 : 33 UT. At this time, the split-band structure was clearly seen, albeit for only a few minutes, in the

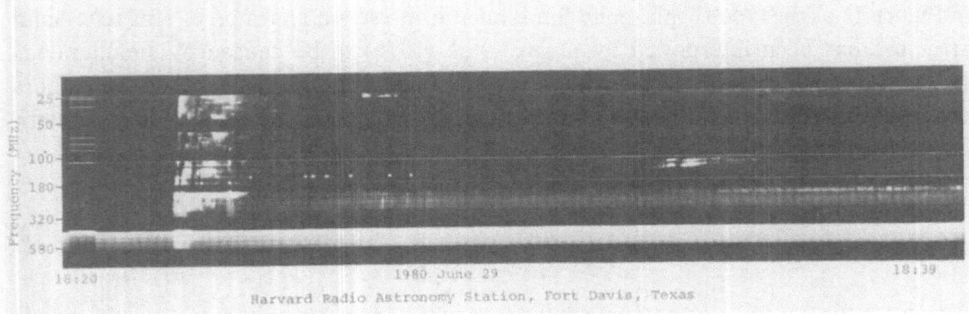

Fig. 12.   Radio dynamic spectra for the flare on 29 June, 1980, 18 : 21 UT (A. Maxwell, private communication, 1981). Note that the shock wave was formed late in the event and dissipated after only a few minutes.

second harmonic. A very faint, split-band fundamental was also seen at the same time interval after which the type II emission disappeared, presumably due to shock attenuation. As discussed by Wu *et al.* (1982c), a coronal transient from the flare was observed by both the NCAR/HAO Mark III *K*-coronagraph at Mauna Loa as well as by the coronagraph/polarimeter on SMM. It is difficult to state with any certainty that a permanent shock wave had been formed, the loss of signal in Figure 12 mitigating against it. In this case, however, the earlier, homologous, energetic flares may have temporarily depleted the corona to such an extent that the shock, while possibly present beyond the ∼ 18 : 36 UT disappearance, travelled outward through a highly-rarefied medium that precluded the plasma emission mechanism. As discussed later, speculation of this kind can be avoided by an extension of the dynamic spectrum to lower frequencies made possible by radio emission reception at spacecraft that operate without the disadvantage of ionospheric cut-off (at ∼ 20 MHz).

An increasing amount of observational evidence is presently being accumulated to support the suggestion (Figures 8 and 9) that the shock wave, when it does indeed form as a result of 'breaking' of the nonlinear high amplitude MHD fast mode wave, is ahead of the leading edge of the white light coronal transient. Gergely *et al.* (1981) recently combined Clark Lake radioheliograph observations of a Type II radio burst with simultaneous observations of a Skylab-observed coronal transient on 27 October, 1973. They estimated the plane-of-sky shock velocity to be 4900 km s^{-1}, six to seven times faster than the velocity of the transient which lagged behind. They also recalled the range of magnetic field strength at $2R_\odot$ as suggested by Dulk and McLean (1978) to be between 1–10 G; as a result they concluded that for their shock, the possibility must be considered that it was indeed a strong shock having a Alfvén Mach number of 7–8, i.e., substantially larger than those considered by Smerd, Sheridan and Stewart (1975). It is necessary to point out that the plane-of-sky radioheliograph results may be affected somewhat by coronal refraction and ionospheric refraction to some extent. Nevertheless the possible correction to the shock velocity would still result in a high value.

More such evidence of a 'shock-ahead-of-transient' is offered in the case of a Mark III
K-coronagraph observation following a class 2 flare near the west limb on 25 March,
1981 (Friend et al., 1981). Following an initial increase of mass within the field of view,
a total net change in the polarized brightness of the corona over the active region
(following image subtraction) amounted to a net decrease of 35%. The coronal transient
moved with a constant velocity of 670 km s^{-1}. This observation is apparently a common
feature (following limb flares) observed above the K-coronagraph's occulting disk at
1.24$R_\odot$ and appears to confirm the MHD model's prediction, qualitatively, shown in
Figure 8 for the rarefaction region. A type II shock was also observed at Fort Davis to
move ahead of the transient (assuming a 2 × Newkirk coronal density model (1961),
which is equivalent to the 10 × Saito and 10 × Baumbach–Allen models) at about
1200 km s^{-1} (A. Maxwell, private communication, 1982). An example of the spatial
separation at a given moment for this case was the 1.25$R_\odot$ location of the coronal
transient's leading edge, following the shock at 1.35$R_\odot$. This result, although model-
dependent, is supported by the Clark Lake radioheliograph result discussed in the
preceding paragraph.

Additional evidence of a 'permanent' flare-generated shock wave has recently been
obtained during the interdisciplinary SCOSTEP/STIP Interval VII (August–September 1979). During this period, Croft (1982) described the procession, at/near superior
conjunction, of Voyager 1, Voyager 2, Pioneer 11, and Pioneer-Venus as shown in
Figure 13. Croft (1982) subsequently found a series of strong Doppler shifts in the Deep
Space Network telemetry signals during this epoch. He plotted the ratio of the Doppler
noise at both low and high frequencies as shown in Figure 14. Each sudden depression
apparently is caused by a transient. Following a solar flare just behind the east limb at
~ 14 : 01 UT, 18 August 1979, (Kane et al., 1981), radio scattering observations consisting of spectral broadening, mean phase and amplitude scintillations following the
resulting shock wave were reported by Woo and Armstrong (1981) who used the same
spacecraft signals as Croft's from the Deep Space Network. This event is the only one
examined in detail thus far among the likely candidates suggested by the Doppler noise

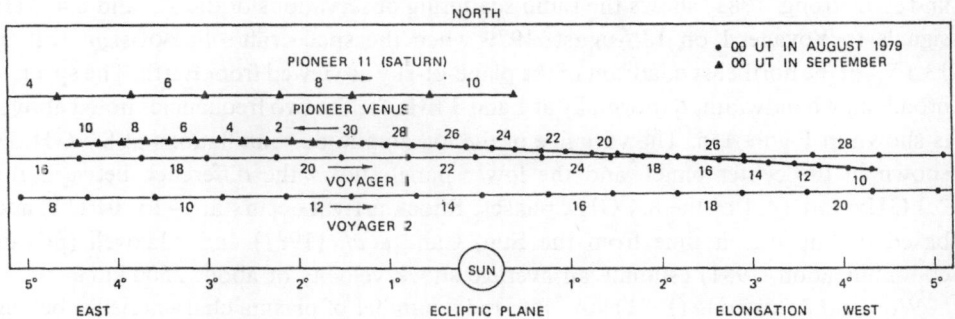

Fig. 13.   The spacecraft 'parade' during STIP INTERVAL VII (August-September, 1979). Deep Space
Network telemetry at both S- and X-bands (2.3 and 8.4 GHz, respectively) was used to detect solar
flare-generated shock waves and coronal disturbances as described in the text. (From Croft, 1982).

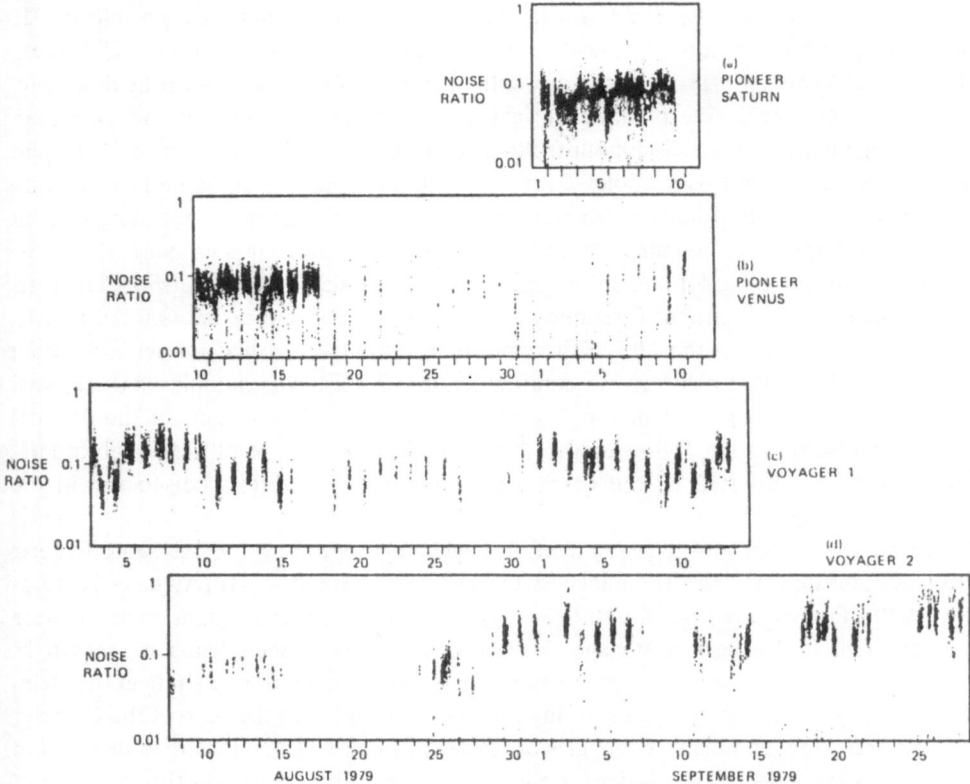

Fig. 14. Doppler noise ratio (low frequency noise divided by high frequency noise) from signals received from the four spacecraft shown in Figure 13 during available periods of simultaneous tracking by the NASA Deep Space Network (Croft, 1982). The depressions are caused by coronal transient activity.

ratio depressions (i.e., the high frequency noise is enhanced during a transient's passage across the line of sight from Earth to the spacecraft shown in Figure 13). Figure 15 (Woo and Armstrong, 1981) shows the radio scattering observations of the 2.3 and 8.4 GHz signals to Voyager 1 on 18 August, 1979 when the spacecraft's line-of-sight was at $13.1 R_\odot$ in the northeast quadrant of the plane-of-sky as viewed from Earth. The spectral broadening bandwidth, B (normally at 1 and 3 kHz for the two frequencies noted above) is shown in Figure 15a. The variance of the log amplitude scintillations at 8.4 GHz is shown in the center panel, and the lower panel shows the difference between the 2.3 GHz and 3/11 of the 8.4 GHz phases. Shock arrival occurs at $\sim 15:01$ UT, and based on the transit time from the Sun, Cane *et al.* (1981) and Maxwell (private communication, 1981) estimate an average shock velocity of about 2500 km s^{-1}.

Woo and Armstrong (1981) have inferred a number of plasma characteristics behind the shock wave as shown in Figure 16. The top panel (Figure 16a) provides a measure of plasma turbulence as indicated by the refractive index, c_n, and electron density fluctuations, c_{n_e}, at 8.4 GHz. The inferred electron density and plasma velocity are

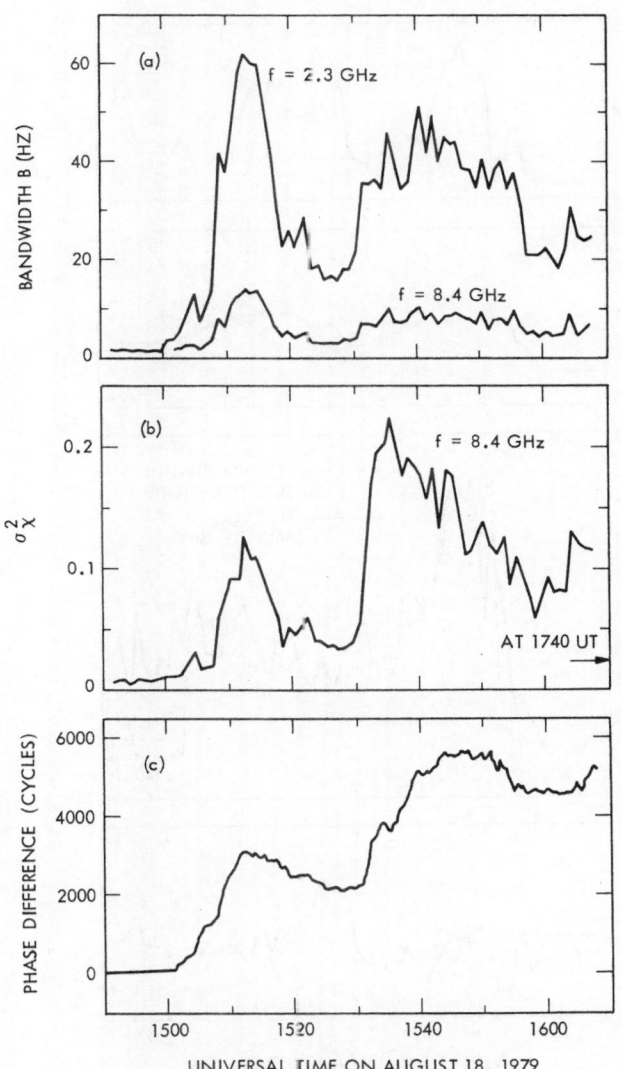

Fig. 15. Voyager 1 radio scattering observations on 18 August, 1979. (Woo and Armstrong, 1981). Shock arrival at ~ 15:01 UT is clearly indicated.

shown, respectively, in Figures 16b and 16c, the latter being estimated by two independent methods. The substantial rise in solar wind velocity about 10–15 min into the post-shock flow to 2600 km s^{-1}, from its pre-shock value of 200 km s^{-1}, is extraordinary! Using the classical Rankine-Hugoniot shock jump relations, Woo and Armstrong (1981) estimated an *in situ* shock velocity of 3500 km s^{-1}, thereby suggesting (Cane *et al.*, 1981) that the shock actually accelerated enroute from the flare site to the line-of-sight to Voyager 1. The bottom panel in Figure 16 shows the ratio (in percent) of the r.m.s. electron density fluctuations, for scale sizes smaller than the Fresnel size

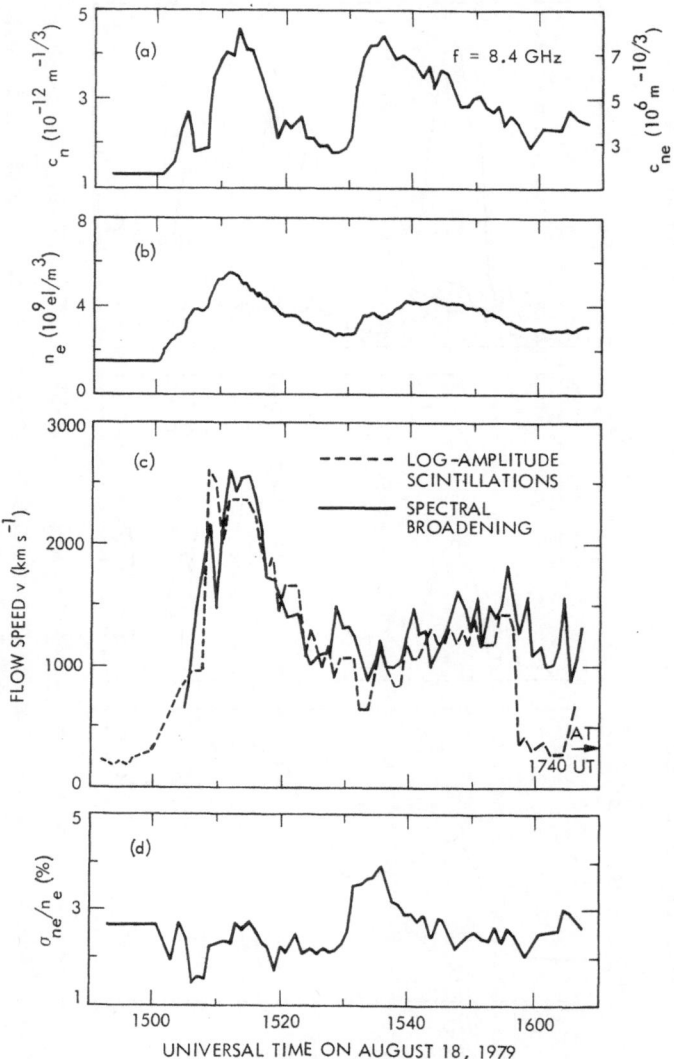

Fig. 16.   Measurements of the shocked solar wind plasma as inferred from the radio scattering observations
of 18 August, 1979 as shown in Figure 15. (Woo and Armstrong, 1981).

(73 km) at 8.4 GHz, to the electron density. The notable increase in the fluctuations at
~ 15 : 35 UT represents the start of a second region of enhanced amplitude scintillations.
Woo and Armstrong (1981) suggest that this position may correspond to the turbulent
'driver gas' of the shock wave. In terms of the previous discussion, this may indeed mark
the position of the contact surface (or piston) where localized turbulence may manifest
itself as anomalous resistivity (Dryer, 1975). This turbulent layer may persist throughout
its outward transit in the interplanetary medium. The question of identification of the

flare's contact surface, together with its implications for flare ejecta identification, is one of the unsolved problems in solar/interplanetary physics.

Substantial progress, however, is being made on the propagation of the shock wave itself. An unambiguous tracking of the same shock wave of 18 August, 1979 was made possible by the low frequency radio type II burst record from ISEE-3. Cane *et al.* (1982) have reported the observations of kilometric wavelength detection of the same shock observed by Woo and Armstrong (1981). Figure 17, from Cane *et al.* (1982), shows an intense type III burst that was initiated in the 1 MHz level at ~ 14 : 15 UT, about 14 min after the X-ray burst from the east limb flare. The type II burst also appeared at the high end of the radiometers' band (2 MHz to 30 kHz, with one radiometer having a 10 kHz bandwidth, the second with a 3 kHz bandwidth) at approximately 17 : 00 UT. The dynamic spectra in Figure 17 were generated by 23 discrete observing frequencies within the range just noted. Revealed therein is a nearly continuous type II slow drift radio burst until its arrival at ISEE-3's position (~ 235 earth radii in front of our planet) at 05 : 58 UT on 20 August, 1979. Part of the difficulty in analyzing data of this type (see, also, Malitson *et al.*, 1976; Boischot *et al.*, 1980) is illustrated by the occasional coalescence of the type II burst with a low frequency continuum. This latter noise band results from the direct coupling of the ISEE-3 antennae to the local ambient plasma. It is, therefore, apparent that earlier (pre-flare) electron density enhancements occasionally engulfed the spacecraft with the unavoidable, but temporary, loss of the type II slow drift. Note, for example, the sudden increase in the low frequency cut-off at

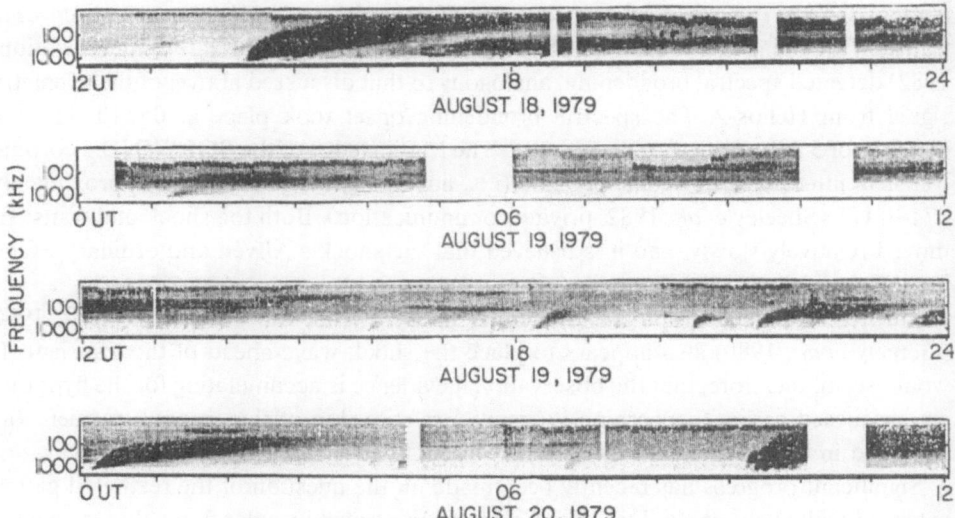

Fig. 17. Dynamic spectra from the ISEE-3 radiometers that are driven by a 90 m dipole antenna in the spacecraft spin plane. The type III radio burst at ~ 14 : 15 UT on 18 August, 1979 is followed by a type II burst from the shock wave at ~ 17 : 00 UT (or even earlier). This shock, the same one detected by Woo and Armstrong (1981), was tracked almost all the way to ISEE-3's position at 05 : 58 UT, 20 August, 1979, where its broadband detection was independently detected by the plasma analyzer and magnetometer on the spacecraft. (Cane *et al.*, 1982)

07 : 40 UT on 19 August, 1979 (second panel in Figure 17). The ambient plasma frequency (proportional to the square root of the plasma density) remained high for about 1 h. Finally, after the shock passed the spacecraft at 05 : 58 UT the next day, it is seen that the local plasma frequency remained above the lowest observing frequency (30 kHz), thereby indicating at least a four-fold increase in density for at least four additional hours. No attempt to date has been made to identify the contact surface. Indeed, it would be remarkable if it were to be identified since the parent flare was more than 90° displaced in heliolongitude! It would seem, without emphasizing the obvious, that such a search should be made closer to a flare's central meridian. Also it should not be surprising to observe occasional coalescence of the desired type II drift (that is being 'tracked') with locally enhanced density structures in the spacecraft's vicinity. It was, for example, pointed out by Davis and Feynman (1977) that variations should be expected from the average inverse-square solar wind density dependency on helioradius as a result of streams that have a habit of temporarily negating this relationship. Thus, Davis and Feynman (1977) used a more realistic hydrodynamic stream interaction model to improve the analysis of a similar type II radio burst (Malitson et al., 1976) for the flare-generated shock wave of 7–8 August, 1972.

There was no unambiguous observation (i.e., a series of images) of the white light transient associated with the 18 August, 1979 flare. A P78-1 SOLWIND image, obtained some time after Woo and Armstrong's observation, compared with a pre-flare coronal image, indicated a brightness change, thus suggesting progress toward a fairly-rapid (hours) return to the pre-event coronal state. Fortunately, another opportunity for comparison of a radio scattering observation of a shock wave with its associated coronal transient took place on 24 October, 1979. On this date, R. Woo (private communication, 1982) detected spectral broadening, analogous to that discussed above, of the telemetry signal from Helios-2. The spectral broadening onset took place at 05 : 10 UT at a distance of $5.2R_\odot$ from the Sun's center. The leading edge of the SOLWIND's coronal transient intersected this same position (i.e., line of sight to Helios-2) at approximately 07 : 30 UT (Sheeley et al., 1982, private communication). Both the shock and transient moved relatively slowly, and it is believed that the shock's Alfvén and ordinary Mach numbers were only slightly greater than one.

Further radio heliographic and coronal imagery from the event of 9 April, 1980 (Gergely et al., 1980) also appears to place the shock wave ahead of the transient. It would seem, therefore, that the observational evidence is accumulating for the hypothesis, discussed earlier, concerning the spatial relationship of the coronal transient (as observed in white light) and its associated shock wave.

Significant progress has recently been made on the question of the restricted global extent of white light coronal transients that are generated by solar flares. In a sequence of subtracted images, Michels et al. (1982) detected, unambiguously, a 360° density enhancement that encircled the occulting disk on the P78-1's SOLWIND coronagraph after a central meridian flare on 27 November, 1979 at ∼ 06 : 45 UT. This enhancement continued to increase (in radius) as seen in Figure 18. The implication that this 'head-on' transient was expanding toward the earth is unavoidable; in fact, the implied head-on

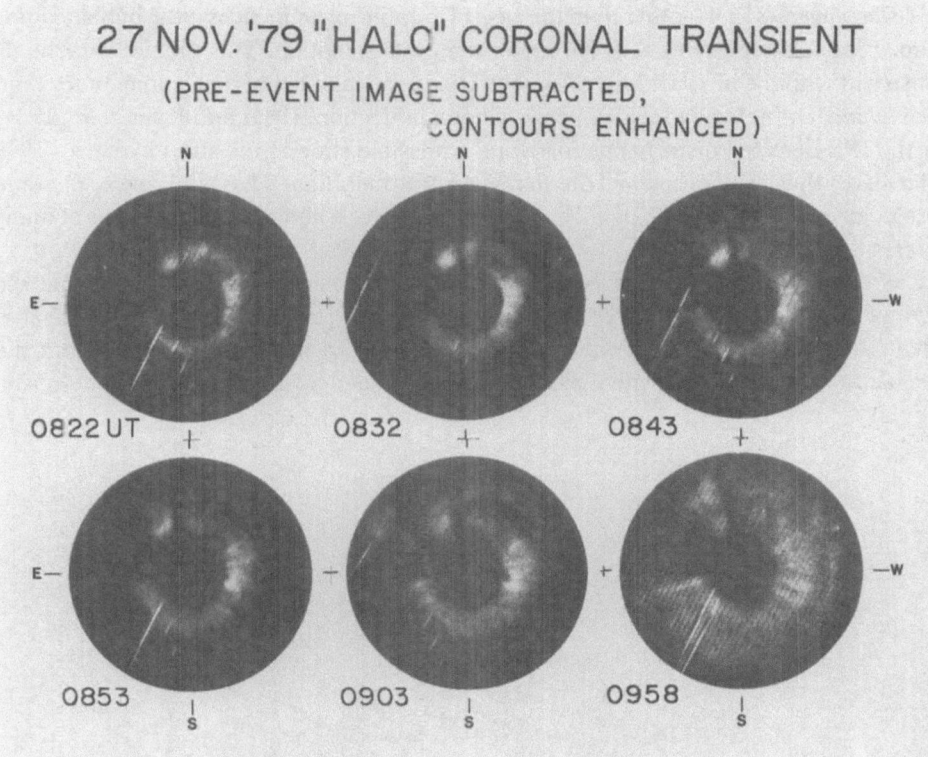

27 NOV.'79 "HALO" CORONAL TRANSIENT
(PRE-EVENT IMAGE SUBTRACTED, CONTOURS ENHANCED)

Fig. 18   'Head-on' or 'Halo' white light coronal transient observed by P78-1's SOLWIND experiment on 27 November, 1979. Note that the gap in intensity to the ENE is caused by the shadow of the occulter's post on the instrument. Also, the field-of-view has been cropped here to just over $8R_\odot$. This first optical observation of a coronal transient directed toward the earth had not been reported at the time of the Lindau Meeting, but was subsequently announced at the Boulder Meeting in January 1982 of the American Astronomical Society, while this review was in preparation for publication. It is included here because of its important bearing on the subject matter. (Michels *et al.*, 1982).

velocity toward Earth was $\sim 1180\ \mathrm{km\ s^{-1}}$. Indeed, a shock wave was observed at ISEE-3 as well as a sudden commencement at Earth. The average shock velocity (Sun to ISEE-3) was, using onset at the flare, about $570\ \mathrm{km\ s^{-1}}$. Analysis of this event is currently in progress.

Related studies concerning the question of the solid angle subtended by transients have been conducted by F. Crifo (private communication, 1981) who used polarization measurements of a Skylab transient on 10 August, 1973 (Crifo *et al.*, 1982) and a SMM C/P transient on 7 April, 1980. Her conclusion, that some particular transients are more bubble-like rather than looplike, appears to be confirmed by this recent SOLWIND result. Of course, we may note that there is a spectrum of geometrical shapes among which loop-like transients (especially the unambiguous eruptive prominences discussed earlier) are still admissible. The SOLWIND observation of Figure 18, however, suggests that – other than the case of eruptive prominences – the bubble/dome shape probably represents the rule rather than the exception. This picture is certainly consistent with the observations of shocks over a range of heliographic longitudes (and even *latitudes* as noted by Jackson, private communication, 1981; and Kane *et al.*, 1981; for the IPS-detected disturbance following a possible shock from the 14 August, 1979 solar flare) that extend beyond the flares' central meridian. The solid angle, in some cases, can extend to beyond 2π sr centered, more-or-less, about the central axes of many flares.

It is now clear, from *in situ* as well as from remote sensing observatories, that the coronal transients' shock waves are asymmetrical but still quite large in their spatial extent. It is therefore appropriate to direct attention to the shock structure and the plasma physics associated with it along the lines pursued by magnetospheric physicists.

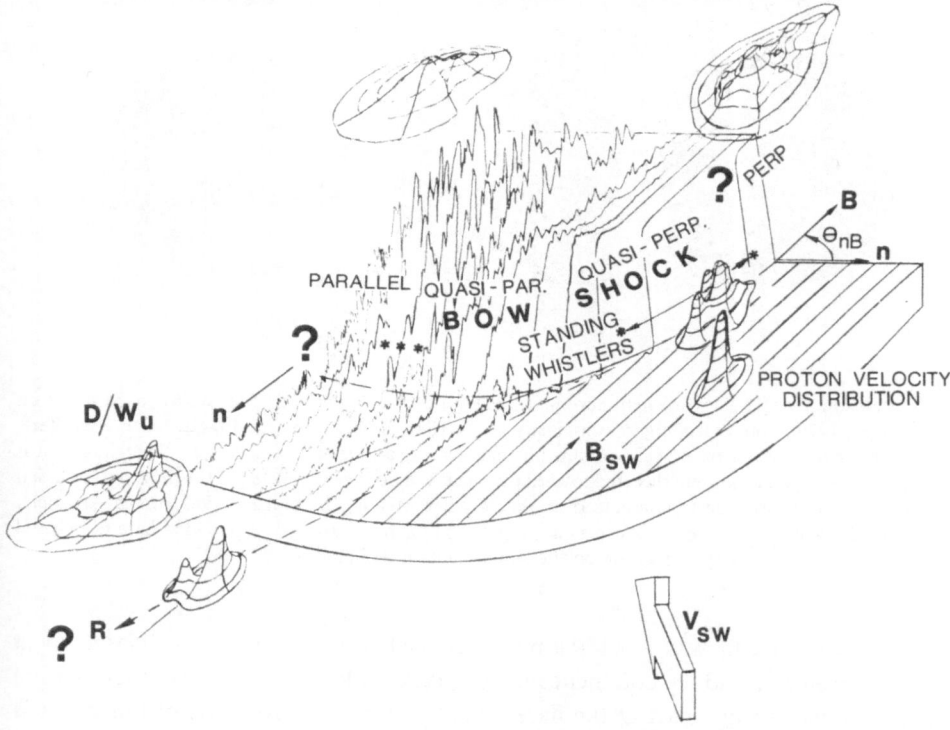

Fig. 19. Macrophenomenology of the Earth's collisionless bow shock that is probably also appropriate for flare-generated interplanetary shock waves. (After Greenstadt and Fredricks, 1979). The upstream interplanetary magnetic field direction $\mathbf{B}_{sw}$ is indicated on the foreground field 'platform'. The solar wind velocity upstream of the bow shock is $\mathbf{V}_{sw}$. For the interplanetary shock, this vector would be the local relative velocity as measured in the shock frame. The field magnitude is plotted vertically. The superimposed three-dimensional sketches represent solar wind proton thermal properties as number distributions in velocity space. A representative distribution of protons in a reflected beam, R, is shown emanating from the region along the shock where the angle θ_{nB} begins to deviate from 90° that is, when the shock is *nearly* a perpendicular shock. As the shock becomes quasi-paralled ($0 < \theta_{nB} < 45°$, say), the region D/W_u represents the spatial extent where diffuse, accelerated particles are detected together with upstream ULF waves.

In their study of the earth's bow shock, Greenstadt and Fredericks (1979) noted that the structure is determined in a fundamental way, not only by the magnitude of the solar wind (that is, upstream) Alfvén Mach number and plasma beta, but particularly by the orientation θ_{nB}, of the upstream magnetic field vector relative to the normal vector at each point on the asymmetrical bow shock. Figure 19 (E. Greenstadt, private communication, 1981) is a sketch that summarizes the macrophenomenology of the magnetic field behavior around the earth's bow shock as well as of the richness of the various proton distribution functions associated with various locations relative to θ_{nB}. The region, D/W_{u}, indicating diffuse ions and upstream ULF waves is probably analogous to the

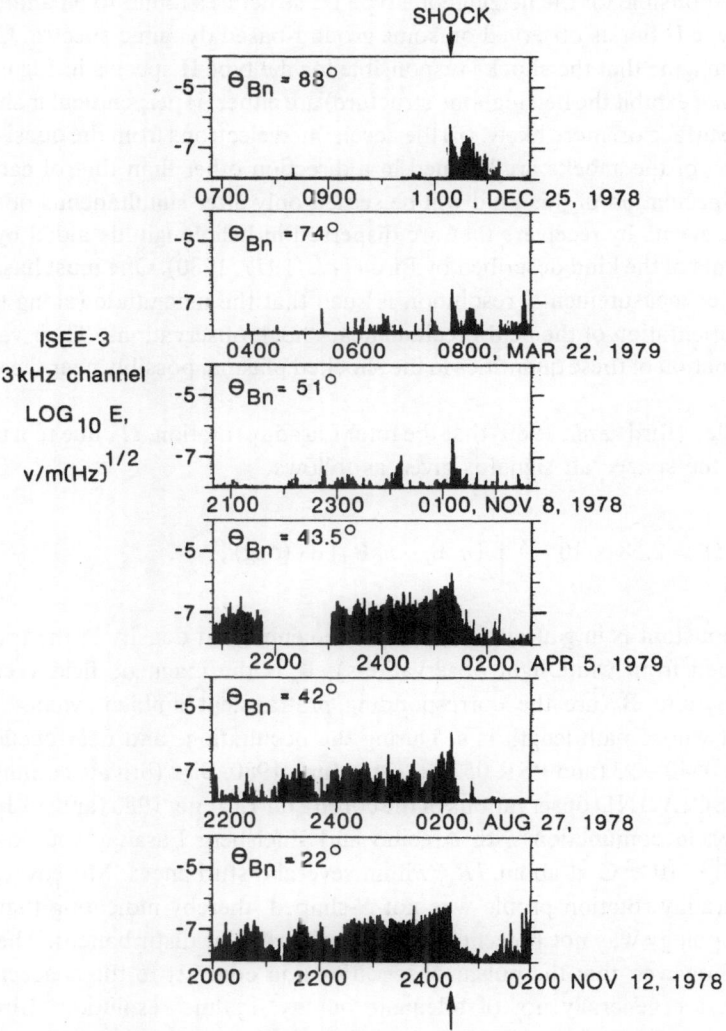

Fig. 20.   ISEE-3 electric field noise upstream of, at, and following six interplanetary shock waves. The 3.16 kHz channel shown here displays upstream noise for the quasi-parallel shocks ($\theta_{nB} < 51°$), whereas it appears downstream of the quasi-perpendicular shocks (Kennel *et al.*, 1982).

locations vis-á-vis the interplanetary shock wave, where precursor plasma waves (possibly ion acoustic waves) are detected by spacecraft that are located upstream of the shock wave. These precursor waves have also been observed prior to the arrival of interplanetary shocks (c.f., Scarf *et al.*, 1981). Examples of the plasma waves observed prior to arrival of interplanetary shock waves (that have a range of θ_{nB} from quasi-parallel to quasi-perpendicular) are shown in Figure 20 (Kennel *et al.*, 1981). It is unknown at this time as to the source of the type II radio bursts; but it is believed that the source of accelerated electrons, responsible for the type III radio bursts, is probably traceable back to the shock along the dashed line labeled R in Figure 19. These electrons are probably responsible for the herringbone type III structures found to emanate from the slow drift type II bursts observed on some ground-based dynamic spectra. Conversely one might imagine that the shocks responsible for the type II spectra in Figures 11 and 12 (that do *not* exhibit the herringbone structure) are either (i) perpendicular shocks over their entire surface or, more likely, (ii) the accelerated electrons from the quasi-perpendicular portion of the shocks are beamed in a direction other than that of earth's radio receivers. Speculation of this kind will be settled only after simultaneous observations of the same events by receivers that are dispersed in heliolongitude aided by Faraday measurements of the kind described by Bird *et al.* (1977, 1980). One must hasten to add that the latter measurements' resolution is such that the magnitude (along the line of sight) and orientation of the field preclude near-shock observations. Their value lies in the determination of these quantities in the shocked plasma, possibly near the white light transient.

It is recalled (Bird *et al.*, 1980) that the total Faraday rotation, Ω_t, due to a transient's crossing of the spacecraft signal is given as follows:

$$\Omega_t = 2.58 \times 10^{-13} \int_t [n_t \, \mathbf{B}_t - n_0 \mathbf{B}_0] \, \mathrm{d}s \; (\deg) , \tag{5}$$

where the constant is in gaussian units; n_t is the enhanced density in the transient (to be determined from white light observations); $\mathbf{B}_t$ is the magnetic field vector in the transient; n_0 and $\mathbf{B}_0$ are the corresponding pretransient ambient values along the line-of-sight whose path length is s. During the occultation (and near-occultation) of pulsar PSR 0540 + 23 (and PSR 0525 + 21) in June 1980, Bird (private communication, 1981) used SOLWIND observations of the corona on 15 June, 1980 (and 13 June, 1980, respectively), in conjunction with Arecibo and Effelsberg Faraday rotation data, to estimate $|\mathbf{B}| \sim 10^{-2}$ G at about $7R_{\odot}$ within several disturbances. Moreover, he noted that the Faraday rotation profile was not S-shaped, thereby indicating that a helical magnetic topology was not present within these particular disturbances. The reader is cautioned, however, that the pulsar observations, in contrast to the spacecraft occultation data, are generally not of adequate quality or time resolution (Bird, private communication, 1982) for studies of coronal dynamics.

In this section I have attempted to summarize a number of observations relevant to coronal transient phenomena, particularly those cases where simultaneous radio and optical imagery data were found in some recent cases. In Section 5, I will try to summarize some of the recent theoretical and modeling work relevant to the understanding of coronal transients.

5. Theory and Modeling of Coronal Transients

At the present time there are three distinct approaches to the theoretical exploration of transients, their initiation and motion.

The first approach assumes that the loop-shaped brightness enhancements (including the outermost rims as well as the obvious eruptive prominences) delineate magnetic flux tubes. Given this assumption, this approach is further divided into three kinematical classes: (i) a model in which the driving force for the transient is derived from the gradient of the magnetic pressure of a self-induced magnetic field which is produced by the current in the loop without consideration of external plasma or surrounding magnetic fields; (ii) a model in which a loop is driven by externally upwelling of magnetic flux that produces, thereby, an imbalance of magnetic pressure on the loop; and (iii) a model in which an extraneous body immersed in an ambient conducting medium will be acted upon by a MHD buoyancy force which results from the inhomogeneity of thermal and magnetic stresses that act on the compressible extraneous body. Model (i) has been outlined by Mouschovias and Poland (1978); Model (ii), by Pneumann (1980); and Model (iii), by Yeh (1982a, b) and Yeh and Dryer (1981a, b). Within this approach only Model (iii) provides an explanation (Yeh, 1982a) for the initial ejection of a coronal loop as a result of abrupt surface activity which is followed by magnetic unwinding and peripheral expansion of a stationary loop. The increase of the loop's volume will enhance the MHD buoyancy force which, having exceeded the gravitational force, provides the initial ejection and continued acceleration, expansion of the loop and eventual blending with the surrounding ambient solar wind. There are two premises included in Models (i) and (ii), namely, that the thermal force is non-essential, and the net force produced by the magnetic gradient is adequate to determine translational motion. It was argued (Dulk *et al.*, 1976; Mouschovias and Poland, 1978) that the thermal force could be ignored on the grounds that the plasma beta is generally small (see also, MacQueen, 1980). However, it was noted (Yeh and Dryer, 1981b) that plasma beta is a measure of the relative magnitudes between thermal and magnetic energy *densities*, whereas forces are the result of *gradients* of energy densities; hence, a greater energy density will have a greater gradient only if the scale lengths for the two energy densities are comparable in magnitude. It follows that the slenderness of transient loops clearly indicates (Yeh, 1982a) that the pertinent scale lengths are quite different from those for the ambient medium; thus a small value (however small) of the plasma beta for the ambient medium or even for the loop material should not be taken as justification for ignoring the thermal forces. As for the second premise of Models (i) and (ii), the gradient of the magnetic forces is a necessary but *not sufficient* condition to determine translational motion of a compressible body. The motion of a compressible flux loop must be studied by examining the force densities at individual mass elements. Hence the thermal force is essential to transmit the ambient stress to the loop's interior so that MHD buoyancy force will affect every mass element thereby providing the inertial force for the outward motion common to all of the elements. A rigorous critique is presented for each of these Models in Yeh's most recent paper (Yeh, 1982a). He concluded that the heliocentrifugal motion of the loop as a whole is driven by the magnetohydrodynamic buoyancy force, whereas the expansionary motion relative to the axis of the loop is driven by the self-induced hydromagnetic force.

To summarize, the physics of the ejection of a coronal loop (such as a prominence) proceeds in the following sense:

(1) Magnetic activity at the solar surface takes place such as flux emergence, translational and/or rotational motion of the loop's footpoints, etc.

(2) The helical field lines in the coronal loop start to unwind (i.e., to decrease their pitch).

(3) The expansive magnetic force due to the axial field within the loop exceeds the contractive magnetic force due to the azimuthal field.

(4) As the loop's volume increases, its mass density decreases.

(5) The upward buoyancy force exceeds the downward gravitational force.

(6) The loop then moves away from the sun (heliocentrifugal motion) accompanied with loop diameter expansion, magnetic field unwinding, and field line stretching (see, for example, Figures 4 and 5).

The second approach considers the quasi-steady evolution of coronal magnetic fields in response to gradual photospheric changes. Self-consistent dynamical motion is not considered. Instead, as pointed out by Low (1981), these studies led to the suggestion that the evolution can terminate abruptly at a point beyond which no equilibrium solution exists. The interpretation is that a magnetic field, evolving toward the terminal configuration, will lose equilibrium and transit into a dynamical state. The problem of energy transport in the early development of this model was avoided by assuming the atmosphere to be isothermal and the field-plasma interaction to be confined spatially to a loop-shaped current sheet. In Low's quasi-steady model, the slow, gradual evolution through static equilibria, toward an untenable state, is brought about by very slight (highly plausible, as in Yeh's model discussed above) photospheric footpoint motions. The result is a ballooning-upward motion of the magnetic field configuration. For equilibrium across the assumed current sheet, the total pressure (plasma plus field) is conserved. A thermal pressure difference, $\Delta p \simeq (p_{out} - p_{in})$, exists across the current sheet where p_{in} and p_{out} are, respectively, constant pressures inside and outside the loop's footpoints at $Y = \pm L$. As the footpoints are displaced, the total flux of the bipolar field, F_0, is also conserved. Low (1981) has solved an illustrative case in which the parameter $C = -F_0/(\Delta p)^{1/2}$ plays an important role in the suggestion of the bipolar field ballooning-outward process. He suggests that, under certain circumstances, the top of the loop must approach the Alfvén speed where a fast dynamical phase must take over, basically as a magnetic buoyancy effect. At this point, he speculates, the external pressure above the current sheet can no longer contain the bipolar field, and a coronal transient takes place. In this regard, Low et al. (1982) applied this model to the observations of a coronal transient by Fisher *et al.* (1981) in which a rarefaction preceded the density enhancement (that started first along the legs, then built up to form a complete loop enhancement of brightness in the subtracted images). This class of non-impulsively-triggered coronal transients is frequently seen in the absence of solar flares on the limb by the Mark III *K*-coronograph (Fisher and Poland, 1981). This model assumes that the prominence (often observed within a transient as discussed earlier) is embedded in a coronal cavity of low density. The cavity is also assumed to be permeated by enhanced magnetic fields that compensate for the reduced pressure of the low density plasma. As the cavity evolves, according to this scenario, the rising dark arch system in the *K*-coronagraph images represents the low density presumed to have been present prior to the footpoint motion that precipates the loss of equilibrium noted above. By a judicious adjustment of the parameter, *C*, Low *et al.* (1982) are able to match the profile (in time) of the height and presumed footpoint separation distance.

The above second approach to the modeling of coronal transients has been extended to include the time domain explicitly by incorporating MHD selfsimilar analysis (Low, 1982). (Application of this technique has most recently been applied to the solar wind double-shock ensembles as reviewed by Dryer, 1975). As in its use in the solar wind, self-similar solutions must be regarded as asymptotic forms of non-self-similar solutions in space and time displaced from the influence of the boundary (here, the Sun) such that its 'memory' of the initial conditions has been essentially lost. Thus, for example, Dryer (1975) has discussed the distant evolution of double-shock ensembles (forward-reverse MHD shock waves), based on classical similarity theory, wherein it is irrelevant (that is, not known) whether such 'corotating interaction regions' originated from a solar flare or from interacting high/low solar wind streams. In the present context of expanding supernova remnants or coronal transients, Low (1982) assumes the existence of a polytrope (limited to the singular value of the polytropic exponent $4/3$), neglects the presence of one or more shock waves, and considers an axisymmetric geometry. Possibly the most attractive feature of this model is its ability to include a non-potential magnetic field that may thus contain substantial free energy above that contained in the potential state. Low's work also demonstrated that, under the assumption of self-similarity, an MHD problem for axially-symmetric geometry can be reduced to a purely magnetostatic problem. Using this same time-similarity procedure, Osherovich (1982) recently proposed a new transient model. Instead of introducing a current sheet (thus, cutting the magnetic field on the moving boundary surface), he required the magnetic energy to be finite with the magnetic field **B** to be continuous everywhere. Osherovich's approach leads to an eigenvalue problem that can be analytically treated. The basic equation of his theory is a Schrödinger-type equation.

The third approach to the theory and modeling of coronal transients (as well as supernova remnants as a logical extension) is to consider the full set of magnetohydrodynamic equations (the usual three-moment magnetofluid-dynamic equations plus Maxwell's equations). This approach provides the thermodynamic properties, velocities and magnetic fields in a totally selfconsistent way, once the initial and boundary conditions are rigorously specified. This approach is fruitful in the determination of the dynamic physical process. Complete data analyses, used in conjunction with such dynamic models, are crucial for understanding cause-and-effect relationships. If the simulations (computer-generated, in view of the highly non-linear nature of the equations that are not susceptible to approximations as discussed above) can be made to reproduce, in a reasonable way, the observed physical properties in some portions of the coronal transient, we can reasonably obtain information on the physics of the solar event and the transient itself for which observations are unavailable. Rigorous specification of the initial and boundary conditions, as mentioned above, is essential. A carefully-crafted program that starts with 'canonical' magnetic topologies and thermodynamic equilibrium (such as potential magnetic fields, single-fluid, dissipationless, two-dimensional, adiabatic processes) is necessary in order to focus attention on the more important physical processes. It is tempting, of course, to initiate such studies by incorporating complex atomic physics, multi-fluid and nonequilibrium processes with various forms of dissipation in, perhaps, three dimensional geometries. In the present writer's opinion, this approach should be exercised with caution so as not to gloss over the essential physics and to mask basic processes in an over-zealous effort to simulate some particular diagnostic such as intensities and profiles in spectral lines.

The third approach was initiated, for an ideal ionized gas in a 2-D planar geometry (a slice, possibly,through a more complex, 3-D transient) by Nakagawa *et al.* (1978), Wu *et al.* (1978), and Steinolfson *et al.* (1978). Subsequently, the solar wind background was incorporated in a simple radial magnetic field by Wu *et al.* (1981) who found that the solar wind (as discussed in the Introductory Remarks) does not significantly affect the general dynamic characteristics of the mass motion. The ambient solar wind, as expected, increases the velocity of the mass motion and its shock wave, as well as only a moderate change in the thermodynamic properties of the coronal plasma. In recent work, Wu (1982a, b) discussed the potential implications of the simulated density enhancements within the context of type IV radiation via either the plasma or gyrosynchroton emission mechanisms. Also, Dryer and Wu (1981) discussed the ramifications of the shock formation and characterization (in terms of θ_{nB} as discussed above) in the lower corona within the context of secondary stage cosmic ray particle acceleration. Wu *et al.* (1982b) applied this model to representative potential and non-force-free magnetic topologies and demonstrated the additional physics of coupled fast, slow and Alfvén wave modes; the result is a clear simulation of twisting magnetic field loops. In a fundamental work, Nakagawa *et al.* (1981) extended their earlier planar 2-D analysis to include the third components of the velocity and magnetic field vectors. The result is a nonplanar 2-D model with either a potential ($\nabla \times \mathbf{B} = 0$) or force-free [$(\nabla \times \mathbf{B}) \times \mathbf{B} = 0$] magnetic field. (This axi-symmetric model, in which all partial derivatives with respect to the azimuthal ordinate are set equal to zero, is referred to, in fusion research, as a quasi-3-D or 2−1/2-D model). Wu *et al.* (1982c) applied the 2-D version of this model to SMM-observed post-solar-flare densities and temperatures, together with a velocity pulse (i.e., a mass ejection pulse) as a test of the 29 June, 1980, 18 : 21 UT, coronal transient that was observed by the Mark III *K*-coronograph. The simulation for this case was not entirely satisfactory; for example, the observations indicated a rarefaction that was not reproduced by the chosen observed pressure pulse. It is believed that the long-lasting X-ray emission measures probably do not adequately represent the shorter time-scale deposition of energy by the flare; that is, the rarefaction would have been successfully simulated, as in earlier studies (cf. Figure 8), had the pulse been shorter in duration. Also, it is possible that the density increase in the flare's pulse was confined in the flare's neighborhood and, therefore, need not have been ejected upward as done by Wu *et al.* (1982c). In another 2-D application of the nonplanar code, Wu *et al.* (1982d) used an injection of cool, dense plasma up both sides of a closed potential field loop in an attempt to simulate the major features of the 8 May, 1979 coronal transient observed by SOLWIND as shown in Figure 2. A plot of the fractional density change, $(n - n_0)/n_0$, as a function of latitude and heliocentric radius is shown in Figure 21 several hours after the west limb flare. In addition to the fast MHD shock ahead of the transient, this simulation demonstrates the creation of two additional shocks within the coronal loop, together with a substantially larger density jump close to the sun. This density increase was accompanied by relatively low temperatures – a combination appropriate for a long duration observation of some excitation of neutral hydrogen in Hα at these low altitudes. A similar mass injection (as the initial perturbation at the photospheric level) was used in a separate study by Wu *et al.* (1982a). The result,

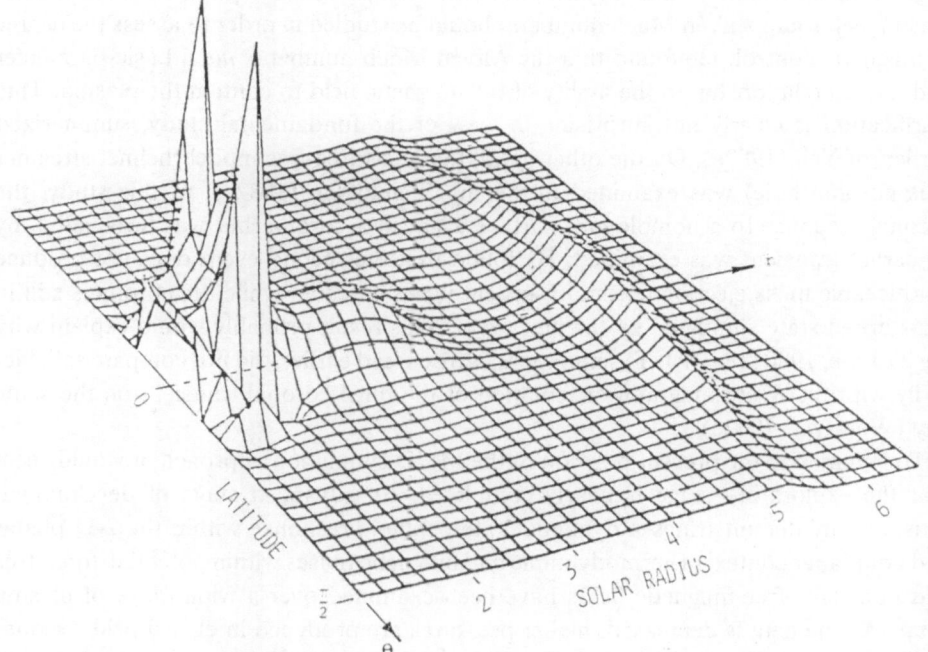

Fig. 21.   Simulated fractional density change, $(n-n_0)/n_0$, as a function of radius and latitude due to injection of cool, dense plasma up both side of a closed magnetic loop. The two spikes represent shocked plasma within the loop; the fast MHD shock at $\sim 5R_\odot$ leads the simulated 'coronal transient'.
(Wu *et al.*, 1982d).

a steepening of a fast MHD wave into a shock (eventually), was suggested as a triggering mechanism for any neighboring region that might be in a neutrally-stable equilibrium. One might speculate that either another flare might be triggered, or that a nearby loop might be caused to erupt as discussed by Low (1981) and Yeh (1982a). The completely nonplanar MHD model has been coupled with mass injection into the loop by Steinolfson (1982a) in order to achieve a quiescent-prominence model.

Returning to the planar 2-D MHD configuration, Steinolfson, Suess and Wu (1982) produced a non-force-free magnetic topology, simultaneously with an outward flowing solar wind, by using the time-relaxation method of numerical computations. The result is the first demonstration of a self-consistent coronal streamer at the equator together with a large coronal hole centered at the pole. This work was followed (Steinolfson and Wu, 1980; Steinolfson, 1928b) by the introduction of pressure pulses (mostly via a relatively-large temperature pulse) similar to those used in the early phases of these MHD studies. They also used a wide range of plasma betas from 0.05 to 0.5; the earlier work had considered values from 0.1 to as high as 1.0 as well as infinity. These workers found that the enhanced density structure was very similar to those observed by the Skylab coronagraph (Hildner, 1977); that is, the coronal transients' legs had little meridional motion, and the enhanced density loop within the closed helmet streamer expanded predominently in the radial direction. Steinolfson (1928b) also examined the suggestion by Maxwell and Dryer (1981) that temporal and spatial distributions of plasma betas and Alfvén Mach numbers should be studied in order to assess the degree of magnetic control. He found that the Alfvén Mach number is *not* a basic parameter and has no relationship to the ability of the magnetic field to contain the plasma. This clarification is clearly not surprising in view of the fundamental study, summarized earlier, of Yeh (1982a). On the other hand, the non-force-free model (helmet streamer plus coronal hole) was examined further by Steinolfson (1982c). In this study, the coronal response to a homologous surface event in a corona that was 'evacuated' by an earlier transient was examined. He found that the second event does not produce a noticeable mass ejection coronal transient when it occurs while the corona is still in a disturbed state simulated by the first event. This result probably would explain why the 29 June, 1980 (18 : 21 UT) simulation, as discussed earlier, did not compare satisfactorily with the *K*-coronagraph observation of the third coronal transient on the same day (Wu *et al.*, 1982c).

To summarize the present position of the MHD simulation approach, it would seem that the exploratory stage is nearly completed. Its advanced stage of development satisfactorily demonstrates a dynamical atmospheric response within the 2-D planar and nonplanar context. Thermodynamic and magnetic pulses within potential, force-free and non-force-free magnetic fields have been examined over a wide range of plasma betas. As the beta is decreased, higher pressures are produced in closed field regions; the flow velocities are higher; and, not unexpectedly, the MHD waves and shocks travel more quickly. Future work, it is suggested, should consider the newer forms of 'pulses': mass injection from the photospheric/chromospheric levels, non-planar perturbations such as loop twisting or footprint spreading, etc. Future work must also consider dissipation (other than at shocks) that should be added in the form of finite resistivity and thermal conduction. While radiation is undoubtedly important very close to the Sun, its importance within the context of coronal transients is probably of a higher order. Finally, to anticipate the identification of transients by *in situ* spacecraft, the MHD models should incorporate more than one species, non-equilibrium effects, and (eventually) coupled fluid-kinetic effects. These are all difficult problems but, nevertheless, solvable with numerical computer techniques.

The MHD modeling of coronal transients must eventually be directly coupled with their progeny, the interplanetary disturbances. Since this approach is limited primarily by computer storage (and cost), it is natural that the two domains have been explored separately (see, for example, the reviews by Wu, 1980, 1982a). The most recent work in the interplanetary domain has been performed by Wu *et al.* (1982) who have utilized the nonplanar 2-D approach in the solar equatorial plane.* In a canonical example, they assumed the steady-state existence of an Archimedean interplanetary magnetic field (with a small southward component) embedded within a representative solar wind from 0.08 to 1.1 AU. A shock wave with a velocity of 1000 km s^{-1} was introduced, over a heliographic longitude width of 24° and maintained (via a trapezoidal time profile) for $1-\frac{1}{2}$ hr at the lower boundary. This pulse, of course, is a fairly coarse attempt to simulate the shock wave from a greatly-restricted geometrical extent of a coronal transient's shock wave. Its justification, of course, is in providing insight for future, more sophisticated pulses.

Part of the result of this interplanetary disturbance is shown in Figure 22. Shown here are the fractional changes in density, temperature and pressure at $t = 60.4$ hr after initiation of the shock pulse. The restricted, but ever-expanding, global extent of the shock wave (and post-shock flow) is clearly seen. The half-tone scales next to each panel show the attenuation (as, for example, in standing planetary bow shocks) of the interplanetary shock in the heliolongitudinal direction. The simulation also demonstrates the widening extent of the perturbed solar wind to an eventual 360° width. Also shown in Figure 22 is the deformed interplanetary magnetic field at the 60.4 hr time marker. The half-plane of the equatorial plane is shown for this purpose while only a quadrant is shown for the thermodynamic variables. The reader is asked to direct attention, for the moment, to post-shock flow in the general vicinity of the flare's central meridian where most of the magnetic field kinking takes place. By examining the density and temperature in, roughly, the same (somewhat generous) regions in radius and longitude, one can see that depressed temperatures and low densities are found within the domain of strong magnetic field kinking**. This result suggests that the 'magnetic cloud' observations (see, for example, Burlaga *et al.*, 1981) can be understood in terms of the temporal MHD explanation proposed here. That is, the demonstrated field-line kinking (to angles equal

---

* The physics is, of course, also equally appropriate for the ecliptic plane or even an idealized, non-warped, heliospheric equatorial plane.

** This magnetic field kinking is accentuated for stronger shock velocity pulses. In fact, the polarity is even reversed from 'outward' (as shown in Figure 22) to 'inward' until the entire disturbance is attenuated with time.

TIME= 60.4 HOURS PRESSURE (P/P0-1)

TIME= 60.4 HOURS MAG. FIELD LINES
FROM .08 TO1.1AU   SCALE=0.1 AU/DIV

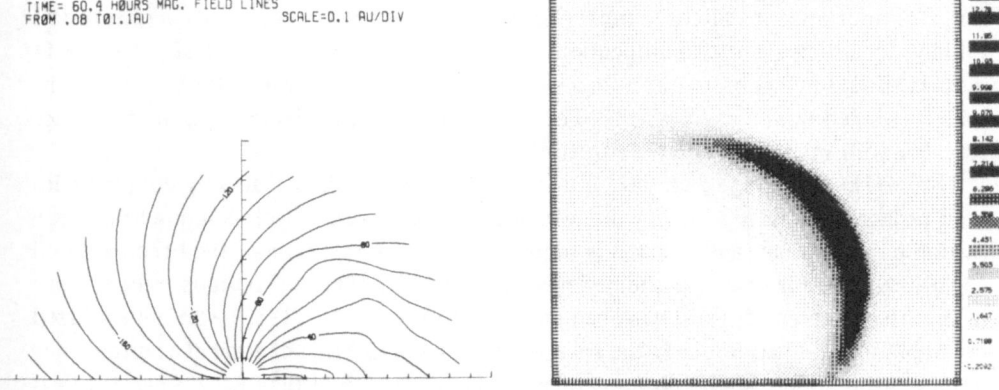

FROM .08 TO1.1AU

TIME= 60.4 HOURS DENSITY (D/D0-1)   TIME= 60.4 HOURS TEMPER. (T/T0-1)

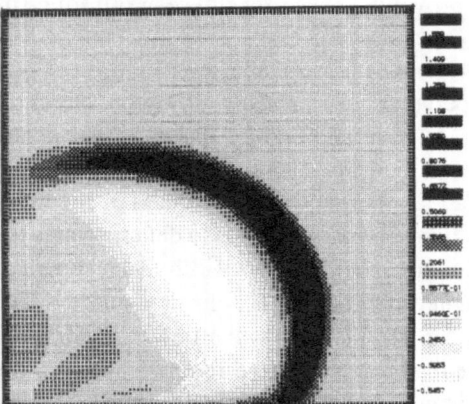   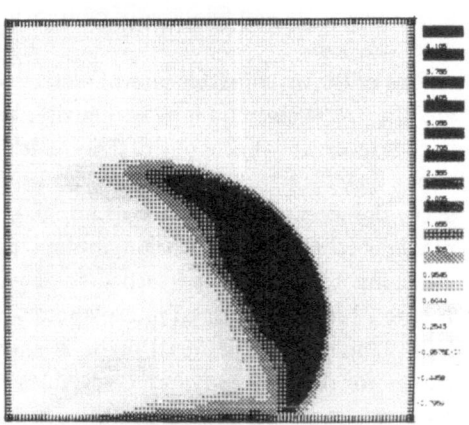

FROM .08 TO1.1AU   FROM .08 TO1.1AU

Fig. 22.   Fractional density, temperature and pressure changes in one quadrant of the solar equatorial plane at a time 60.4 hr after introduction of the simulated solar flare-generated shock wave at 0.08 AU. The maximum (minimum) fractional density change is $(n-n_0)/n_0 = +1.56(-0.55)$; for the fractional temperature change, $(T-T_0)/T_0 = +4.1(-0.80)$; and for the pressure change, $(p-p_0)/p_0 = +12.8(-0.21)$. The scale, as viewed from above the equatorial plane, is from 0.08 to 1.1 AU. The deformed interplanetary magnetic field is shown at the same time in the equatorial (or ecliptic) half-plane. (After Wu *et al.*, 1982). The center of the 24° (in heliolongitudinal width) shock pulse is at 30° from the horizontal axis. It is suggested in the text that the simultaneous post-shock field-line kinking, juxtaposed upon the depressed temperatures and low densities (in the general vicinity of the 'flare's' central meridian) is appropriate for a necessary and sufficient explanation of 'magnetic clouds' observed by Burlaga *et al.* (1981) and others. The rotation of the interplanetary field, shown projected into the equatorial plane in the top left panel, also occurs in planes perpendicular to this plane.

to and greater than 90° to the radial direction) is appropriate for the bi-directional streaming of electrons observed by several groups. The demonstrated simultaneous depressed temperatures, low densities, and elevated magnetic fields are also appropriate for these same observations as reported, for example by Burlaga *et al.* (1981). Although the MHD simulation has, as yet, not incorporated finite resistivity (in order to demonstrate reconnection and production of 'bubbles'), it has shown that a necessary and sufficient condition for the demonstration of observed interplanetary quantities is the temporal behavior of a dissipationless plasma (except at shocks) in response to a pulse near the Sun. Stated in another way, reconnection of the field lines on the sunward side of a deformed magnetic configuration is a sufficient condition for the formation of a 'bubble' but it is *not a necessary condition*. This study is continuing with a view toward incorporating additional physics (such as multi-fluids, thermal conductivity, resistivity, etc.) as well as the previously-mentioned requirement to connect the lower coronal activity with its consequences in interplanetary space.

6. Concluding remarks

The observational phase of coronal transient phenomena, observed first via optical imagery, has advanced to the stage of multi-coronagraph and coronagraph/radio comparisons. The link with interplanetary observations has been made via reasonable tracking of a shock wave by spacecraft at several heliocentric radii as well as by the remote sensing of type II radio bursts.

Insight concerning the dynamic physical processes within the transients is being provided by sophisticated MHD modeling, complemented, in special cases, by analytical methods. Further comparison of major features of specific events with the models, especially for the well-observed Solar Maximum Year events, promises a rich harvest of physical understanding of the origin, structure and interplanetary heritage of coronal transients.

Acknowledgements

It is with pleasure and gratitude that I thank a large number of colleagues for their advice and encouragement during the preparation of this overview paper. Many of them have given permission to refer to work in progress and to unpublished observations. I take full responsibility for any views and interpretations expressed here that may not be in accord with their own positions, preliminary or otherwise. I would like to thank, especially, the scientists at the Naval Research Laboratory (D. J. Michels, R. A. Howard, M. J. Koomen, and N. R. Sheeley, Jr.); Harvard Radio Astronomy Station, Fort Davis (A. Maxwell); University of California, Los Angeles (C. T. Russell); NCAR/High Altitude Observatory (L. L. House, W. J. Wagner, C. Sawyer, R. Fisher, D. G. Sime, and R. Illing); Jet Propulsion Laboratory (R. Woo); TRW Space Systems, Inc. (E. Greenstadt and F. Scarf); SRI International (T. Croft); Universität Bonn (M. K. Bird); Observatoire de Meudon (F. Crifo); University of Maryland (T. Gergely);

NASA Goddard Space Flight Center (R. G. Stone, A. Poland, and H. V. Cane); and NASA Marshall Space Flight Center (E. Tandberg-Hansen and E. Hildner) for their helpful advice, encouragement and counsel concerning the experimental observations and comparison with theory and simulations. Last, but not least, I wish to acknowledge and thank my colleagues in the theoretical and modeling area who have each taught, argued with and counseled me with infinite patience: University of Tel-Aviv (S. Cuperman); University of Alabama at Huntsville (S. T. Wu and S. M. Han); University of Alaska (S.-I. Akasofu and G. Gislason); NOAA Space Environment Laboratory (T. Yeh, S. T. Suess, and V. Osherovich); University of California at Irvine (R. S. Steinolfson); Chiba Institute of Technology (Y. Nakagawa); NCAR/High Altitude Observatory (B. C. Low); and Chinese Institute of Science and Technology (Y. Q. Hu and S. Wang). I am grateful to S. Cuperman, V. Osherovich, S. T. Suess, S. T. Wu, and T. Yeh for reading and commenting on various aspects in the manuscript and to L. Conner for its rapid and efficient preparation. This work was supported primarily by NOAA and, in part, by my participation in the Guest Investigator programs of NASA's Solar Maximum Mission and Pioneer-Venus-Orbiter Mission.

References

Anzer, U.: 1980, in M. Dryer and E. Tandberg-Hansen (eds.), 'Solar and Interplanetary Dynamics', *IAU Symp.* **91**, 263.

Armstrong, J. W. and Woo, R.: 1981, *Astron. Astrophys.* **103**, 415.

Barnes, A. and Hartle, R. E.: 1972, in C. P. Sonnett, P. J. Coleman, and J. M. Wilcox (eds.), *Solar Wind*, NASA SP-308.

Billings, D. E.: 1966, *A Guide to Solar Corona*, Academic Press, New York.

Bird, M. K., Volland, H., Stelzried, C. T., Levy, G. S., and Seidel, B. L.: 1977, in M. S. Shea, D. F. Smart, and S. T. Wu (eds.), *Contributed Papers to the Study of Travelling Interplanetary Phenomena*, Air Force Geophysics Laboratory Report AFGL-TR-77-0309, p. 63.

Bird, M. K., Volland, H., Seidel, B. L., and Stelzried, C. T.: 1980, in M. Dryer and E. Tandberg-Hanssen (eds.), 'Solar and Interplanetary Dynamics', *IAU Symp.* **91**, 475.

Boischot, A., Riddle, A. C., Pearce, J. B., and Warwick, J. W.: 1980, *Solar Phys.* **65**, 397.

Bruzek, A. and DeMastus, H. L.: 1970, *Solar Phys.* **12**, 447.

Burlaga, L., Sittler, E., Mariani, F. and Schwenn, R., 1981, *J. Geophys. Res.* **86**, 6673.

Cane, H. V., Stone, R. G., Fainberg, J., Steinberg, J.-L., and Hoang, S.: 1982, *Solar Phys.* **78**, 187.

Cane, H. V., Stone, R. G., and Woo, R.: 1981, *EOS* **62**, 984 (abstract).

Couturier, P., Mangeney, A., and Souffrin, P.: 1980, in M. Dryer and E. Tandberg-Hanssen (eds.), 'Solar and Interplanetary Dynamics', *IAU Symp.* **91**, 127.

Crifo, F., Picat, J. P., and Cailloux, M.: 1982, *Solar Phys.* (submitted).

Croft, T.: 1982, *STIP VII: Four Neighboring Radio Paths Traverse the Northern Corona in 1979*, Air Force Geophysics Laboratory Report TR−82−0035, Bedford, Massachusetts.

Cuperman, S.: 1980, *Space Sci. Rev.* **26**, 277.

Cuperman, S. and Harten, A.: 1970a, *Cosmic Electrodyn.* **1**, 205.

Cuperman, S. and Harten, A.: 1970b, *Astrophys. J.* **162**, 315.

Cuperman, S. and Harten, A.:1971, *Astrophys. J.* **163**, 383.

Davis, W. D. and Feynman, J.: 1977, *J. Geophys. Res.* **82**, 4699.

Dryer, M.: 1975, *Space Sci. Rev.* **17**, 277.

Dryer, M.: 1981, in H. Rosenbauer (ed.), *Solar Wind IV*, Max-Planck-Institute für Aeronomie, Report No. MPAE-W-100-81-31, Katlenburg-Lindau, pp. 199–210.

Dryer, M. and Cuperman, S.: 1972, in P. S. McIntosh and M. Dryer (eds.), *Solar Activity Observations and Predictions*, M.I.T. Press, Cambridge, p. 197.

Dryer, M. and Maxwell, A.: 1979, *Astrophys. J* **231**, 945.

Dryer, M. and Wu, S. T.: 1981, *Adv. Space Res.* **1**, 85.

Dryer, M., Wu, S. T., Steinolfson, R. S., and Wilson, R. M., 1979, *Astrophys. J.* **227**, 1059.

Dryer, M., Pérez-de-Tejada, H., Taylor, H. A., Jr., Intriligator, D. S., Mihalov, J. D., and Rompolt, B.: 1982, *J. Geophys. Res.*, (in press).

Dulk, G. A.: 1980, in M. Kundu and T. E. Gergely (eds.), 'Radiophysics of the Sun', *IAU Symp.* **86**, 419.

Dulk, G. A. and McLean, D. J.: 1978, *Solar Phys.* **57**, 279.

Dulk, G. A., Smerd, S. F., MacQueen, R. M., Gosling, J. T., Magnum, A., Stewart, R. T., Sheridan, K. V., Robinson, R. D., and Jacques, S., 1976, *Solar Phys.* **49**, 369.

Ekers, R. D. and Little, L. T.: 1971, *Astron. Astrophys.* **10**, 310.

Feldman, W. C., Asbridge, J. R., Bame, S. J., and Gosling, J. T.: 1977, in O. R. White (ed.), *The Solar Output and Its Variations*, Colorado Assoc. University Press, Boulder, p. 351.

Fisher, R. R., Garcia, C. J., and Seagraves, P.: 1981, *Astrophys. J. Letters* **246**, L161.

Fisher, R. R. and Poland, A. I., 1981, *Astrophys. J.* **246**, 1004.

Friend, D., Munro, R. H., Fisher, R. R., and McCabe, M. K.: 1981, *Bull. Am. Astron. Soc.* **13**, 890 (abstract).

Gergely, T. E., Kundu, M. R., Erskine, F. T., Sawyer, C. B., Wagner, W. J., Illing, R. M. E., House, L. L., McCabe, M. K., Stewart, R. T., and Nelson, G. J.: 1980, *Bull. Am. Astron. Soc.* **12**, 900 (abstract).

Gergely, T. E., Kundu, M. R., and Hildner, E.: 1981, *Astrophys. J.* (submitted).

Glass, I. I.: 1977, *Prog. Aerospace Sci.* **17**, 269.

Greenstadt, E. W. and Fredericks, R. W.: 1979, in L. J. Lanzerotti, C. F. Kennel, and E. N. Parker (eds.), *Solar System Plasma Physics*, Vol. III, Chapter III. 1.1, North-Holland Publ. Co.

Hansen, R. T., Garcia, C. J., Hansen, S. F., and Yasukawa, E.: 1974, *Publ. Astron. Soc. Pacific* **86**, 500.

Hildner, E., 1977, in M. A. Shea, D. F. Smart, and S. T. Wu (eds.), *Study of Travelling Interplanetary Phenomena/1977*, D. Reidel Publ. Co., Dordrecht, Holland, p. 3.

Hildner, E., Illing, R. M. E., Wagner, W. J., House, L. L., Sawyer, C. B., and Hyder, C. L.: 1981, *Bull. Am. Astron. Soc.* **13**, 861 (abstract).

Hirshberg, J.: 1968, *J. Geophys. Res.* **16**, 309.

Hollweg, J. V.: 1978, *Rev. Geophys. Space Phys.* **16**, 698.

House, L. L., Wagner, W. J., Hildner, E., Sawyer, C. B., and Schmidt, H.: 1981a, *Astrophys. J. Letters* **244**, L117.

House, L. L., Illing, R. M. E., Sawyer, C. B., and Wagner, W. J.: 1981b, *Bull. Am. Astron. Soc.* **13**, 862 (abstract).

Howard, R. A., Koomen, M. J., Michels, D. J., Tousey, R., Detwiler, C. R., Roberts, D. E., Seal, R. T., Whitney, J. D., Hansen, R. T., Hansen, S. F., Garcia, C. J., and Yasukawa, E., 1975, in WDC-A Report UAG-48, NOAA Boulder.

Howard, R. A., Sheeley, N. R., Michels, D. J., and Koomen, M. J., 1981, (abstract), Max-Planck-Institut für Aeronomie IX-th Lindau Workshop on *The Source Region of the Solar Wind*, (2–6 November, 1981).

Intriligator, D. S.: 1980, in M. Dryer and E. Tandberg-Hanssen (eds.), 'Solar and Interplanetary Dynamics', *IAU Symp.* **91**, 357.

Jackson, B. V.: 1981, *Solar Phys.* **73**, 133.

Jackson, B. V. and Hildner, E.: 1978, *Solar Phys.* **60**, 155.

James, J. C.: 1966, *Astrophys. J.* **146**, 356.

Kane, S. R., *et al.*: 1981, reprint.

Karpen, J. T., Oran, E. S., Boris, J. P., Mariska, J. T., and Brueckner, G. E.: 1981, *Bull. Am. Astron. Soc.* **13**, 913 (abstract); also *EOS* **62**, 1008 (abstract).

Kennel, C. F., Scarf, F. L., Coroniti, F. V., Smith, E. J., and Gurnett, D. A.: 1982, *J. Geophys. Res.* **87**, 17.

Kopp, R. A. and Holzer, T. E.: 1976, *Solar Phys.* **49**, 43.

Low, B. C.: 1981, *Astrophys. J.* **251**, 352.

Low, B. C.: 1982, *Astrophys. J.* (in press).

Low, B. C., Munro, R. H., and Fisher, R. R.: 1982, *Astrophys. J.* **254**, 335.

MacQueen, R. M.: 1980, *Phil. Trans. Roy. Soc. London*, A **297**, 605.

Malitson, H. H., Fainberg, J., and Stone, R. G.: 1976, *Space Sci. Rev.* **19**, 511.

Martin, S. F. and Ramsey, H. E.: 1972, in P. S. McIntosh and M. Dryer (eds.), *Solar Activity Observations and Predictions*, MIT Press, Cambridge, Mass., pp. 371–388.

Maxwell, A. and Dryer, M.: 1981, *Solar Phys.* **73**, 313.

Maxwell, A. and Dryer, M.: 1982, *Space Sci. Rev.* **32**, 11.

Michels, D. J., Howard, R. A., Koomen, M. J., Sheeley, N. R. Jr., and Rompolt, B.: 1980, in M. Dryer and E. Tandberg-Hanssen (eds.), 'Solar and Interplanetary Dynamics', *IAU Symp.* **91**, 382.
Michels, D. J., Howard, R. A., Koomen, M. J., and Sheeley, N. R. Jr.: 1982, *Bull. Am. Astron. Soc.* **14**, 572 (abstract).
Mouschovias, T. and Poland, A. I.: 1978, *Astrophys. J.*, **220**, 675.
Munro, R. H.: 1977, *Bull. Am. Astron. Soc.* **9**, 371, (abstract).
Munro, R. H., Gosling, J. T., Hildner, E., MacQueen, R. M., Poland, A. I., and Ross, C. L.: 1979, *Solar Phys.* **61**, 201.
Munro, R. H. and Jackson, B. V.: 1977, *Astrophys. J.* **213**, 874.
Nakagawa, Y.: 1976, *Space Sci. Rev.* **19**, 459.
Nakagawa, Y., Wu, S. T., and Han, S. M.: 1978, *Astrophys. J.* **219**, 314.
Nakagawa, Y., Wu, S. T., and Han, S. M.: 1980, in M. Dryer and E. Tandberg-Hanssen (eds.), 'Solar and Interplanetary Dynamics', *IAU Symp.* **91**, 495.
Nakagawa, Y., Wu, S. T., and Han, S. M.: 1981, *Astrophys. J.* **244**, 331.
Newkirk, G., Jr.: 1967, in *Annual Review of Astronomy and Astrophysics*, Vol. 5, Annual Reviews, Inc., Palo Alto, p. 213.
Osherovich, V.: 1982, *Astrophys. Space Sci.* **86**, 453.
Parker, E. N.: 1958, *Astrophys. J.* **128**, 664.
Parker, E. N.: 1963, *Interplanetary Dynamical Processes*, Intersciences Publishers, New York.
Parker, E. N.: 1977, in G. Burbridge, D. Layzer, and J. G. Phillips (eds.), *Annual Review of Astronomy and Astrophysics*, Annual Reviews, Inc., Palo Alto, p. 45.
Pneumann, G. W.: 1980, *Solar Phys.* **65**, 369.
Rottman, G. J., Orall, F. Q., and Klimchuck, J. A.: 1981, *Astrophys. J. Letters* **247**, L135.
Rottman, G. J., Orrall, F. Q., and Klimchuck, J. A.: 1982, *Astrophys. J.* **260**, (in press).
Scarf, F. L., Gurnett, D. A., and Kurth, W. S.: 1981, in H. Rosenbauer (ed.), *Solar Wind Four*, Max-Planck Institut für Aeronomie Report MPAE-W-100-81-31, Lindau, F.R.G., pp. 305–316.
Sheeley, N. R. Jr., Michels, D. J., Howard, R. A., and Koomen, M. J.: 1980, *Astrophys. J. Letters* **237**, L97.
Sheeley, N. R., Jr., Howard, R. A., Koomen, M. J., Michels, D. J., Schwenn, R., Rosenbauer, H., Mühlhauser, K., Neubauer, F., Mihalov, J. D.: and Rompolt, B.: 1982, (in preparation).
Smerd, S. F., Sheridan, K. V., and Stewart, R. T.: 1975, *Astrophys. Letters* **16**, 23.
Steinolfson, R. S.: 1982a, *Astron. Astrophys.* (submitted).
Steinolfson, R. S.: 1982b, 'MPI für Physik and Astrophysik Report Astro 289', *Astron. Astrophys.*, (in press).
Steinolfson, R. S.: 1982c, 'MPI für Physik and Astrophysik Report Astro 290', *Astron. Astrophys.*, (in press).
Steinolfson, R. S. and Dryer, M.: 1982, *Astrophys. Space Sci.* (submitted).
Steinolfson, R. S. and Wu, S. T.: 1980, in M. Dryer and E. Tandberg-Hanssen (eds.), 'Solar and Interplanetary Dynamics', *IAU Symp.* **91**, 483.
Steinolfson, R. S. and Wu, S. T.: 1980, Univ. of Alabama, Huntsville, Research Report No. 240.
Steinolfson, R. S., Wu, S. T., Dryer, M., and Tandberg-Hanssen, E.: 1978, *Astrophys. J.* **225**, 259.
Steinolfson, R. S., Suess, S. T., and Wu, S. T.: 1982, *Astrophys. J.* **255**, 730.
Stewart, R. T., 1980, in M. Dryer and E. Tandberg-Hanssen (eds.), 'Solar and Interplanetary Dynamics', *IAU Symp.* **91**, 333.
Stewart, R. T., Dulk, G. A., Sheridan, K. V., House, L. L., Wagner, W. J., Sawyer, C. B., and Illing, R. M. E.: 1982, *Astron. Astrophys.*, (in press).
Suess, S. T.: 1979, *Space Sci. Rev.* **23**, 159.
Suess, S. T.: 1981, *EOS* **62**, 1008 (abstract).
Suess, S. T., Richter, A. K., Winge, C. R., and Nerney, S. F.: *Astrophys. J.* **217**, 296.
Tanaka, K. and Nakagawa, Y.: 1973, *Solar Phys.* **33**, 187.
Tyler, G. L., Vesecky, J. F., Plume, M. A., Howard, H. T., and Barnes, A.: 1981, *Astrophys. J.* **249**, 318.
Wagner, W. J., Hildner, E., House, L. L., Sawyer, C. B., Sheridan, K. V., and Dulk, G. A.: 1981, *Astrophys. J. Letters* **244**, L123.
Wagner, W. J., Sawyer, C. B., Illing, R. M. E., House, L. L., Querfeld, C. W., Sheeley, N. R., Jr., Howard, R. A., Koomen, M. J., Michels, D. J., and Smartt, R. N.: 1980, *Bull. Am. Astron. Soc.* **12**, 902 (abstract).
Withbroe, G. L., Kohl, J. L., Weiser, H., Noli, G., and Munro, R. H.: 1982, *Astrophys. J.* **254**, 361.
Wolff, L. C., Brandt, J. C., and Southwick, G. R.: 1971, *Astrophys. J.* **165**, 181.
Woo, R.: 1978, *Astrophys. J.* **219**, 727.

Woo, R. and Armstrong, J. W.: 1981, *Nature* **292**, 608.

Wu, S. T.: 1980, in M. Dryer and E. Tandberg-Hanssen (eds.), 'Solar and Interplanetary Dynamics', *IAU Symp.* **91**, 443.

Wu, S. T.: 1982a, *Space Sci. Rev.* **32**, 83.

Wu, S. T.: 1982b, in *STIP Symposium on Solar Radio Astronomy, Interplanetary Scintillations, and Coordination with Spacecraft*, AFGL Rept., Bedford, Mass , (in press).

Wu, S. T., Dryer, M., and Han, S. M.: 1976, *Solar Phys.* **49**, 187.

Wu, S. T., Dryer, M., and Han, S. M.: 1982, *Solar Phys.* (submitted).

Wu, S. T., Dryer, M., Nakagawa, Y., and Han, S. M., 1978, *Astrophys. J.* **219**, 325.

Wu, S. T., Steinolfson, R. S., Dryer, M., and Tandberg-Hanssen, E.: 1981, *Astrophys. J.* **243**, 641.

Wu, S. T., Hu, Y. Q., Wang, S., Dryer, M., and Tandberg-Hanssen, E.: 1982a, *Astrophys. Space Sci.* **83**, 189.

Wu, S. T., Nakagawa, Y., Han, S. M., and Dryer, M.: 1982b, *Astrophys. J.*, (in press).

Wu, S. T., Wang, S., Dryer, M., Poland, A. I., Sime, D., Wolfson, C. J., Orwig, L. E., and Maxwell, A.: 1982c, (in preparation).

Wu, S. T., Wang, S., Michels, D. J., Howard, R. A., Koomen, M. J., and Sheeley, N. R.: 1982d, (in preparation).

Yakovlev, O. I., Efimov, A. I., Razmanov, V. M., and Shtyrkov, V. K.: 1980, *Acta Astron* **7**, 235.

Yeh, T.: 1982a, *Solar Phys.* **78**, 287.

Yeh, T.: 1982b, *Astrophys. J.* **264**, (in press).

Yeh, T. and Dryer, M.: 1981, *Astrophys. J.* **245**, 704.

Yeh, T. and Dryer, M.: 1981b, *Solar Phys.* **71**, 141.

Zirin, H. and Tanaka, K.: 1981, *Astrophys. J.* **250**, 791.